About Island Press

Island Press is the only nonprofit organization in the United States whose principal purpose is the publication of books on environmental issues and natural resource management. We provide solutions-oriented information to professionals, public officials, business and community leaders, and concerned citizens who are shaping responses to environmental problems.

In 2000, Island Press celebrates its sixteenth anniversary as the leading provider of timely and practical books that take a multidisciplinary approach to critical environmental concerns. Our growing list of titles reflects our commitment to bringing the best of an expanding body of literature to the environmental community throughout North America and the world.

Support for Island Press is provided by The Jenifer Altman Foundation, The Bullitt Foundation, The Mary Flagler Cary Charitable Trust, The Nathan Cummings Foundation, The Geraldine R. Dodge Foundation, The Charles Engelhard Foundation, The Ford Foundation, The Vira I. Heinz Endowment, The W. Alton Jones Foundation, The John D. and Catherine T. MacArthur Foundation, The Andrew W. Mellon Foundation, The Charles Stewart Mott Foundation, The Curtis and Edith Munson Foundation, The National Fish and Wildlife Foundation, The National Science Foundation, The New-Land Foundation, The David and Lucile Packard Foundation, The Pew Charitable Trusts, The Surdna Foundation, The Winslow Foundation, and individual donors.

About World Wildlife Fund

Known worldwide by its panda logo, World Wildlife Fund is dedicated to protecting the world's wildlife and the rich biological diversity that we all need to survive. The leading privately supported international conservation organization in the world, WWF has sponsored more than 2,000 projects in 116 countries and has more than 1 million members in the United States.

Through its Living Planet Campaign, WWF aims to make the remaining days of this century a turning point in the worldwide struggle to preserve species and habitats. The campaign calls on governments, corporations, and others to take significant actions to help preserve the world's endangered spaces—places we call the Global 200; to protect endangered species; and to address the global threats that put all living things in harm's way. To learn more about the Living Planet Campaign, visit the WWF Web site at www.worldwildlife.org or write to us.

Freshwater Ecoregions
of North America

Freshwater Ecoregions
of North America

A CONSERVATION ASSESSMENT

Robin A. Abell
David M. Olson
Eric Dinerstein
Patrick T. Hurley
James T. Diggs
William Eichbaum
Steven Walters
Wesley Wettengel
Tom Allnutt
Colby J. Loucks
and Prashant Hedao

WORLD WILDLIFE FUND—
UNITED STATES

ISLAND PRESS

WASHINGTON, D.C. • COVELO, CALIFORNIA

Library of Congress Cataloging-in-Publication Data

    Freshwater ecoregions of North America : a conservation assessment/
Robin A. Abell . . . [et al.] (World Wildlife Fund-United States).
      p.   cm.
    Includes bibliographical references and index.
    ISBN 1-55963-734-X
    1. Biological diversity conservation—North America. 2. Biotic communities—
North America 3. Freshwater ecology—North America.
    I. Abell, Robin A. II. World Wildlife Fund (U.S.)
    QH77.N56F69   2000
    577.6—dc21   99–16796
                                                   CIP

Contents

LIST OF SPECIAL ESSAYS   ix

LIST OF FIGURES   xi

LIST OF TABLES   xiii

ACRONYMS   xv

FOREWORD   xvii

PREFACE   xix

ACKNOWLEDGMENTS   xxi

1. INTRODUCTION   1
　　Assessment Overview   2
　　Structure of the Book   6
　　The Challenge   6

2. APPROACH   9
　　Delineation of Ecoregions and Geographic Scope of the
　　　　Study   9
　　Assignment of Major Habitat Types   12
　　Discriminators   14
　　Biological Distinctiveness Index (BDI)—Overview   15
　　Conservation Status—Overview   17
　　Integrating Biological Distinctiveness and Conservation
　　　　Status   22

3. BIOLOGICAL DISTINCTIVENESS OF NORTH AMERICAN
　　ECOREGIONS   25
　　Species Richness   25
　　Endemism   39
　　Rare Ecological or Evolutionary Phenomena   46
　　Rare Habitat Type   48
　　Synthesis of Biological Distinctiveness Data   48

4. Conservation Status of North American
   Ecoregions   59
   Snapshot Conservation Status   59
   Conservation Snapshot Criteria   62
   Threat Assessment   70
   Final Conservation Status   70
   Additional Conservation Status Data   70

5. Setting the Conservation Agenda: Integrating
   Biological Distinctiveness and Conservation
   Status   87

6. Recommendations   101
   North America's Global Responsibilities for Biodiversity
      Conservation   101
   Targets Requiring Urgent Action   103

7. Site-Specific Conservation   109
   Dams   112
   Additional Site-Selection Tools   119

Appendixes
   A. Methods for Assessing the Biological Distinctiveness
      of Freshwater Ecoregions   121
   B. Methods for Assessing the Conservation Status of
      Freshwater Ecoregions   127
   C. Biological Distinctiveness Data and Scores   131
   D. Conservation Status Assessment and Scores   141
   E. Statistical Analysis of Biological Distinctiveness and
      Conservation Status Data   149
   F. Integration Matrices for the Eight Major Habitat
      Types   155
   G. Ecoregion Descriptions   165
   H. Conservation Partner Contact Information   265

Glossary   281

Literature Cited and Consulted   289

Authors   305

Contributors   307

Index   309

List of Special Essays

Essay 1.   Diversity, Levels of Imperilment, and Cryptic Fishes in the Southeastern United States   30
N.M. Burkhead and H.L. Jelks

Essay 2.   The Mississippi River Basin: Its Megafauna and Hydrological Modifications   33
B.M. Burr and J.B. Ladonski

Essay 3.   Preserving North America's Unique Crayfish Fauna   36
C.A. Taylor

Essay 4.   Freshwater Mussels: A Complicated Resource to Conserve   37
G.T. Watters

Essay 5.   Subterranean Freshwater Biodiversity in Northeastern Mexico and Texas   41
D.A. Hendrickson and J. Krejca

Essay 6.   Connecting the Land to the Sea: Anadromous Fish   46
P.B. Moyle

Essay 7.   Western Springs: Their Faunas and Threats to Their Existence   52
W.L. Minckley and P.J. Unmack

Essay 8.   California Vernal Pools   53
B. Vlamis

Essay 9.   Prairie Potholes in Decline   55
M.H. Stolt

Essay 10.   Invasive Nonindigenous Species: A Major Threat to Freshwater Biodiversity   67
J.D. Williams and G.K. Meffe

Essay 11.   Categorizing Imperiled and Vulnerable Species   73
L.L. Master

Essay 12.   Canadian Freshwater Biodiversity   83
D. McAllister

Essay 13. The Valley of Cuatro Ciénegas, Coahuila: Its Biota and Its Future   94
S. Contreras-Balderas

Essay 14. North America's Freshwater Invertebrates: A Research Priority   104
D.L. Strayer

Essay 15. Conservation of Aquatic Karst Biotas: Shedding Light on Troubled Waters   106
S.J. Walsh

Essay 16. Reoperation and Decommissioning of Hydropower Dams: An Opportunity for River Rehabilitation   117
A. Fahlund

List of Special Essays

List of Figures

2.1   Freshwater Ecoregions of North America   10

2.2   Hierarchy of Spatial Units Used in Conservation Assessment Framework   12

2.3   Freshwater Major Habitat Types of North America   14

2.4   Weightings for Conservation Status Indicators, by Major Habitat Type   18

2.5   Integration Matrix for Priority Setting   22

3.1   Total Species Richness   26

3.2   Total Number of Endemic Species   26

3.3   Total Percentage of Endemic Species (Average)   26

3.4a   Fish Richness   27

3.4b   Number of Endemic Fish Species   27

3.4c   Percentage of Endemic Fish Species   27

3.5   Levels of Imperilment of Darters (family Percidae) in the Southeastern United States   32

3.6a   Crayfish Richness   35

3.6b   Number of Endemic Crayfish Species   35

3.6c   Percentage of Endemic Crayfish Species   35

3.7a   Unionid Mussel Richness   38

3.7b   Number of Endemic Unionid Mussel Species   38

3.7c   Percentage of Endemic Unionid Mussel Species   38

3.8a   Herpetofauna Richness   40

3.8b   Number of Endemic Herpetofauna Species   40

3.8c   Percentage of Endemic Herpetofauna Species   40

3.9   Rare Ecological or Evolutionary Phenomena   49

3.10 Biological Distinctiveness Category, Based on Richness and Endemism   49

3.11 Final Biological Distinctiveness   57

4.1   Snapshot Conservation Status   60

4.2   Land-Cover (Catchment) Alteration (Estimated)   63

4.3   Surface Water Quality Degradation (Estimated)   63

4.4   Alteration of Hydrographic Integrity (Estimated)   63

4.5   Degree of Habitat Fragmentation   65

4.6   Effects of Introduced Species   65

4.7   Direct Species Exploitation   65

4.8   Number of Nonindigenous Fishes Introduced into Inland Waters of the United States, Presented by Watershed (USGS Hydrologic Units, top) and by State (bottom)   69

4.9   Threat Assessment   71

4.10 Final Conservation Status   72

4.11a Percentage of Fish Species That Are Imperiled   76

4.11b Percentage of Endemic Fish Species That Are Imperiled   76

4.12a Percentage of Herpetofauna Species That Are Imperiled   77

4.12b Percentage of Endemic Herpetofauna Species That Are Imperiled   77

4.13 U.S. National Watershed Characterization—Water Quality   80

4.14 U.S. National Watershed Characterization—Compiled by Ecoregion   82

5.1a Priority Classes of Ecoregions Using Snapshot Conservation Status   89

5.1b Priority Classes of Ecoregions Using Final Conservation Status   90

6.1   The Global 200 Freshwater Ecoregions   102

7.1a Important Sites for the Conservation of Freshwater Biodiversity in North America—Eastern United States   113

7.1b Important Sites for the Conservation of Freshwater Biodiversity in North America—Western United States and Canada   114

7.1c Important Sites for the Conservation of Freshwater Biodiversity in North America—Southwestern United States and Mexico   115

7.2   Dams Proposed for Decommissioning or Modification   116

A.1   Steps for Evaluating Biological Distinctiveness of Ecoregions   123

A.2   Example of Threshold Determination for Species Richness in the Temperate Coastal Rivers and Lakes MHT   125

B.1   Steps for Evaluating Conservation Status of Ecoregions   128

List of Tables

2.1   Assignment of Ecoregions to Major Habitat Types   13

2.2   Biological Distinctiveness and Conservation Status
        Criteria   15

3.1   Biological Distinctiveness Rankings   28

3.2   Mean Biodiversity Indicators of MHTs, and Associated
        Ranks   45

3.3   Rare Ecological and Evolutionary Phenomena   50

4.1   Distribution of Ecoregions by MHT for the Snapshot (1997)
        Conservation Status Index   60

4.2   Conservation Status Rankings   61

4.3   Distribution of Ecoregions by MHT for the Final Conserva-
        tion Status Index   72

4.4   Average Faunal Imperilment of MHTs   78

5.1a   Distribution of Ecoregions within Integration Matrix, Using
        Snapshot Assessment   88

5.1b   Distribution of Ecoregions within Integration Matrix, Using
        Final Conservation Status   88

5.2a   Distribution of Ecoregions by Priority Class, Using Snapshot
        Assessment   91

5.2b   Distribution of Ecoregions by Priority Class, Using Final
        Conservation Status   91

5.3   Distribution of Ecoregions by Priority Class   91

5.4a   Distribution of Priority Classes by MHT, Using Snapshot
        Assessment   97

5.4b   Distribution of Priority Classes by MHT, Using Final Con-
        servation Status   97

6.1   Global 200 Freshwater Ecoregions in North America   103

7.1   Important Sites for the Conservation of Freshwater Biodiver-
        sity in North America   110

A.1   Data Sources for Biological Distinctiveness Indicators   122

E.1   Area and Latitude of Ecoregions   151

Acronyms

ARL	Arctic Rivers and Lakes
ANOVA	Analysis of Variance
BDI	Biological Distinctiveness Index
CONABIO	Comisión Nacional Para el Conocimiento y Uso de la Biodiversidad
CRP	Conservation Reserve Program
CSI	Conservation Status Index
EPA	U.S. Environmental Protection Agency
ERBC	Ecoregion-Based Conservation
ERLS	Endorheic Rivers, Lakes, and Springs
FERC	Federal Energy Regulatory Commission
GAP	Gap Analysis Approach
IUCN	International Union for the Conservation of Nature and Natural Resources (now called World Conservation Union)
IWI	Index of Watershed Indicators
LTL	Large Temperate Lakes
LTR	Large Temperate Rivers
MHT	Major Habitat Type
NGO	Nongovernmental Organization
SCRL	Subtropical Coastal Rivers and Lakes
TCRL	Temperate Coastal Rivers and Lakes
THL	Temperate Headwaters and Lakes
TNC	The Nature Conservancy
TVA	Tennessee Valley Authority
USDA	United States Department of Agriculture
USFS	United States Forest Service
USFWS	United States Fish and Wildlife Service
USGS	United States Geological Survey
WWF	World Wildlife Fund
XRLS	Xeric-Region Rivers, Lakes, and Springs

Foreword

With this report, WWF-US provides a frame of reference for action to conserve biodiversity in the United States, Canada, and Mexico. Across North America, the demands we make for land and water resources crowd out species, degrade their habitats, and stress the underlying ecological processes that are the lifeblood of natural systems. Conservationists are responding to these threats in many ways. By conducting an ecoregion-based assessment of freshwater biodiversity, WWF aims to speed up conservation planning and action across our continent. At a minimum, this will help to focus our own efforts as an organization. We hope it will also help knit together the work of others working to safeguard North American biodiversity, regardless of how they are approaching this challenge. For example, our principal funding partner for this project, the U.S. Environmental Protection Agency (EPA), has an important role to play as it sets its national priorities for freshwater conservation over the short and long term.

The key message we hope readers will take away is one of urgency. In some ecoregions, this urgency is due to the losses already incurred, and the need to save what remains or carry out major efforts at restoration. In other ecoregions, the situation is urgent because the opportunity remains to prevent similar losses from occurring in the first place.

Given this urgency, what is WWF doing? Increasingly, all our conservation programs are designed to achieve ecoregion-based conservation by means of a two-pronged strategy: establishing protected areas and achieving sustainable management of the lands and waters outside protected areas. In the United States, our efforts are now concentrated on urgent conservation issues in five globally outstanding and endangered ecoregions or ecoregion complexes: the Klamath-Siskiyou; southern Florida, including the Everglades; the Chihuahuan Desert; the Bering Sea; and the rivers and streams of the southeastern United States. All of these ecoregions harbor important freshwater biodiversity, or, in the case of the Bering Sea, support migratory species that move into and out of freshwater systems.

By preparing this assessment of North America's fresh-water biodiversity, we now have a road map for conservation action and a yardstick for judging the success or failure of WWF's own conservation work, as well as that of others. We look forward to cooperating with all those who share our mission so that, together, we will succeed.

James Leape
Executive Vice President
WWF-US

Preface

This book is part of a long-term effort undertaken by the Conservation Science Program of World Wildlife Fund-US to conduct conservation assessments of terrestrial, freshwater, and marine ecoregions around the world. This project began with a terrestrial biodiversity assessment of the Russian Federation (Krever et al. 1994), followed by Latin America and the Caribbean (Dinerstein et al. 1995), mangrove ecoregions of Latin America (Olson et al. 1996), freshwater ecoregions of Latin America and the Caribbean (Olson et al. 1997), and terrestrial ecoregions of North America north of Mexico (Ricketts et al. 1999a). Forthcoming in this series are conservation assessments of marine ecoregions of North America, and of terrestrial ecoregions of Africa and Asia. Ecoregions in these analyses evaluated as globally outstanding in biodiversity value were elevated to a map of the Global 200 ecoregions recently published by WWF (Olson and Dinerstein 1998).

Acknowledgments

First and foremost we thank the participants in our expert workshop who contributed expertise, years of experience, and valuable time to the project. Several of these participants, in addition to other experts, were invited to write the essays included in this book. Their contributions have expanded its breadth enormously.

Within WWF, many people not noted in either category helped in innumerable ways at every stage of the project. Marlar Oo, Kim McCrary, Quinn McKew, Michele Thieme, Meghan McKnight, and Belaine Lehman provided assistance at crucial junctures, and the entire staff of the Conservation Science Program found time to help in vital ways when needed, despite other commitments. Additionally, Carla Langeveld located many of the literature sources crucial to this study.

Peter Moyle, Christopher Taylor, W.L. Minckley, and Don McAllister conducted a peer review of this book, and their comments have resulted in many improvements. David Propst and Paul Loiselle provided important contributions to our ecoregion descriptions, and Kevin Kavanagh of WWF-Canada made time to review and add to the descriptions covering Canada.

We thank James Maxwell and Clayton Edwards for generously supplying their map of zoogeographic zones, on which our freshwater ecoregion map is largely based. Our ability to use their map as a solid starting point allowed us to marshal our resources for other time-intensive parts of the assessment.

Christopher Taylor, James Williams, and Larry Master provided crayfish and mussel distribution data, making our analysis of biological distinctiveness far more comprehensive than if we had relied solely on published data. Additionally, Larry Master supplied global rank data from The Nature Conservancy (TNC), allowing us to include an analysis of species imperilment.

This freshwater report benefited enormously from the exhaustive effort expended on the companion assessment of North American terrestrial ecoregions. The terrestrial report provided a template for this project, and many of the comments offered by reviewers of the first assessment were incorporated into this one.

Both CONABIO and WWF-México provided important information for this assessment, particularly for the identification of important sites in Mexico. Similarly, this study benefited from the results of a joint WWF-US, WWF-México, and TNC ecoregion-based conservation project in the Chihuahuan Desert.

Likewise, WWF-Canada assisted by contributing information and evaluating our assessment results.

We thank the many authors, cartographers, biogeographers, and conservation biologists on whose work this assessment is based. They are cited in the book but deserve thanks here.

Finally, we would like to thank and acknowledge organizations and individuals providing support for this conservation effort, including Environmental Systems Research Institute, Inc. (ESRI) for its software donations and Hewlett-Packard Co. for its generous hardware donations. In particular, Tom Born and his colleagues at the EPA have given this project their full backing and encouragement, spearheading a move to elevate freshwater conservation to the high priority that its urgency mandates.

Freshwater Ecoregions
of North America

CHAPTER 1

Introduction

North America's freshwater habitats—its lakes, springs, streams, and rivers—support some of the most extraordinary biotic assemblages in the world. When compared with similar habitats across the globe, many of North America's communities are virtually unrivaled in their highly distinctive fish, mussel, crayfish, amphibian, and aquatic reptilian faunas (Olson and Dinerstein 1998).

North America's freshwater environments have the unfortunate distinction of also being among the most threatened (Moyle 1994). Flow alteration, habitat degradation and fragmentation, introduced species, and overall land-use change have already taken a heavy toll on the continent's freshwater biota, with many remaining species and communities in serious decline (Miller et al. 1989; The Nature Conservancy 1996a). For instance, in the United States, only 2 percent of the nation's 5.1 million kilometers of rivers and streams remain free flowing and undeveloped, with over 75,000 large dams and 2.5 million small dams on them (Langner and Flather 1994; Naiman et al. 1995a; McAllister et al. 1997). The cumulative impact of all forms of disturbance to aquatic systems is staggering. Within the United States alone, 67 percent of freshwater mussels and 65 percent of crayfish species are rare or imperiled; 37 percent of freshwater fish species are at risk of extinction; and 35 percent of amphibians that depend on aquatic habitats are rare or imperiled (The Nature Conservancy 1996c). These numbers do not include the twenty-seven species of freshwater fish and ten species of mussels that are known to have gone extinct in North America in the last 100 years (Miller et al. 1989; The Nature Conservancy 1996c).

While the global conservation community has mobilized to conserve the biological wealth of tropical rain forests and coral reefs, the freshwater realm has until recently gone virtually unnoticed (Allan and Flecker 1993; Olson and Dinerstein 1998). This is as true for North America as for lesser-known parts of the world. For instance, at the United States' 1986 National Forum on Biodiversity, not a single speaker focused on freshwater issues (Wilson 1988; Blockstein 1992). Since then, this oversight has given way to numerous calls of alarm, as tallies of imperiled species increasingly indicate that freshwater taxa are far more threatened than are terrestrial species (Neves 1992; Allan and Flecker 1993; Williams et al. 1993; McAllister et al. 1997). Such inventories can only account for described species, and even within well-known groups such as fish, species are apparently going extinct before they can be classified (N. Burkhead, pers. comm.; McAllister et al. 1985).

The gravity of threats to freshwater species and habitats has begun to garner attention among policy makers and the public, and support for conservation initiatives is growing. With nearly every freshwater system suffering from some degree of degradation, and with conservation resources limited, there is an urgent and practical need for priority-setting. Recognizing this, World Wildlife Fund–United States, with support from the U.S. EPA, has conducted a conservation assessment of freshwater ecoregions as an initial step in identifying those areas where protective and restorative measures should be implemented first.

The goals of this assessment are as follows: (1) to identify freshwater ecoregions that support globally outstanding biological diversity and to emphasize our global responsibility to protect and restore them; (2) to assess the types and immediacy of threats to North American ecoregions; (3) to begin to identify specific sites within ecoregions where conservation activities may result in substantial benefits to biodiversity; (4) to identify important gaps in information that hamper an accurate evaluation of biodiversity; and (5) to provide a broad-scale framework so that conservation agencies and other groups can position their activities within a continental and global context, resulting in more effective allocation of conservation resources.

These objectives are similar to those guiding companion conservation assessments for terrestrial, marine, and freshwater ecoregions in North America, Latin America and the Caribbean, Africa, and Asia (Dinerstein et al. 1995; Olson et al. 1997; Ricketts et al. 1999a; Wikramanayake et al. in press; Olson et al. in prep.; Ford in prep.). Conservation assessments have been conducted for each of these regions because the state of emergency facing so many habitats and species requires priority-setting that is both rapid and based on accurate information. North American freshwater systems are no exception, and it is hoped this assessment will provide important information needed to address their urgent conservation needs.

Assessment Overview

Recognizing the urgent need to conserve freshwater biodiversity, conservation biologists are faced with a conundrum: how can we conserve biodiversity if the biodiversity has yet to be quantified adequately at a landscape scale? One answer is to take a habitat representation approach. Gaining acceptance by a growing number of conservationists, this approach emphasizes the importance of conserving a full representation of diverse habitats and ecosystems (Noss and Cooperider 1994). At the continental scale, representation can be achieved by using ecoregions.

Ecoregions are defined as relatively large areas of land or water that contain a geographically distinct assemblage of natural communities. These communities share a large majority of their species, dynamics, and environmental conditions and function together effectively as a conservation unit. For this assessment,

North America, defined here as Canada, the continental United States, and Mexico, was divided into seventy-six freshwater ecoregions (see figure 2.1). These span the Nearctic and northern section of the Neotropical biogeographic realms. The freshwater ecoregions, in most cases, comprise aggregations of catchments, also known as watersheds or drainage basins. A catchment includes all of the land draining into a particular river (or lake, in the case of closed-basin systems without exterior drainage).

This study uses an ecoregion-based methodology specifically tailored to freshwater systems. Instead of forcing freshwater species into a terrestrial framework, we adopt an ecoregion-based scheme that more accurately reflects species patterns across the aquatic landscape (see box 1). The conservation of freshwater and terrestrial biodiversity should not occur separately, as the two realms are intimately connected, but we must use the best tools available for each.

Box 1

The North American Freshwater Biota and the Justification for Freshwater Ecoregions

Compared with other temperate-zone continents, North America harbors an impressive array of freshwater life. With more than 1200 fish species, it is second only to Asia (with 1,500 described species; Briggs 1986). Worldwide, North America is home to the largest number of margaritiferid and unionid mussel species, with 281 species and 16 subspecies (Williams et al. 1993). The continent also contains 77 percent of all crayfishes, including 99 percent of all species in the family Cambaridae (Hobbs 1988). Within the largely undescribed groups of freshwater invertebrates, there are more than 10,000 species in North America (see essay 14); other studies estimate that there are 5,000 freshwater Coleoptera (beetle) species, 20,000 Chironomids (midges), 4,000 Crustacea, and 1,350 Trichoptera (caddisflies) (McAllister et al. 1997). These numbers for invertebrate diversity, while rough estimates, offer an idea of the magnitude of species that may be at risk from threats to freshwater systems.

Data on species numbers at the continental scale are interesting but not particularly useful to the conservationist, as they give no indication of biogeography or ecological or evolutionary processes. Conservationists searching for viable solutions first need to know how species are aggregated across the landscape so that they can target their efforts most effectively. The inability of the U.S. Endangered Species Act to do more than rescue individual species on the brink of extinction has become obvious, prompting approaches that target landscape-scale processes and species groupings for conservation (Moyle 1994; Angermeier and Schlosser 1995). Angermeier and Schlosser (1995) note that species assemblages, "distinguished directly on the basis of biotic factors and indirectly on the basis of physical factors," would be good units for conservation because they allow consideration of ecological and evolutionary factors in addition to an accounting of species numbers.

Freshwater species assemblages, like the catchments they inhabit, can be defined and described within a range of spatial scales. Many continent-wide assessments of terrestrial biodiversity have adopted the ecoregion as the unit of choice, as ecoregions incorporate broad biogeographic patterns without losing many of the finer details of individual species distributions (Dinerstein et al. 1995; Ricketts et al. 1999a). Several ecoregion-based schemes, based largely on climate, land-surface form, potential natural vegetation, and soils are in wide use (Bailey 1976; Omernik 1987; Dinerstein et al. 1995). Although the character of aquatic systems is determined largely by the terrestrial landscape (Lyons 1989; Bryce and Clarke 1996), none of these ecoregional schemes is sufficient for describing the distribution and distinctiveness of freshwater communities (see Hughes et al. 1994 for review). The natural distribution limits of species confined to freshwater systems generally correspond to lake margins, spring locations, and, in most cases, larger drainage areas. This contrasts with terrestrial species distributions, which are tied more closely to vegetation, landscape features such as soil type, and topography. An ecoregion-based approach to freshwater biodiversity conservation, then, requires a separate set of freshwater ecoregions.

If our goal is to identify ecoregions with distinct and functionally interactive biotas, it makes sense to construct ecoregions around patterns of species assemblages. Maxwell et al. (1995) derived such an ecoregion map using the distributions of native freshwater fish. Their preliminary map, which identified sixty-one "subregions" in Canada, the continental United States, and most of Mexico, provides a strong foundation for our work. Incorporating additional faunal groups and more detailed information for Mexico, we modified Maxwell et al.'s map to better reflect geographic patterns. To our knowledge, this is the first time a conservation assessment of freshwater biodiversity has been conducted using ecoregions derived specifically for freshwater species. It is also the first time that all of Mexico has been included in a North American assessment.

Delineation of appropriate ecoregions is only the first step in a conservation assessment. The final step is to prioritize ecoregions for conservation action. To identify those ecoregions that should be targeted, we must decide what we want to save or restore. Our conservation analyses are aimed at achieving the fundamental goals of biodiversity conservation: (1) representation, (2) sustaining viable populations, (3) maintaining ecological processes, and (4) responsiveness to short- and long-term change (Noss 1991).

The goal of biological conservation is to retain biodiversity, but the term biodiversity is variously defined (Angermeier and Schlosser 1995). Most often, it is used as a synonym for species richness. However, species richness takes into account neither the distinctiveness of species, communities, and habitat types, diversity at higher taxonomic scales (e.g., genus, family), nor the rarity of ecological or evolutionary processes. In this study we adopt an index termed *biological distinctiveness* that incorporates all of these factors, under the assumption that within ecosystems, structure and process are as valuable to protect as strict composition. An evaluation of biological distinctiveness by ecoregions reveals strong differences among them and allows us to rank their biodiversity value on global and continental scales.

Were all ecoregions in a pristine state, we might decide to focus our conservation energies on those with the greatest biological distinctiveness. Unfortunately, nearly all ecoregions are altered to some degree and face additional impending threats. However, habitat degradation and species imperilment are not uniform across the continent. These factors are related to landscape development, water use, and vulnerability of particular ecosystems and communities to disturbance. Some freshwater ecosystems may be so irreversibly transformed from their natural states as to be beyond repair. Others may still retain enough of their original structure, function, and composition to be restored without enormous expenditure. Evaluation of where ecoregions lie on this continuum constitutes an assessment of their *conservation status*.

In order to set priorities for conservation, we must consider both an ecoregion's biological distinctiveness and its conservation status. While it may be obvious that ecoregions with the highest distinctiveness should be targeted first, it is not equally obvious that the most critically imperiled ecoregions should also receive the highest priority. Following the recommendations of experts who assisted with this assessment, we have constructed a matrix that assigns priorities to ecoregions based on the combination of biological distinctiveness and conservation status. We recognize that other conservation groups may choose to set priorities differently, and the matrix easily allows for the reassignment of priorities.

Some of the information used in our assessment is available from published sources (e.g., species distribution data, which we used to produce species richness and endemism indices for each ecoregion). However, for the most part, information on critical variables at the scale of ecoregions is unavailable in published form. For example, quantitative data on habitat loss over entire ecoregions, available in standardized form for the whole continent, simply do not exist. To gather this kind of important information efficiently, we relied heavily on expert assessment. We convened an expert workshop at which ecologists and conservationists collectively assessed the ecoregions for a variety of biological distinctiveness and conservation status criteria. We asked the experts first to critique our ecoregion boundaries and to adjust them if necessary. Regional subgroups then assessed the ecoregions and, based on their collective experience and knowledge, placed them in broad categories for each criterion. The entire process was guided by explicit decision rules and category thresholds to standardize the treatment of all ecoregions. In this manner, we were able to gather an enormous amount of subjective, but quantified, information that is unavailable from published sources. Perhaps most importantly, the decision rules and methods for calculating all criteria are contained in appendices of this book so that the process is as transparent and repeatable as possible.

Studies of this geographic extent are necessarily coarse. Ecoregion boundaries are approximations of what in reality are gradual shifts in ecological communities. The information we have gathered, from published data and the expert workshop, are rough categorical rankings. Nevertheless, this level of data precision is appropriate for the type of analysis we conduct: a broad-scale assessment of biodiversity and the threats facing it. This book is intended to help conservation planners take the first, broad-scale step toward strategically setting priorities in North America; it is not an end in itself. Appropriate conservation policies and activities at finer scales (i.e., within ecoregions or small groups of them) must follow if the value of this study is to be realized.

During the course of the expert workshop, we gathered a wealth of information that should help those interested in taking the more fine-scale approach of within-ecoregion conservation. For each ecoregion, we asked experts to identify important threats, priority actions that would best advance biodiversity conservation, and major conservation organizations operating in the region. We have compiled these details in the ecoregion descriptions in appendix G. Therefore, this book combines a broad-scale framework and inter-ecoregional assessment with more detailed information on each ecoregion to help direct more local initiatives. We hope it will lead to better choices and implementation of local-scale conservation activities by placing them in the context of continental and global priorities.

Structure of the Book

The material that follows is divided into three main sections. In the first we briefly outline the approach used to conduct the assessment. More detailed descriptions of the specific methods and the scoring system for ecoregions are found in appendices (biological distinctiveness index, appendix A; conservation status index, appendix B). In the second section, we present the results of the biological distinctiveness and conservation status assessments and the integration of these indices into a priority-setting matrix. Raw scores for biological distinctiveness criteria (appendix C), conservation status (appendix D), and statistical analyses of these results (appendix E) are provided. The integration of the two indices, conducted separately for each major habitat type, is also illustrated (appendix F).

In the third section, we discuss the major themes that emerged from the analysis. Interspersed throughout the results and discussion are short essays providing supplementary information on topics in North American freshwater conservation. In the last chapters we provide a set of recommendations that, if followed, will go far in enhancing biodiversity conservation in North America. We also place the biodiversity values of North American ecoregions within the context of the recently completed Global 200 Ecoregions analysis (Olson and Dinerstein 1998) to emphasize the global responsibility that North Americans must accept to be leaders in international conservation efforts.

Throughout the text, ecoregion names are capitalized and followed by their unique identification number in brackets (e.g., the Tennessee-Cumberland [35]). We include a glossary of important conservation terms to aid readers unfamiliar with conservation or freshwater terminology. Finally, we provide descriptions of all ecoregions (appendix G); descriptions include each ecoregion's geographic extent, defining biological features, conservation status, analysis of threats, recommended suite of conservation activities, and contacts for conservation partners active in the ecoregion. Contact information for the conservation partners is listed in appendix H.

The Challenge

As citizens of North America, we have a global responsibility to conserve the biodiversity within our own freshwater ecosystems. Unfortunately, this assessment shows that we are not meeting our responsibility. Some freshwater ecoregions may already be irreversibly degraded, but many others offer important opportunities for preserving our natural heritage. This study identifies twelve ecoregions with globally outstanding biodiversity features that could still be protected if we act immediately. The longer we wait, though, the less will remain to be saved.

Ecoregion-based conservation has now been adopted by some of the leading conservation groups in North America. These groups include WWF-US, The Nature Conservancy (TNC), Defenders of Wildlife, the WWF Canada Endangered Spaces Campaign, State GAP analysis programs, and Comisión Nacional Para el Conocimiento y Uso de la Biodiversidad (CONABIO) in Mexico. The combination of broad- and fine-scale analyses provides a valuable framework and strategic planning tool to help ensure the conservation of North America's diverse natural heritage.

This study adds an important dimension to these analyses, as it considers the ways in which freshwater biodiversity conservation complements and diverges from terrestrial-based efforts. Areas of highly distinct freshwater biodiversity are often not distinguished using terrestrial ecoregions, yet in other cases globally outstanding freshwater and terrestrial ecoregions overlap. Furthermore, the threats facing both freshwater and terrestrial systems often stem from the same activities. Well-planned conservation activities may benefit biota inhabiting both realms.

We hope that the results of this study contribute to the successful implementation of ecoregion-based conservation across North America. In particular, we hope that this assessment will bring attention to both the outstanding diversity and extreme imperilment of our continent's freshwater biodiversity. It is not hyperbole to say that time is running out. Freshwater habitats have a tremendous capacity for physical regeneration, but the most special biological elements of these habitats may disappear forever.

CHAPTER 2

Approach

Delineation of Ecoregions and Geographic Scope of the Study

An ecoregion is defined as a relatively large area of land or water that contains a geographically distinct assemblage of natural communities. These communities (1) share a large majority of their species, dynamics, and environmental conditions, and (2) function together effectively as a conservation unit (Dinerstein et al. 1995). This study focused exclusively on freshwater ecoregions, defined as lotic systems (rivers, streams, creeks, and springs) and lentic systems (lakes and ponds). Wetlands are treated in separate terrestrial assessments for the United States and Canada (Ricketts et al. 1999a), and estuaries are included in a forthcoming marine assessment (Ford in prep.).

The seventy-six ecoregions that form the conservation units for this study (figure 2.1) are based largely on a USDA Forest Service mapping project (Maxwell et al. 1995), which used native fish distributions to delineate subregions (renamed ecoregions for our study). Maxwell et al.'s subregion boundaries were modified by biodiversity experts to better represent the distributions of a greater range of freshwater species. The result is a map that reflects patterns of biodiversity across the United States, Canada, and Mexico that can be used for effective large-scale conservation planning, reporting, and monitoring. We have retained Maxwell et al.'s regions and subzones, representing increasingly broader scale delineations of biotic affinity, and renamed them complexes (e.g., Colorado, Mississippi complexes) and bioregions (e.g., Pacific, Arctic-Atlantic bioregions) (see figure 2.1 legend and figure 2.2).

Because fish, mussels, and most crayfish are confined to aquatic environments and have been subject to many of the same forces over evolutionary time, their distributional patterns tend to be more similar than different. More vagile taxa, such as many amphibians, reptiles, and aquatic insects, exhibit distributions that correspond less closely to catchments. Thus, we used the distributions of fish, crayfish, and mussels to guide our delineation of ecoregions. These taxa, as obligate freshwater inhabitants over their entire life cycles, may serve as the best targets for the conservation of entire freshwater systems. Knowledge of additional aquatic invertebrates is far from complete but could be used to revisit ecoregion boundaries when such data become available.

Figure 2.1   Freshwater ecoregions of North America (with legend, next page).

Pacific Bioregion

Coastal Complex

1. North Pacific Coastal
2. Columbia Glaciated
3. Columbia Unglaciated
4. Upper Snake
5. Pacific Mid-Coastal
6. Pacific Central Valley
7. South Pacific Coastal

Great Basin Complex

8. Bonneville
9. Lahontan
10. Oregon Lakes
11. Death Valley

Colorado Complex

12. Colorado
13. Vegas-Virgin
14. Gila

Arctic-Atlantic Bioregion

Rio Grande Complex

15. Upper Rio Grande (Rio Bravo del Norte)
16. Guzmán
17. Rio Conchos
18. Pecos
19. Mapimí
20. Lower Rio Grande (Rio Bravo del Norte)
21. Rio Salado
22. Cuatro Ciénegas
23. Rio San Juan

Mississippi Complex

24. Mississippi
25. Mississippi Embayment
26. Upper Missouri
27. Middle Missouri
28. Central Prairie
29. Ozark Highlands
30. Ouachita Highlands
31. Southern Plains
32. East Texas Gulf
33. West Texas Gulf
34. Teays-Old Ohio
35. Tennessee-Cumberland
36. Mobile Bay
37. Apalachicola
38. Florida Gulf

Atlantic Complex

39. Florida
40. South Atlantic
41. Chesapeake Bay
42. North Atlantic

St. Lawrence Complex

43. Superior
44. Michigan-Huron
45. Erie
46. Ontario
47. Lower St. Lawrence
48. North Atlantic-Ungava

Hudson Bay Complex

49. Canadian Rockies
50. Upper Saskatchewan
51. Lower Saskatchewan
52. English-Winnipeg Lakes
53. South Hudson
54. East Hudson

Arctic Complex

55. Yukon
56. Lower Mackenzie
57. Upper Mackenzie
58. North Arctic
59. East Arctic
60. Arctic Islands

Mexican Transition Bioregion

61. Sonoran
62. Sinaloan Coastal
63. Santiago
64. Manantlan-Ameca
65. Chapala
66. Llanos el Salado
67. Rio Verde Headwaters
68. Tamaulipas-Veracruz
69. Lerma
70. Balsas
71. Papaloapan
72. Catemaco
73. Coatzacoalcos
74. Tehuantepec
75. Grijalva-Usumacinta
76. Yucatán

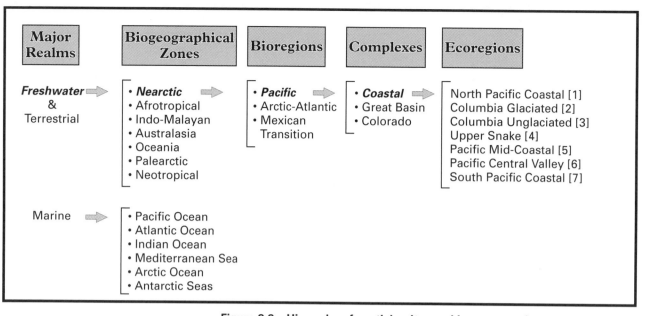

Major Realms	Biogeographical Zones	Bioregions	Complexes	Ecoregions

Figure 2.2   Hierarchy of spatial units used in conservation assessment framework (see glossary for definitions of terms). Terrestrial and freshwater conservation assessments share biogeographical zones but employ different bioregions, complexes, and ecoregions. See Ricketts et al. (1999a) for North American terrestrial ecoregions.

Assignment of Major Habitat Types

Ecological processes, general patterns of biodiversity, and responses to disturbance vary widely in their scale and importance among different habitat types. To address this variation, we grouped ecoregions into eight major habitat types (MHTs), following a global framework applied in other regional analyses (Dinerstein et al. 1995; Olson et al. 1997). These MHTs include arctic rivers and lakes; large temperate lakes; temperate headwaters and lakes; large temperate rivers; endorheic (closed-basin) rivers, lakes, and springs; xeric-region rivers, lakes, and springs; temperate coastal rivers and lakes; and subtropical coastal rivers and lakes (table 2.1 and figure 2.3). MHTs are not geographically defined units; rather they refer to the dynamics of ecological systems and the broad habitat structures and patterns of species diversity that define them.

The conservation biology community concurs that representation of all of the earth's habitat types is critical for conserving the planet's biodiversity. By organizing ecoregions within MHTs and then evaluating each class separately, we can facilitate representation in regional conservation strategies (Noss and Cooperrider 1994; Olson and Dinerstein 1998).

Criteria for analysis of biological distinctiveness and conservation status were tailored to the distinct characteristics of different MHTs. The biological distinctiveness index (BDI), based largely on species richness and endemism, would not allow a fair evaluation of ecoregions if they were grouped together (see appendix A). Biogeographic studies have established that species richness for most taxa increases with decreasing latitude; this has been shown for aquatic vertebrate groups, although it may not be true for aquatic invertebrates (Gee 1991; Allan and Flecker 1993; Ricketts et al. 1999a). For this reason, a direct comparison of species rich-

TABLE 2.1   Assignment of Ecoregions to Major Habitat Types.

Arctic Rivers and Lakes
South Hudson [53]
East Hudson [54]
Yukon [55]
Lower Mackenzie [56]
Upper Mackenzie [57]
North Arctic [58]
East Arctic [59]
Arctic Islands [60]

Large Temperate Lakes
Superior [43]
Michigan-Huron [44]
Erie [45]
Ontario [46]

Temperate Headwaters and Lakes
Central Prairie [28]
Ozark Highlands [29]
Ouachita Highlands [30]
Southern Plains [31]
Teays-Old Ohio [34]
Tennessee-Cumberland [35]
Canadian Rockies [49]
Upper Saskatchewan [50]
Lower Saskatchewan [51]
English-Winnipeg Lakes [52]

Large Temperate Rivers
Colorado [12]
Upper Rio Grande/Bravo [15]
Lower Rio Grande/Bravo [20]
Mississippi [24]

Mississippi Embayment [25]
Upper Missouri [26]
Middle Missouri [27]

Endorheic Rivers, Lakes, and Springs
Bonneville [8]
Lahontan [9]
Oregon Lakes [10]
Death Valley [11]
Guzmán [16]
Mapimí [19]
Llanos el Salado [66]
Lerma [69]

Xeric-Region Rivers, Lakes, and Springs
South Pacific Coastal [7]
Vegas-Virgin [13]
Gila [14]
Rio Conchos [17]
Pecos [18]
Rio Salado [21]
Cuatro Ciénegas [22]
Rio San Juan [23]
Sonoran [61]
Chapala [65]
Rio Verde Headwaters [67]

Temperate Coastal Rivers and Lakes
North Pacific Coastal [1]
Columbia Glaciated [2]

Columbia Unglaciated [3]
Upper Snake [4]
Pacific Mid-Coastal [5]
Pacific Central Valley [6]
East Texas Gulf [32]
Mobile Bay [36]
Apalachicola [37]
Florida Gulf [38]
South Atlantic [40]
Chesapeake Bay [41]
North Atlantic [42]
Lower St. Lawrence [47]
North Atlantic-Ungava [48]

Subtropical Coastal Rivers and Lakes
West Texas Gulf [33]
Florida [39]
Sinaloan Coastal [62]
Santiago [63]
Manantlan-Ameca [64]
Tamaulipas-Veracruz [68]
Balsas [70]
Papaloapan [71]
Catemaco [72]
Coatzacoalcos [73]
Tehuantepec [74]
Grijalva-Usumacinta [75]
Yucatán [76]

ness in arctic versus subtropical habitats would not yield much useful information. Furthermore, the relatively recent glaciation (10,000–15,000 B.P.) of the northern portion of North America has led to fewer opportunities for speciation there, resulting in low numbers of total species and endemics (Briggs 1986). Finally, there is a relationship between fish species richness and the areas of both river basins and lakes; in other words, one would expect to find more fish species in large lakes or rivers than in smaller systems (Allan and Flecker 1993; McAllister et al. 1997). (Ecoregion size, however, was not statistically correlated with the number of species, as shown in appendix E.) To judge the biological distinctiveness of ecoregions fairly, categories of distinctiveness were adopted for each MHT.

Conservation status indicators were also given different weightings according to their perceived impact on species and communities within each MHT. This was necessary because disturbances do not have uniform effects on all habitats. For example, altered hydrologic integrity would likely have a more profound effect within a xeric region than a wet coastal zone. Water quality degradation might affect a lake more severely than a large river.

Figure 2.3 Freshwater major
habitat types of North America.

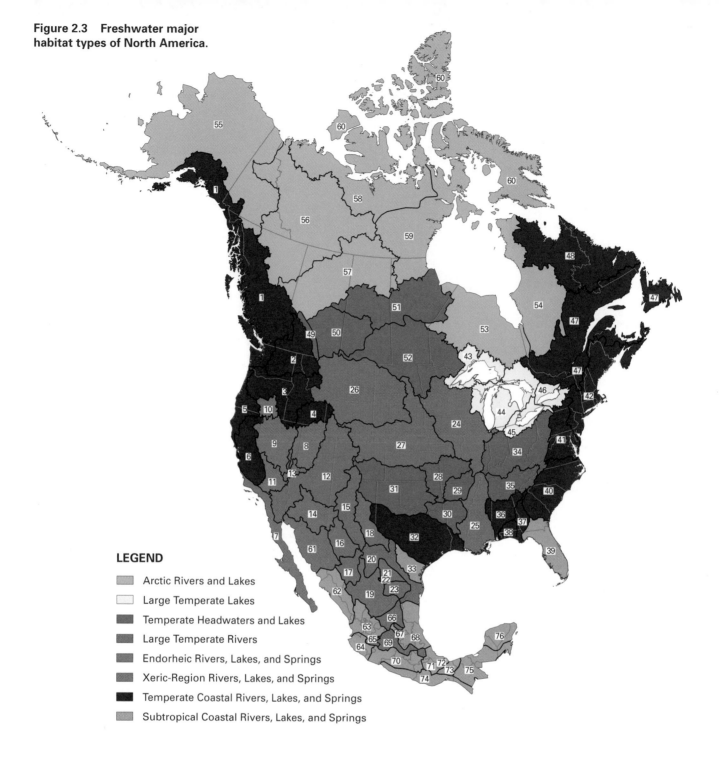

LEGEND

- Arctic Rivers and Lakes
- Large Temperate Lakes
- Temperate Headwaters and Lakes
- Large Temperate Rivers
- Endorheic Rivers, Lakes, and Springs
- Xeric-Region Rivers, Lakes, and Springs
- Temperate Coastal Rivers, Lakes, and Springs
- Subtropical Coastal Rivers, Lakes, and Springs

Discriminators

The biological distinctiveness and conservation status discriminators were first developed by WWF-US for a conservation assessment of terrestrial ecoregions in Latin America and the Caribbean
(Dinerstein et al. 1995). For this study, the conservation status
index (CSI) has been tailored to freshwater systems. Both indices
are evaluated using several criteria (table 2.2; detailed descriptions
are found in appendixes A and B, as are data sources used to evaluate the component criteria.)

TABLE 2.2 Biological Distinctiveness and Conservation Status Criteria.

Biological Distinctiveness	Conservation Status
1. Species richness	1. Degree of land cover (catchment) alteration
2. Species endemism	2. Water quality degradation
3. Rare ecological or evolutionary phenomena	3. Alteration of hydrographic integrity
4. Rare habitat type	4. Degree of habitat fragmentation
	5. Additional losses of original intact habitat
	6. Effects of introduced species
	7. Direct species exploitation

Biological Distinctiveness Index (BDI)—Overview

For this study, we followed Dinerstein et al. (1995) in interpreting the biological importance of an ecoregion as the degree to which its biodiversity is distinctive at different biogeographic scales. We used this scale-dependent assessment to assign each ecoregion to one of four categories: globally outstanding, continentally (i.e., within a biogeographic province) outstanding, bioregionally outstanding, and nationally important.

Biodiversity assessments typically focus at the level of the species. Our use of the term biological distinctiveness invokes the strict definition of biodiversity—besides species, we incorporated ecosystem diversity and ecological processes that sustain biodiversity. Specifically, biological distinctiveness is based on broad measures of species richness, endemism, uniqueness of higher taxa, unusual ecological and evolutionary phenomena, and the global rarity of MHTs.

Criteria and Evaluation

Species richness and endemism were assessed by systematically comparing our ecoregion map with the estimated ranges of more than 2,200 North American species representing five taxonomic groups: fishes, crayfishes, unionid mussels, amphibians that depend on aquatic habitats, and aquatic and semiaquatic reptiles. These groups were chosen because they are important elements of freshwater biotas and the best distribution data are available for them. Sufficient data for mussels in Mexico are lacking, but mussels were nonetheless included in the analysis because they are perhaps the most threatened of all freshwater taxa. We feel that, in aggregate, these five groups can be used with caution as an effective proxy for more numerous and less well-known groups such as aquatic insects, and thus as indicators of overall biodiversity patterns.

Besides species distributions, we included two ecoregion-scale criteria in the BDI: rare ecological or evolutionary phenomena and the global rarity of habitat types. Rare evolutionary phenomena include globally outstanding centers of evolutionary radiation, higher-level taxonomic diversity, and unique species assemblages. An example of rare ecological phenomena is the large-scale migration of fish in large rivers or from marine to freshwater habitats.

Rarity is of two types: original and human-induced. Because this criterion attempts to measure the remaining opportunities for conservation of a certain phenomenon, both kinds of rarity were taken into account. Thus an ecological process that was once widespread but has been disrupted in the majority of its original locations was treated, in its remaining locations, as the equivalent of a process that was originally rare. For example, spawning runs of salmon were once common along the Pacific Coast but are much reduced today. This criterion was designed to capture ecoregion characteristics that are difficult to quantify consistently using other measures. It was scored by expert assessment as being globally outstanding, continentally outstanding, or not appropriate.

We gave additional recognition to ecoregions that contain species assemblages with high levels of beta-diversity. Beta-diversity is a measure of turnover, or replacement of species, with distance or along environmental gradients. In a sense, beta-diversity indicates the biological complexity of an ecoregion and is correlated with pronounced local endemism. It also provides an important indicator of the kind and level of effort needed for conservation in an ecoregion. Typically, ecoregions with very high levels of beta-diversity will require multiple protected areas distributed across the landscape to conserve scattered, highly distinct assemblages of freshwater species. Ecoregions in some MHTs, such as arctic rivers and lakes, show little turnover in species, while ecoregions in xeric-region ecosystems often exhibit high beta-diversity. Cuatro Ciénegas [22], an endorheic ecosystem composed of hundreds of small springs, lakes, and streams and harboring numerous very localized endemic species, provides a prime example of high beta-diversity.

The global rarity of habitat types was also scored by expert assessment using the same three categories as above. An ecoregion was considered globally outstanding if fewer than eight ecoregions worldwide contain the particular habitat type, and continentally outstanding if fewer than three occur in North America. Otherwise it was scored as not applicable. This measure represents the number of opportunities to conserve this habitat type worldwide, and the corresponding importance of a North American ecoregion that contains it.

Combining BDI Criteria

Within each MHT, species richness and endemism were ranked as being either high, medium, or low (see appendix A for a full discussion of methods). Endemism was weighted twice as high, to account for its proportionately greater contribution to biological distinctiveness. These two criteria were combined to place each ecoregion in one of four biological distinctiveness categories: globally outstanding, continentally outstanding, bioregionally outstanding, or nationally important. Ecoregions were also raised to either the globally outstanding or continentally outstanding category if they were judged to have rare ecological or evolutionary phenomena or rare habitat types. At a minimum, then, every ecoregion is nationally important. All ecoregions provide essential ecosystem services, natural resources, or natural recreation opportunities for local human communities, and many nationally important ecoregions contain habitat for rare or imperiled species. However, certain ecoregions within each MHT are so extraordinarily rich or unusual in their biodiversity that conservation efforts within them take on global or continental importance.

Conservation Status—Overview

The second major discriminator, conservation status, was designed to estimate an ecoregion's current and future ability to meet three fundamental goals of biodiversity conservation: maintaining viable species populations and communities, sustaining ecological processes, and responding effectively to short- and long-term environmental change. The snapshot conservation status was based on seven criteria: degree of land-cover (catchment) alteration; water quality degradation; alteration of hydrographic integrity; degree of habitat fragmentation; additional losses of intact original habitat; effects of introduced species; and direct species exploitation. Each ecoregion was ranked on a scale from 0 to 4 (0 for the least degraded, 4 for the most). The scores were weighted according to the ecoregion's MHT (figure 2.4). Scores were then tallied for each ecoregion and divided by the total possible score for that ecoregion to obtain a percentage score. This score was used to assign each ecoregion to one of five categories: critical, endangered, vulnerable, relatively stable, and relatively intact (appendix B).

Criteria

The threats to freshwater systems are well documented, and, in general, multiple threats exist for any given ecosystem (Allan and Flecker 1993). Developing criteria for a conservation assessment is complicated by the fact that habitat quality within a freshwater ecosystem is largely determined by processes occurring elsewhere. Specifically, terrestrial land use within a catchment will affect downstream freshwater systems to one degree or another. The spatial scale over which impacts occur varies according to the activity or disturbance. Loss of riparian vegetation abutting a stream will have direct effects on the temperature, organic matter input, and nutrient and sediment load of the stream habitat, whereas the effects of urbanization and associated development many kilometers away can be equally damaging (e.g., changes in runoff and the pollutants carried by it). It is difficult to distinguish between the effects of different land uses on freshwater habitats, but it is perhaps harder to trace observed changes in species or communities to specific source impacts.

Our choice of criteria was guided by prior studies of freshwater biodiversity conservation. Miller et al. (1989) evaluated North American fish species that have gone extinct over the past century and identified five main causative factors: physical habitat alteration, introduced species, chemical alteration or pollution, hybridization, and overharvesting. Of the twenty-seven species and thirteen subspecies lost, only seven had a single factor as the

Major Habitat Type (MHT)	Conservation Status Indicator						
	Degree of Land Cover (Catchment) Alteration	Water Quality Degradation	Alteration of Hydrographic Integrity	Degree of Habitat Fragmentation	Additional Losses of Original Intact Habitat	Effects of Introduced Species	Direct Species Exploitation
Arctic Rivers and Lakes							x 2
Large Temperate Lakes		x 2				x 2	
Temperate Headwaters and Lakes	x 2		x 2	x 2		x 2	
Large Temperate Rivers	x 2		x 2	x 2			
Endorheic Rivers, Lakes, and Springs	x 2	x 2	x 3	x 2	x 2	x 2	x 2
Xeric-Region Rivers, Lakes, and Springs	x 2	x 3	x 3	x 2	x 2	x 2	x 2
Temperate Coastal Rivers and Lakes	x 2		x 2	x 2			x 2
Subtropical Coastal Rivers and Lakes	x 2		x 2	x 2			x 2

Figure 2.4 Weightings for conservation status indicators, by major habitat type. Certain threats or potential threats have varying impacts.

probable cause of extinction, and the most common contributing cause (73 percent of cases) was habitat loss. Hybridization, the only category not covered in our conservation assessment, was implicated in 38 percent of the cases but was not the single cause in any case.

The Committee on Restoration of Aquatic Ecosystems (1992) lists different classes of impacts for lake and river biota. For lakes, the major classes are excessive inputs of nutrients and organic matter; hydrologic and physical changes; siltation from inadequate erosion control in agricultural and mining activities; introduction of exotic species; acidification from atmospheric sources and acid mine drainage; and contamination by toxic or potentially toxic metals and organic compounds. Thus in lakes, various forms of water quality degradation are so important as to be classified separately, whereas hydrologic and physical changes are grouped together. In rivers, on the other hand, the five major classes of factors listed by the authors are changes in energy source (e.g., decreased coarse particulate organic matter, increased algal production), water quality, habitat quality, flow regime, and biotic interactions. In this case, change in energy source is a product of altered land cover within the catchment.

Allan and Flecker (1993) suggest that six factors are critical to biodiversity conservation in lotic habitats: habitat loss and degradation, the spread of exotic species, overexploitation, secondary extinctions (when the removal of one species has cascading effects throughout the species assemblage of the area), chemical and organic pollution, and climate change. Secondary extinctions are difficult to quantify and tend to result in local population losses rather than global extinctions. Climate change, though perhaps already having impacts on habitats and communities, is considered a threat over the long term more than a current threat. The remaining four factors are included in our conservation assessment criteria.

In their conservation status assessment of South African rivers, O'Keeffe et al. (1987) used forty-one indicators, twenty-six of which were threats (as opposed to positive attributes, such as percentage of river unregulated). Each of the forty-one indicators, categorized as being related to "the river," "the catchment," or "the biota," was highly specific; in many cases, several of the indicators were similar enough to be grouped together under a more general category, such as "water quality degradation."

When considered together, these attempts to categorize the major threats to freshwater systems tend to identify the same factors but group them somewhat differently. We retained most of the major threat categories and modified them to ensure that the variety of threats to different freshwater habitats would be covered. Our criteria overlap with each other in some cases (the effects of catchment land-cover alteration will be manifested in a number of other categories), but we chose to err on the side of overcounting threats. This decision was based on the assumption that each expert scoring the criteria would tend not to count an impact more than once. To ensure comprehensiveness, we also added the fifth criterion, additional losses of intact original habitat, to capture any impacts that were not otherwise included. Following the categorization of O'Keeffe et al. (1987), one of our criteria is related to the catchment, four are related to the river, lake, or spring (or, generally, habitat), and two are related to the biota.

Unlike some other conservation assessments, ours did not include species imperilment as a criterion, because imperilment may be the consequence of the various threats included as other indicators. However, we did use information regarding species imperilment to help validate the assessment results (Natural Heritage Central Databases 1997; see results and appendix E for complete details).

Although we evaluated conservation status at the ecoregion scale, the processes that affect species occur at a variety of spatial and temporal scales. With the exception of the first criterion, catchment land-cover alteration, scales are not specified for evaluation of the remaining indicators. We assumed that experts would score the criteria by considering the appropriate scale for the given habitat type of each ecoregion.

Detailed descriptions of how each criterion was scored are presented in appendix B.

Assessment of Criteria

Several national and continental data sets and maps exist that are potentially useful for quantitatively assessing some of the seven criteria (U.S. EPA 1997; Natural Heritage Central Databases 1997; Federal Emergency Management Agency 1996). However, we found none to be of sufficient spatial resolution or consistency to be relied upon completely for the analysis. To gather high-quality information in a timely fashion, we convened a workshop of biologists and conservationists with regional expertise to assess each ecoregion in terms of the seven landscape-level features (see list of contributors). Workshop participants were provided with standardized decision rules and data sheets with which to score the ecoregions (see appendix B for detailed description).

Snapshot Conservation Status

The seven criteria were weighted and combined into a single index, from which five categories of conservation status were derived: critical, endangered, vulnerable, relatively stable, and relatively intact. These categories, following those of Dinerstein et al. (1995), were developed in the tradition of the IUCN Red Data Book series (IUCN 1988; Collar et al. 1992). The IUCN Red Data Books list species under various levels of threat, to call attention to species and populations on trajectories toward extinction. WWF has adopted similar categorical schemes to describe ecoregions and to assess their conservation status (Dinerstein et al. 1995). The rationale for using these criteria is simple: almost 90 percent of all species found in the Red Data Books are listed as endangered because of loss of habitat (Wilcove et al. 1996). Because many more species that share those same ecoregions are either undescribed or unlikely ever to be officially listed, we applied Red Data Book criteria directly to ecoregions to determine where overall species loss or declines are most likely to occur.

Methods, weighting, and thresholds used in deriving the conservation status index categories are detailed in appendix B. Below we provide a generalized and qualitative description of each category in terms of the ecoregion's habitat integrity and predicted ecological impacts. Not all factors listed below need to occur to warrant a given status. The descriptions reflect how, with increasing habitat loss, degradation, and fragmentation, ecological processes cease to function naturally, populations no longer occur within the natural range of variation, and major components of biodiversity are eroded (adapted from Dinerstein et al. 1995).

- Critical: The remaining intact habitat is restricted to isolated areas or stream segments that have low probabilities of persistence over the next 5–10 years without immediate or continuing protection and restoration. Remaining habitat does not meet the minimum requirements for maintaining viable populations of many species and ecological processes. Surrounding land-use practices are incompatible with maintaining aquatic habitat structure and function. Hydrographic integrity has been severely modified by permanent, large-scale structures. Established exotic species seriously threaten native species populations. Consistently poor water quality excludes all but the hardiest species from large portions of remaining habitat. Many species are already extirpated or extinct.

- Endangered: The remaining intact habitat is restricted to isolated areas or segments of varying size or length (a few larger areas or reaches may be present) that have medium to low probabilities of persistence over the next 10–15 years without immediate or continuing protection or restoration. Remaining habitat does not meet the minimum requirements for many species populations and large-scale ecological processes. Surrounding land-use practices are largely incompatible with maintaining aquatic habitat structure and function. Hydrographic integrity has been modified by structures of varying size and permanence. Spread of exotic species poses a potentially serious threat to native species populations. Poor water quality excludes many species from remaining habitat. Some species are already extirpated or extinct.

- Vulnerable: The remaining intact habitat occurs in blocks or segments ranging from large to small; many intact areas will likely persist over the next 10–20 years, especially if given adequate protection and moderate restoration. Some remaining habitat meets minimum requirements for species populations and large-scale ecological processes. Surrounding land-use practices are sometimes compatible with maintaining aquatic habitat structure and function. Hydrographic integrity may be restored in some areas by implementing moderate changes. Established exotic species may be controllable. Some species may already be extirpated or extinct.

- Relatively Stable: Natural communities have been altered in certain areas, causing local declines in some populations and disruption of ecosystem processes. These disturbed areas can be extensive but are still patchily distributed relative to the area of intact habitats. Ecological linkages among intact habitats are still largely functional. Sensitive species are still present but at diminished densities. Hydrographic integrity, if altered, could be restored by implementing minor changes. Surrounding land-use practices do not impair aquatic habitats, or could be easily modified to minimize impacts. Exotic species pose little or no threat to natives. A nearly full complement of native species still exists.

- Relatively Intact: Native communities within an ecoregion are largely intact with species, populations, and ecosystem processes occurring within their natural ranges of variation. Populations of sensitive species are not diminished. Biota move and disperse naturally within the ecoregion. Ecological processes fluctuate naturally throughout largely contiguous natural habitats. Hydrographic integrity is unmodified, and surrounding land use does not impair aquatic habitat. Maintenance of current conditions will conserve native species over both the short and long term.

Final (Threat-Modified) Conservation Status

The snapshot conservation status of each ecoregion was modified according to the degree of expected future threat, as assessed at the expert workshop. Using this measure we looked beyond the ecological threats implicit in habitat loss, fragmentation, and other existing forces to evaluate the future trajectories of these phenomena. Experts estimated the cumulative impact of all threats on species persistence over the next 10–20 years to categorize ecoregions by level of threat: high, medium, or low. An ecoregion with high threat was promoted to the next-highest conservation status category to arrive at its modified conservation status. (For example, an endangered ecoregion with high threat was promoted to critical.) The conservation status for an ecoregion with moderate or low threat remained unchanged (see appendix B).

Integrating Biological Distinctiveness and Conservation Status

The biological distinctiveness and conservation status indices are two essential discriminators for biodiversity conservation planning at large scales. They combine an evaluation of the relative biological importance of ecoregions with a measure of anthropogenic impacts, both current and projected, facing each ecoregion. Considered together, the two indices provide a powerful tool for indicating appropriate conservation activities within ecoregions and for

Figure 2.5   Integration matrix for priority setting.

Biological Distinctiveness	Final Conservation Status				
	Critical	Endangered	Vulnerable	Relatively Stable	Relatively Intact
Globally Outstanding	II	I	I	II	II
Continentally Outstanding	III	II	II	III	IV
Bioregionally Outstanding	IV	III	III	IV	V
Nationally Important	V	IV	IV	V	V

setting regional priorities when limited resources require careful and strategic planning.

To integrate the two discriminators, we adapted a matrix developed by Dinerstein et al. (1995). The biological distinctiveness categories lie along the vertical axis, with the conservation status categories along the horizontal axis (figure 2.5). Ecoregions fall into one of the twenty cells of the matrix based on their categories for both indices. The entire assessment process, including this integration step, is carried out independently for each MHT to ensure its representation in the final analysis. Following suggestions of freshwater experts, the twenty cells were organized into five classes, which reflect the nature and extent of the management activities likely to be required for effective biodiversity conservation.

Experts agreed that the most critically imperiled freshwater ecoregions are threatened by effects that will be virtually irreversible without both enormous expenditure and a complete shift in land- and water-use practices. Consequently, they recommended that critical ecoregions *not* receive the highest priority for conservation. They agreed that the limited funds available for freshwater conservation should be targeted first at sites within ecoregions where conservation gains could realistically be achieved. Endangered and vulnerable ecoregions ranked globally outstanding were therefore given the highest priority. The experts, however, were not willing to abandon all conservation efforts within globally outstanding ecoregions that are critically imperiled; these ecoregions received secondary priority, along with relatively stable and relatively intact globally outstanding ecoregions.

This integration matrix follows a "triage" logic at each level of biological distinctiveness, with endangered and vulnerable ecoregions receiving the same priority as both more and less imperiled ecoregions of greater distinctiveness. This priority-setting should not discount the fact that biodiversity conservation is important in every ecoregion, as naturally functioning ecosystems provide many services to animal, plant, and human communities. Conservation of natural areas in all ecoregions ensures preservation of distinct species and communities as well as the genetic and functional diversity of populations across species ranges. Flood control, groundwater recharge, freshwater purification, and innumerable recreational opportunities are examples of local ecosystem services that must be protected within each ecoregion.

However, some ecoregions around the world warrant more immediate attention from conservationists because they are of global importance biologically and are at imminent risk from anthropogenic threats (Olson and Dinerstein 1998). With limited resources and time available for conservation, it is important to be strategic in the allocation and timing of conservation effort and funds. The framework we have devised is designed to assist in this process and to highlight the extraordinary nature of North American biodiversity. We recognize that others might choose a different priority-setting scheme, placing greater emphasis on cells other than those we have highlighted. We include in this report all data generated in the assessment so that others might set their own priorities.

Identification of Important Sites for Conservation

This study sets priorities among ecoregions, but actual conservation occurs at the scale of the individual site. Recognizing this, experts listed priority sites or activities for conservation within each ecoregion, and provided contact information for important people or groups working toward biodiversity conservation. The sites are mapped in chapter 7, and conservation partners for each ecoregion are listed in appendix G.

CHAPTER 3

Biological Distinctiveness of North American Ecoregions

Species Richness

The highest number of species occurs in the Southeast and mid-Atlantic (figure 3.1). This geographic pattern is largely the result of the distribution of MHTs across North America. When species richness data are grouped by MHT, certain MHTs routinely support more species than others (see appendix E for full statistical discussion). For fish, temperate headwaters and lakes, large temperate lakes, and large temperate river ecoregions support, on average, far more species than other MHTs, particularly xeric, endorheic, and arctic ecoregions. For crayfish, temperate headwaters and lakes ecoregions contain substantially more species than most other MHTs, and for herpetofauna, subtropical ecoregions support, on average, nearly twice as many species as any other MHT.

Within some MHTs, certain ecoregions consistently stand out for their high species richness (table 3.1). Within the large temperate river MHT, for instance, the Mississippi Embayment [25] ecoregion has the highest species richness in all faunal groups measured. Among temperate headwaters and lakes, the Tennessee-Cumberland [35] and the Teays-Old Ohio [34] ecoregions have the highest richness values. Richness in such ecoregions often dwarfs that in other ecoregions of the same MHT, revealing that there truly are important differences in biological distinctiveness.

Ecoregions with the highest fish richness in North America are all within the Mississippi drainage of the Arctic-Atlantic bioregion; they are the Mississippi Embayment [25], the Teays-Old Ohio [34], and the Tennessee-Cumberland [35] (figure 3.4a; see essays 1 and 2). Each of these ecoregions includes more than 200 fish species. Fish species richness tends to decline in ecoregions with distance away from these foci. The Mississippi [24], Ozark Highlands [29], Mobile Bay [36], and South Atlantic [40] ecoregions have between 150 and 200 species, and an additional 11 ecoregions have between 100 and 150 species. Arctic, xeric, and endorheic ecoregions have relatively low species richness, as do those draining to the Pacific Coast. The most species-poor ecoregions are the Arctic Islands [60], the Oregon Lakes [10], Death Valley [11], and the Llanos el Salado [66] ecoregions, all with nine species; the latter three ecoregions are endorheic systems.

Crayfish richness follows zoogeographic patterns similar to that of fish, though with far fewer species (figure 3.6a; see essay 3).

Totals

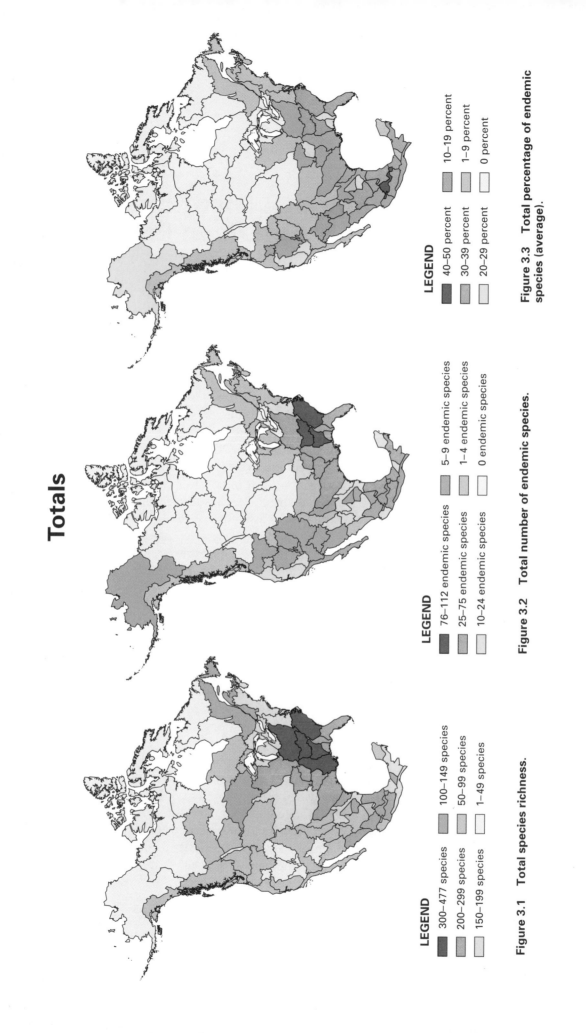

LEGEND

- 300–477 species
- 200–299 species
- 150–199 species
- 100–149 species
- 50–99 species
- 1–49 species

Figure 3.1 Total species richness.

LEGEND

- 76–112 endemic species
- 25–75 endemic species
- 10–24 endemic species
- 5–9 endemic species
- 1–4 endemic species
- 0 endemic species

Figure 3.2 Total number of endemic species.

LEGEND

- 40–50 percent
- 30–39 percent
- 20–29 percent
- 10–19 percent
- 1–9 percent
- 0 percent

Figure 3.3 Total percentage of endemic species (average).

Fishes

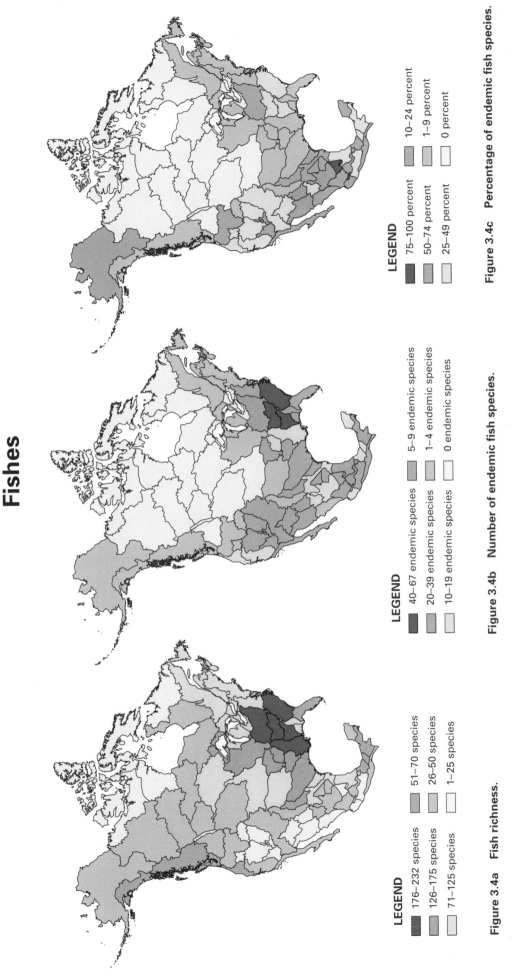

LEGEND

- 176–232 species
- 126–175 species
- 51–70 species
- 26–50 species
- 71–125 species
- 1–25 species

Figure 3.4a   Fish richness.

LEGEND

- 40–67 endemic species
- 20–39 endemic species
- 10–19 endemic species
- 5–9 endemic species
- 1–4 endemic species
- 0 endemic species

Figure 3.4b   Number of endemic fish species.

LEGEND

- 75–100 percent
- 50–74 percent
- 25–49 percent
- 10–24 percent
- 1–9 percent
- 0 percent

Figure 3.4c   Percentage of endemic fish species.

TABLE 3.1 Biological Distinctiveness Rankings.

Ecoregion	Major Habitat Type (MHT)	Richness Points	Endemism Points	Total of Richness & Endemism Points	Biological Distinctiveness Category	Phenomenon Category	Final Biological Distinctiveness Category
South Hudson [53]	ARL	3	0	3	NI		NI
East Hudson [54]	ARL	2	0	2	NI		NI
Yukon [55]	ARL	2	6	8	CO		CO
Lower Mackenzie [56]	ARL	2	0	2	NI		NI
Upper Mackenzie [57]	ARL	3	0	3	NI		NI
North Arctic [58]	ARL	1	0	1	NI	CO	CO
East Arctic [59]	ARL	1	0	1	NI		NI
Arctic Islands [60]	ARL	1	0	1	NI	CO	CO
Superior [43]	LTL	1	0	1	NI	CO	CO
Michigan-Huron [44]	LTL	3	2	5	BO	CO	CO
Erie [45]	LTL	3	0	3	NI		NI
Ontario [46]	LTL	2	0	2	NI		NI
Central Prairie [28]	THL	2	4	6	BO		BO
Ozark Highlands [29]	THL	2	4	6	BO		BO
Ouachita Highlands [30]	THL	2	4	6	BO		BO
Southern Plains [31]	THL	1	2	3	NI		NI
Teays-Old Ohio [34]	THL	3	6	9	GO		GO
Tennessee-Cumberland [35]	THL	3	6	9	GO		GO
Canadian Rockies [49]	THL	1	0	1	NI	CO	CO
Upper Saskatchewan [50]	THL	1	0	1	NI		NI
Lower Saskatchewan [51]	THL	1	0	1	NI	CO	CO
English-Winnipeg Lakes [52]	THL	1	0	1	NI	CO	CO
Colorado [12]	LTR	1	4	5	BO	CO	CO
Upper Rio Grande/Bravo [15]	LTR	1	2	3	NI	CO	CO
Lower Rio Grande/Bravo [20]	LTR	2	6	8	CO	CO	CO
Mississippi [24]	LTR	3	2	5	BO		BO
Mississippi Embayment [25]	LTR	3	6	9	GO	CO	GO
Upper Missouri [26]	LTR	1	0	1	NI		NI
Middle Missouri [27]	LTR	2	0	2	NI		NI
Bonneville [8]	ERLS	1	4	5	BO	CO	CO
Lahontan [9]	ERLS	1	4	5	BO	CO	CO
Oregon Lakes [10]	ERLS	1	2	3	NI		NI
Death Valley [11]	ERLS	1	6	7	CO	CO	CO
Guzmán [16]	ERLS	2	6	8	CO	CO	CO
Mapimí [19]	ERLS	2	6	8	CO	CO	CO
Llanos el Salado [66]	ERLS	2	6	8	CO	CO	CO
Lerma [69]	ERLS	3	6	9	GO	GO	GO
South Pacific Coastal [7]	XRLS	3	2	5	BO	CO	CO

Vegas-Virgin [13]	XRLS	1	4	5	BO		CO
Gila [14]	XRLS	1	4	5	BO	CO	CO
Rio Conchos [17]	XRLS	3	6	9	GO	CO	GO
Pecos [18]	XRLS	3	4	7	CO	CO	CO
Rio Salado [21]	XRLS	3	2	5	BO	GO	CO
Cuatro Ciénegas [22]	XRLS	2	4	6	BO	CO	GO
Rio San Juan [23]	XRLS	3	6	9	GO	GO	GO
Chapala [65]	XRLS	2	6	8	CO	CO	GO
Rio Verde Headwaters [67]	XRLS	1	4	5	BO	CO	GO
Sonoran [61]	XRLS	2	4	6	BO	CO	CO
North Pacific Coastal [1]	TCRL	1	2	3	NI		CO
Columbia Glaciated [2]	TCRL	1	0	1	NI		NI
Columbia Unglaciated [3]	TCRL	1	4	5	BO	CO	CO
Upper Snake [4]	TCRL	1	2	3	NI		NI
Pacific Mid-Coastal [5]	TCRL	1	4	5	BO	CO	CO
Pacific Central Valley [6]	TCRL	1	6	7	CO	GO	GO
East Texas Gulf [32]	TCRL	2	6	8	CO	CO	CO
Mobile Bay [36]	TCRL	3	6	9	GO		GO
Apalachicola [37]	TCRL	2	6	8	CO		CO
Florida Gulf [38]	TCRL	2	6	8	CO		CO
South Atlantic [40]	TCRL	3	6	9	GO	CO	GO
Chesapeake Bay [41]	TCRL	2	4	6	BO	GO	NI
North Atlantic [42]	TCRL	2	2	4	NI		CO
Lower St. Lawrence [47]	TCRL	2	2	4	NI	CO	NI
North Atlantic-Ungava [48]	TCRL	1	0	1	NI		CO
West Texas Gulf [33]	SCRL	1	4	5	BO		CO
Florida [39]	SCRL	3	6	9	GO	CO	GO
Sinaloan Coastal [62]	SCRL	1	4	5	BO	CO	CO
Santiago [63]	SCRL	1	4	5	BO	CO	CO
Manantlan-Ameca [64]	SCRL	1	4	5	BO	CO	GO
Tamaulipas-Veracruz [68]	SCRL	3	6	9	GO	GO	GO
Balsas [70]	SCRL	1	6	7	CO	CO	CO
Papaloapan [71]	SCRL	2	6	8	CO	CO	CO
Catemaco [72]	SCRL	1	4	5	BO	GO	GO
Coatzacoalcos [73]	SCRL	2	4	6	BO	CO	CO
Tehuantepec [74]	SCRL	2	6	8	CO		CO
Grijalva-Usumacinta [75]	SCRL	2	6	8	CO	CO	CO
Yucatán [76]	SCRL	1	4	5	BO	CO	CO

Diversity, Levels of Imperilment, and Cryptic Fishes in the Southeastern United States

Noel M. Burkhead and Howard L. Jelks

About 800 species of freshwater fish reside in North America north of Mexico (Page and Burr 1991). This fauna is unequally distributed across the continent: the majority of the species, about 500, occur in the southeastern United States (Warren et al. 1997). Most of these are upland fishes inhabiting southern Appalachia (349 species, our data). Altogether, these species represent a panoply of evolutionary history and diversity, from ancient fishes that swam in rivers when dinosaurs roamed the landscape, to modern families represented by diverse lineages, many with brilliant spawning colors that delight the eye. The range of forms and adaptations include members of marine families. For example, the shads or Clupeidae, slab-sided, anadromous fishes, formerly migrated hundreds of miles by the thousands to scatter their eggs and milt in the rivers of the Blue Ridge Mountains (Jenkins and Burkhead 1994). Conversely, the small trispot darter (*Etheostoma trisella*), a tiny perch, also journeys in spring, but instead of large rivers it ascends tiny rivulets seeping from ephemeral ponds in fields to attach its few eggs to submerged blades of grass (Ryon 1986). Between these extremes of adaptations and lineages lives such diversity—from genes to elaborate, ritualized behaviors—that students of fishes, even after 265 years of study, are still discovering not just new species but a whole host of biodiversity that fascinates, captivates, and challenges the intellect. The reasons underlying the high level of fish diversity are threefold: (1) the Southeast escaped the last ice-age glaciation; (2) southeastern river systems drain landscapes having diverse geologic, physiographic, and climate elements, so habitats in turn are quite diverse; and (3) the Southeast is a region with a rich history of zoogeographic and evolutionary processes in freshwater fishes (Hocutt and Wiley 1986; Mayden 1988).

Game fishes represent only a small fraction of this fauna (about 6 percent); therefore, most of the fauna is almost unknown to the public. In many respects the fauna is orphaned. Unlike songbirds, nongame freshwater fishes lack a public following with the associated support. The southeastern United States is represented by no less than twenty-eight families of native fishes, but most of the diversity is represented by only five families: minnows (Cyprinidae); suckers (Catostomidae); catfishes (Ictaluridae), particularly madtoms; basses and sunfishes (Centrarchidae); and perches (Percidae). The remaining families represent a few to a dozen or so species, and some families are represented by only one species (monotypic). The pirate perch (*Aphredoderus sayanus*) is an example of a monotypic species. A relatively small, lowland denizen of swamps, marshes, creeks, and river backwaters, it is a fish with an anatomical oddity—its anus, during early development, migrates forward and is located just behind the head in juveniles and adults. Why? We have no idea. Also monotypic, at least in North America (a cousin lives in China), is the paddlefish (*Polyodon spathula*), a large species with a paddle-shaped snout like a swimming Pinocchio. This structure is loaded with tiny organs that probably detect minute electrical impulses emanating from microscopic planktonic prey (Jenkins and Burkhead 1994).

In the midst of this diversity are alarming patterns of faunal decline. The number of federally listed threatened and endangered freshwater fishes from the southeastern United States nearly doubled—from twelve to twenty-two species—in a recent decade (Federal Register 1986, 1995). This relevant metric of faunal decline is corroborated by other analyses (Williams et al. 1989; Warren and Burr 1994; Warren et al. 1997). The vast majority of southeastern fishes are rheophillic, exhibiting an array of adaptations to living in, or adjacent to, current. These fishes and other aquatic faunas (especially mollusks) face escalating imperilment primarily due to widespread decline and loss of lotic (flowing-water) habitats across this region, which harbors the foremost temperate aquatic faunal diversity in the world (Lydeard and Mayden 1995; Warren and Burr 1994; Walsh et al. 1995; Taylor et al. 1996; Neves et al. 1997). The salient causes of lotic habitat loss in the southeastern United States are well known: pollution, impoundment (on both large and small scales), channelization, excessive

An example of cryptic darter species from the Southeast. The holiday darter, *Etheostoma brevirostrum*, described in 1991 (left photo), and its highly similar relative, *Etheostoma* sp. cf. *brevirostrum* (right photo). Since 1990, 26 new darter species have been described from North America. Of the 154 darter species described to date, at least one-fourth are considered imperiled. Photos courtesy of Noel Burkhead.

sedimentation, and urbanization (Moyle and Leidy 1992; Warren and Burr 1994; Lydeard and Mayden 1994; Burkhead et al. 1997; Richter et al. 1997). Deterioration of lotic habitats in the southeastern United States is a by-product of rapid human population growth (Noss and Peters 1995). The broad impact of regional growth on aquatic biodiversity is the basis for recognition of the Southeast as a global freshwater priority (Master et al. 1998; Olson and Dinerstein 1998).

A symptomatic indicator of faunal decline is population fragmentation: the breaking up of a contiguous population of a species into separate populations isolated by impoundments, or by river reaches degraded by pollution or excessive sedimentation (Burkhead et al. 1997). Because isolated populations are more vulnerable to perturbations, the probability that one or more of these units will become extirpated constantly increases because of pressure from human population growth. Given the lack of recent, intensive population survey data for most of the southeastern United States, detecting fragmentation is problematic. Our knowledge of species distributions is sometimes based on data collected more than twenty years ago (e.g., many distributions depicted in Lee et al. 1980). Unless serious attention is directed to stream ecosystem conservation, populations of fishes and other imperiled aquatic organisms will continue to decline, some certainly to extinction.

As presently understood, levels of imperilment are likely underestimated. Even though the southeastern fish fauna is one of the most intensively studied in the world, new species are still being discovered. Many of the recently discovered fishes are cryptic species. Generally, cryptic species appear very similar to known species; hence some are simply not recognized until a more thorough faunal or species-group assessment is made. Cryptic species are frequently localized—some even restricted to single creek systems—thus diminishing their chance of discovery or long-term survival (Boschung et al. 1992; Page et al. 1992). Because degradation of aquatic habitats is virtually ubiquitous across the southeastern landscape, some cryptic species with small ranges may become extinct before they are discovered (or described). The majority of the southeastern cryptic fauna are small fishes (e.g., darters—small perches, family Percidae), but examples include relatively large, riverine fishes, such as the impressive robust redhorse sucker (*Moxostoma robustum*) (Hendricks 1998) and archaic fishes such as the Alabama sturgeon (*Scaphirhynchus suttkusi*) (Williams and Clemmer 1991).

Darters provide the most numerous examples of the cryptic species, and, because the level of imperilment is disproportionately high in darters (figure 3.5), there is obvious reason for concern for species not yet documented. Darters occur in all fluviatile habitats in the southeastern United States, from cascading mountain brooks to large Coastal Plain rivers, and several species are adapted to swamps (Page 1983). To date, 166 species of darters have been described, 154 of which reside entirely or partially in the southeastern United States (Ceas and Page 1997; Warren et al. 1997, our counts). Many of the recently described species are cryptic: recognition of the new forms was facilitated by comparing male nuptial coloration patterns among the different populations (Boschung et al. 1992; Page et al. 1992; Wood and Mayden 1993; Bauer et al. 1995). As a

Figure 3.5 Levels of imperilment of darters (family Percidae) in the southeastern United States.

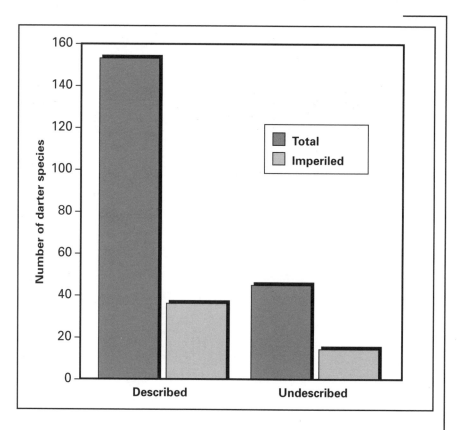

group, darters are disproportionately imperiled because they share attributes that render them more vulnerable to degradation of stream substrata and alteration of streams by impoundment: small body size, relatively low fecundity, benthic specialization, and rheophillic (current-loving) orientation (Berkman and Rabeni 1987; Angermeier 1995).

Since 1990, a remarkable twenty-six new species of darters have been described from the southeastern United States (Suttkus and Etnier 1991; Boschung et al. 1992; Page et al. 1992; Wood and Mayden 1993; Jenkins and Burkhead 1994; Suttkus et al. 1994; Bauer et al. 1995; Thompson 1995; Ceas and Page 1997). Of these twenty-six species, thirteen were discovered relatively recently; the remaining species have long been known to ichthyologists (but were only recently described). Of great conservation concern is the estimated level of imperilment among the new putative taxa (figure 3.5). Of the 154 described species, 37 are considered imperiled (24 percent) (Warren et al. 1997). However, among the forty-three putative new darter taxa (our count based on personal communications with nine ichthyologists), we estimate that fifteen, or 35 percent of the undescribed cryptic fauna, are imperiled. Based on our data, some of these undescribed darters likely merit listing as endangered species. We emphasize that these counts are estimates; however, given the description of thirteen new species in this decade, we are likely underestimating actual species diversity.

Faunal decline can be slowed, even reversed. Rivers are remarkable systems that will rebound if given the opportunity to do so. Abatement of major threats to these systems can become part of intelligent development, if we so choose. One bit of inherited myopia hinders younger southeastern citizens: they have grown up observing their rivers as turbid, muddy systems. But except for the dark-water Coastal Plain rivers, virtually all southeastern rivers used to flow clear, right to the bottom. These rivers used to have a "postcard" clarity that would entice passersby to stop for a moment, transfixed by their swirling beauty. Folks used to let river sensations—an eddy line melting into the current, light performing tricks on the water, the staccato of a startled kingfisher—reawaken in them something special. Declining aquatic biodiversity echoes not only habitat losses but intangibles as well, and one cannot help thinking, inasmuch as the land molds the American character, that all of us are poorer.

The Mississippi River Basin: Its Megafauna and Hydrological Modifications

Brooks M. Burr and James B. Ladonski

The third-longest river in the world and the largest river in North America, the Mississippi River flows 3,731 km from its source in Minnesota south to the Head-of-Passes, Louisiana, where it splits into several distributaries and extends another 32 km to the Gulf of Mexico (Fremling et al. 1989). It drains a basin of 4,759,049 km^2, or about one-eighth the area of North America, including all or parts of thirty-one states and two Canadian provinces (Keown et al. 1981). Immensely productive for fishes during times of maximum flooding when pre-development, flood-pulse conditions are manifested, the river is now severely regulated for transportation and flood control; consequently, it has lost much of its historical and natural character.

The Mississippi Basin has by far the richest fish fauna in North America, including at least 375 species in 31 families (Burr and Mayden 1992). Thirty-four percent of these are endemic to the basin, the majority restricted to highland areas—the Ouachitas, Ozarks, and Appalachians. Comparable, but relative, numbers of crayfishes and mussels fit the same pattern of richness and endemism in the basin. The mainstem river has just over 240 fish species (Fremling et al. 1989), several of which (e.g., the pallid sturgeon, *Scaphirhynchus albus*, and the sturgeon chub, *Macrhybopsis gelida*) occur only in the main channel and mainstems of the largest (i.e., Missouri River) tributaries. The lower reaches and coastal wetlands of the mainstem support seasonal populations of otherwise marine or estuarine fishes (e.g., the bay anchovy, *Anchoa mitchelli*; the Atlantic needlefish, *Strongylura marina*) and invertebrates.

The basin is a globally outstanding example of fish evolution. It has been a refuge during times of glaciation and a refuge for representatives of ancient fishes and other aquatic or semiaquatic vertebrates. For example, the Mississippi Basin and the Yangtse Basin of China harbor the two extant species of paddlefish (Polyodontidae), the ancestral catostomid suckers (*Cycleptus* and *Myxocyprinus*), the genus *Alligator*, and the giant aquatic salamanders (Cryptobranchidae; Zhao and Adler 1993). The Mississippi fauna is clearly the "mother" fauna of North America, and much of the diversity of surrounding ecoregions can be accounted for by spillover from this cradle of temperate fish diversity.

Significant tributary systems with highly differentiated faunas include the Tennessee, Cumberland, and Green Rivers, with approximately fifty endemic fishes, forty-three crayfishes, and a comparable number of mussels. The Mississippi Embayment [25] has thirty-three endemic crayfishes. The Ozark and Ouachita Highlands (ecoregions [29] and [30], respectively) have twenty-six endemic fishes, thirty-six endemic crayfishes, and about seven endemic mussel species (Oesch 1984). Unusual aquatic organisms restricted or nearly restricted to these regions include the cavefishes (Amblyopsidae), the crayfish genera *Barbicambarus* and *Bouchardia*, and numerous species of mussels. The spectacular darter (Percidae) and minnow (Cyprinidae) fauna dominate the diversity of this region.

Exceptional aquatic habitats include some of the largest springs and sources of groundwater in the world; large, free-flowing, unimpounded rivers such as the Wabash River; outstanding karst topography; and headwater mountain streams of high water quality. Under natural conditions the Mississippi River was one of only two great temperate rivers that provided backwaters important to fish production and nursery habitats, and a concomitant river fishery unparalleled in North America.

The Mississippi River has undergone extreme modification within the last 200 years for navigation and flood control. Channelization of the upper portion of the river for navigation (from Minneapolis, Minnesota, to the mouth of the Ohio River) began in 1878 (Fremling et al. 1989). Today the river is controlled by forty locks and/or dams from the headwaters at Lake Itascan, Minnesota, to St. Louis, Missouri (Fremling et al. 1989). In addition, much of the river, particularly the middle and lower reaches, has been buffered by levees for flood control. Increased water flow below dams has resulted in channel deepening, which in turn draws water out of side channels and backwater areas, causing these critical habitats to dry up (Sheehan and Rasmussen 1993). In some cases impoundments have actually increased fish populations because the flooding of adjacent areas, caused by dam construction, increases the amount of available habitat for lentic, shallow-water species such as sunfishes (Centrarchidae). Such benefits, however, are usually short lived because of the accompanying increase in sedimentation rates. Most of the newly created backwaters are filled in within 50 to 100 years after an impoundment is created (Sheehan and Rasmussen 1993).

The most severe and relatively recent threat to the Mississippi River system is the introduction of exotic species. Although intentional stocking of fishes for sport or commercial use began in the late 1800s, invasions of exotic organisms continue today at alarming rates. While the occurrence of most of the non-native fish species in the river is due to intentional stocking by management agencies, many other species have reached the basin through accidental introductions. The most infamous of these invaders may be the zebra mussel (*Dreissena polymorpha*). The zebra mussel was first discovered in the Great Lakes region in 1988 and made its way into the Mississippi River Basin in 1992 through the Illinois River (Oesch 1984). Other invaders of the Great Lakes, including the ruffe (*Gymnocephalus cernuus*), tubenose goby *(Proterorhinus marmoratus)*, and round goby (*Neogobius melanostomus*), also may eventually find their way into the basin. Exotic carps (genera *Cyprinus, Ctenopharyngodon,* and *Hypophthalmichthys*), all now firmly established in the basin, account for the greatest biomass of fishes in the mainstem river.

The Mississippi River mainstem and its drainage basin, although highly altered by human activity, are still a unique and valuable natural resource. The overall biotic diversity, coupled with the large percentages of endemic species of megafauna, and including the sport and commercial fishery, clearly demonstrates that the Mississippi River is an invaluable resource by any measurement. The outlook for the next ten years is encouraging: recently enacted, powerful environmental legislation requires mitigation for loss of fish and wildlife habitat, as well as rehabilitation of areas already degraded. Private-sector funding has resulted in the creation of several aquatic bioreserves (e.g., the Green River bioreserve) that emphasize watershed-wide management of all aquatic resources.

Twenty ecoregions support no native crayfish, and an additional thirteen have only a single species. With three-quarters of all ecoregions containing fewer than ten crayfish species, the high numbers supported by the Tennessee-Cumberland [35] (N = 65), Mobile Bay [36] (N = 60), Mississippi Embayment [25] (N = 57), and South Atlantic [40] (N = 56) ecoregions are globally outstanding. As with fish, crayfish richness tends to decrease outward from this center, and arctic, xeric, and endorheic ecoregions have few crayfish. In Mexico, the Tamaulipas-Veracruz [68] ecoregion stands out for its relatively large number of crayfish (N = 20), compared with neighboring units.

Data for mussels, the other group of invertebrates analyzed, were available only for Canada and the United States; therefore, ecoregions located entirely or primarily in Mexico were not scored for mussels (figure 3.7a). As with fish and crayfish, the highest mussel richness is found in ecoregions of the Mississippi Basin, with the Tennessee-Cumberland [35] and the Teays-Old Ohio [34] ecoregions supporting by far the greatest number of species (125 and 122 species, respectively) (see essay 4). Ecoregions adjacent to these two units support between 50 and 100 species of mussels; this includes the South Atlantic [40] ecoregion, which drains to the Atlantic rather than the Gulf of Mexico. Low numbers of mussel species are found in the West and the northern ecoregions of Canada.

Unlike fish, mussels, and crayfish, freshwater herpetofauna richness reaches its highest levels in subtropical Mexico (Neotropical realm), where three ecoregions support more than 100 species

Crayfish

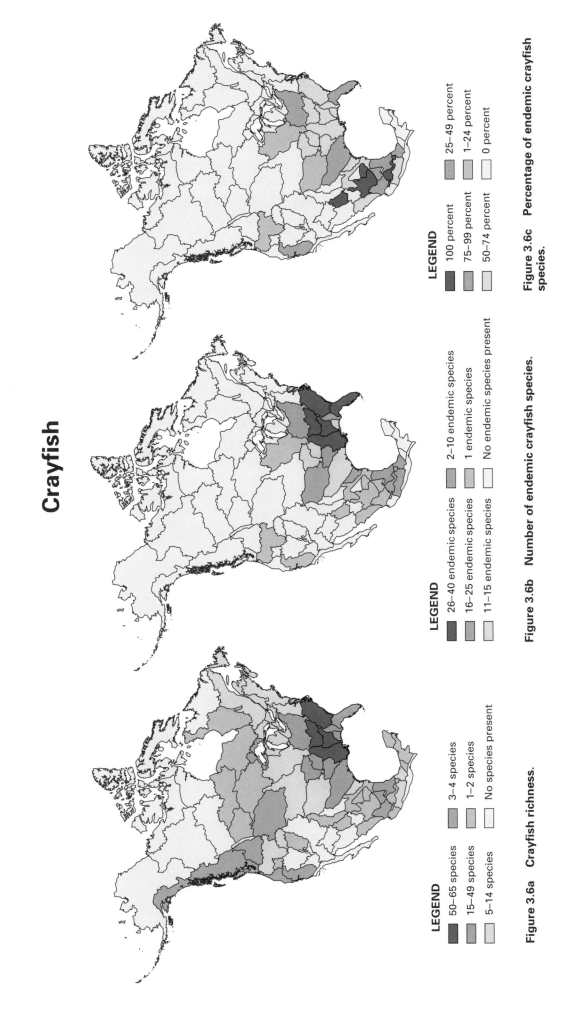

LEGEND

- 50–65 species
- 15–49 species
- 5–14 species
- 3–4 species
- 1–2 species
- No species present

Figure 3.6a   Crayfish richness.

LEGEND

- 26–40 endemic species
- 16–25 endemic species
- 11–15 endemic species
- 2–10 endemic species
- 1 endemic species
- No endemic species present

Figure 3.6b   Number of endemic crayfish species.

LEGEND

- 100 percent
- 75–99 percent
- 50–74 percent
- 25–49 percent
- 1–24 percent
- 0 percent

Figure 3.6c   Percentage of endemic crayfish species.

Preserving North America's Unique Crayfish Fauna

Christopher A. Taylor

Crayfishes, also known as crawfishes or crawdads, are native to freshwater ecosystems on every continent except Africa and Antarctica. Crayfishes are noticeably absent from the biologically diverse tropical regions of northern South America, southern Central America, and southeast Asia (Hobbs 1988). In North America crayfishes have successfully colonized almost every type of aquatic ecosystem including streams, lakes, swamps, and subterranean caves and springs. Burrowing species of crayfishes have evolved life-history strategies that allow them to inhabit semiaquatic habitats such as floodplain forests, wet prairies and savannas, and seasonally flooded ditches.

In addition to its ecological diversity, North America's crayfish fauna is extremely diverse taxonomically (Hobbs 1988; Taylor et al. 1996). It is this taxonomic diversity that sets North America apart from Europe, South America, Australia, and Asia. Of the 506 species and subspecies currently known worldwide, 393 (78 percent) occur in North America. Of these 393 taxa, 346 (68 percent of total known taxa) occur in the United States and Canada, 43 (8 percent) occur in Mexico, and 4 (less than 1 percent) inhabit the island of Cuba. In the United States crayfishes have a distribution pattern similar to those shown by freshwater fishes (Warren and Burr 1994) and mussels (Williams et al. 1993), in that the southeastern United States is home to the highest number of crayfish species and subspecies. Approximately 68 percent of all U. S. species are endemic to an area roughly defined as east and south of (and including) the state of Kentucky. Secondary areas of high endemism and species diversity for crayfishes are the Ozark and Ouachita Highlands of Missouri and Arkansas (ecoregions [29] and [30], respectively) and the Gulf of Mexico coastal plain of Louisiana and Texas (ecoregion [32]).

The population declines of fishes and mussels that have occurred in recent decades are well documented and often are quite clearly the result of human activity (Karr et al. 1985; Parmalee and Hughes 1993; Williams et al. 1992). The extirpation of 5 percent of the U.S. native fish fauna (Warren and Burr 1994) and 7 percent of the native mussel fauna (Williams et al. 1993) has also been reported. Crayfishes, on the other hand, have not been as severely affected by human activity. A recent analysis (Taylor et al. 1996) found 162 species (48 percent of the North American fauna north of Mexico) to be imperiled and two species (less than 1 percent) to be extinct. However, limited natural range, as opposed to anthropogenic activity, was highlighted as a primary cause. Taylor et al. (1996) document eleven species known from a single location and another twenty from five or fewer locations. Using data from Taylor et al. (1996), I find that 43 percent of all crayfishes known from North America north of Mexico are distributed entirely within one state's political boundaries. While organisms that have restricted natural ranges may not meet the criteria for listing at the federal level (sec. 1533 (a), Endangered Species Act of 1973), they should not be denied conservation attention, as species with small ranges are extremely vulnerable to extirpation (Gilpin and Soulé 1986; Rabinowitz et al. 1986). Without proper recognition, entire species or significant portions of species' ranges could be eliminated by relatively small-scale habitat alteration in the form of urban development, drainage of wetlands, or introduction of aggressive non-native crayfishes.

While habitat alteration does threaten North American crayfishes, introduction of non-native crayfishes poses a far more serious threat. In the United States the rapid displacement of native crayfishes by the rusty crayfish, *Orconectes rusticus*, and the signal crayfish, *Pacifastacus leniusculus*, has been well documented (Capelli and Munjal 1982; Light et al. 1995; Taylor and Redmer 1996). Both the signal and rusty crayfishes are native to the United States but displace native species when they are transplanted into river basins outside their native ranges. The avenue of introduction is usually through their use as fishing bait (Page 1985; Taylor and Redmer 1996). Through this practice the rusty crayfish has spread across the United States from New Mexico to Maine. Research has pointed to the aggressive nature and larger body size of the rusty crayfish as factors contributing to its ability to displace other crayfishes (Butler and Stein 1985; Garvey and

Stein 1993). Hybridization between the rusty crayfish and native species has also been suggested as a displacement mechanism (Capelli and Capelli 1980; Page 1985).

Very few states have adequate, enforceable laws that prevent introduction of non-native crayfishes; states that have enacted laws already have well-established populations of non-natives. Federal legislation pertaining to the introduction of non-native species from foreign nations is practically nonexistent. This is particularly troubling given the fact that aquacultural facilities and markets for foreign species are being developed in the United States without an adequate understanding of the potential effects of non-natives on native species. Some foreign species grow considerably larger than do most native North American crayfishes. If we are to protect our native crayfishes, especially the endemics with small ranges, proactive and effective legislation must be implemented at both state and federal levels. We cannot afford to ignore the data showing the negative impacts that non-native species such as the rusty crayfish can have on our native fauna.

Essay 4

Freshwater Mussels: A Complicated Resource to Conserve

G. Thomas Watters

The freshwater realm has been invaded repeatedly by mollusks, so that today we have rather distantly related groups living side by side in our lakes and streams. One of the largest groups, in terms of both diversity and physical size, comprises freshwater mussels in the families Margaritiferidae and Unionidae. This group also is the most imperiled. Within the past 200 years or so, dozens of species have been driven to extinction wholly by the activities of humankind.

The eastern half of North America represents the greatest center of diversity for these animals on earth. Nearly 300 species are recognized from this region, particularly from the rich Mississippi, Ohio, Tennessee, Cumberland, and Mobile Rivers (ecoregions [24] and [25], [34], [35], and [36], respectively). Any relatively intact stream in this region supports more mussel species than all of Australia and Europe combined. Yet more than 75 percent of these species are believed to be declining in numbers, their ranges rapidly shrinking to tiny, distantly isolated populations. Some are extinct, and others are functionally extinct—their numbers too low to support a viable population.

Mussels live by lying imbedded in the lake or river bottom where they filter food from the water with their gills. As adults, most mussels move very little. They may live for a prodigious time—more than 120 years for one species. Their shells may be massive, weighing several pounds and becoming several centimeters thick. The inside of the shell has a lustrous layer known as nacre or mother-of-pearl. Like many other mollusks, mussels protect themselves against foreign material that becomes trapped between their shells by secreting layers of nacre around the object, forming a pearl. Pearls and the nacreous shells themselves have been objects of desire among humans for thousands of years.

Mussels have a complicated life history that involves a parasitic larval stage called a glochidium. Female mussels brood these larvae by the thousands in specialized portions of their gills called marsupia. When mature, the glochidia are released into the water to find hosts—generally fishes, although amphibians may be used to a lesser extent. Mussels show a wide range of host specificity, from generalists capable of using nearly any fish that happens by, to specialists that use only certain types of fishes, such as darters or sculpins. Glochidia attach themselves to the gills, fins, lips, barbels, or other exposed surfaces of their hosts and become encapsulated. Here they remain from days to months before metamorphosing into juvenile mussels, breaking out of their capsules, and settling to the lake or stream bottom. During this parasitic phase mussels become distributed by riding on their highly mobile hosts. This phase is obligatory and without the proper host, no recruitment can take place. It is suspected that some mussel populations consist of aging individuals that

Unionid Mussels

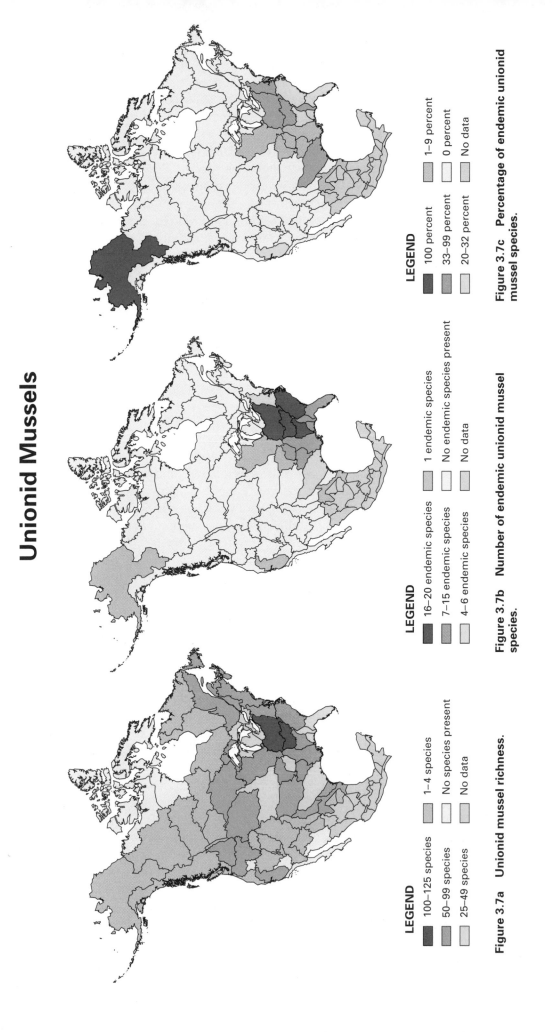

LEGEND

- 100–125 species
- 50–99 species
- 25–49 species
- 1–4 species
- No species present
- No data

Figure 3.7a   Unionid mussel richness.

LEGEND

- 16–20 endemic species
- 7–15 endemic species
- 4–6 endemic species
- 1 endemic species
- No endemic species present
- No data

Figure 3.7b   Number of endemic unionid mussel species.

LEGEND

- 100 percent
- 33–99 percent
- 20–32 percent
- 1–9 percent
- 0 percent
- No data

Figure 3.7c   Percentage of endemic unionid mussel species.

can never reproduce because the host fishes are no longer present. From a conservation and management point of view, this means that it is not sufficient just to protect the habitat of the mussel—the habitat of the host must be protected as well. Because these hosts may engage in migrations or spawning runs, protecting their habitats becomes a serious challenge.

It is safe to say that we have learned more about the basic biology of these animals in the past twenty years than in all previous time. We now know that juveniles, when they leave their hosts, live for several years in the substrate, feeding with ciliary tracts on their mantles and feet. We also know that their parasitic life cycle is not simply a matter of releasing glochidia and relying on a chance encounter with the right host. Many mussels have developed methods to lure hosts to them. Some have highly modified body portions that resemble minnows or worms, to the extent that these lures have "eyes," "fins," and "lateral lines" and undulate in a swimming motion. When a predatory fish such as a bass strikes the lure the female releases a cloud of glochidia in the fish's face. Other mussels package their glochidia in structures called conglutinates, which they release to the water. These conglutinates may mimic insects, worms, fishes, or even fish eggs. Each conglutinate contains thousands of glochidia. When bitten, the conglutinates break apart, releasing the parasitic larvae.

Despite these adaptations, mussels have suffered at the hands of humans, both directly and indirectly. By damming rivers, humans have changed natural habitats, indirectly eliminating both mussels and hosts. Channelization, dredging, and snagging remove substrate, change stream hydrology, and eliminate habitat heterogeneity. Runoff of substrate from agricultural, construction, and residential areas buries the sedentary mussels or smothers their gills. Pollutants in the forms of pesticides, heavy metals, and fertilizers accumulate in their tissues; some are passed up the food web. To confuse matters more, evidence suggests that these indirect impacts affect juveniles and glochidia differently than adults. Humans have directly affected mussels as well. Just as primitive people coveted the nacreous pearl thousands of years ago, so too now are mussels harvested by the thousands of tons to be shipped overseas for the cultured pearl industry. Finally, the introduction of the exotic zebra mussel (*Dreissena polymorpha*) to North America has had a catastrophic impact on native mussels. Once-thriving mussel beds have been all but eliminated as these exotics overwhelm them.

The US Fish and Wildlife Service (USFWS) has stated that freshwater mussels are the most endangered group of animals in North America. Their protection is of paramount importance, but not just because of their intrinsic worth. They are sessile, filter-feeding, long-lived animals with a great range of tolerances to environmental perturbations. As such, they have great potential to act as biological monitors of riverine health—the proverbial canary in the coal mine.

(figure 3.8a). This high richness is largely derived from high numbers of certain groups of frogs and salamanders; in the Tehuantepec [74] ecoregion, for instance, 70 of the 125 aquatic reptile and amphibian species are frogs and 36 are salamanders. In contrast, the largest number of frogs in any temperate ecoregion is thirty-six species in the South Pacific Coastal [7] ecoregion, which includes Baja California. The South Atlantic [40] and Mobile Bay [36] ecoregions stand out in North America north of Mexico for their eighty-two and seventy-seven species of herpetofauna, respectively. Especially in Mexico, where new species are routinely being described, herpetofauna numbers likely underestimate the true number of species.

Endemism

Patterns of endemism across North America are similar to those for richness, with the highest number of endemics generally found

Herpetofauna (Reptiles and Amphibians)

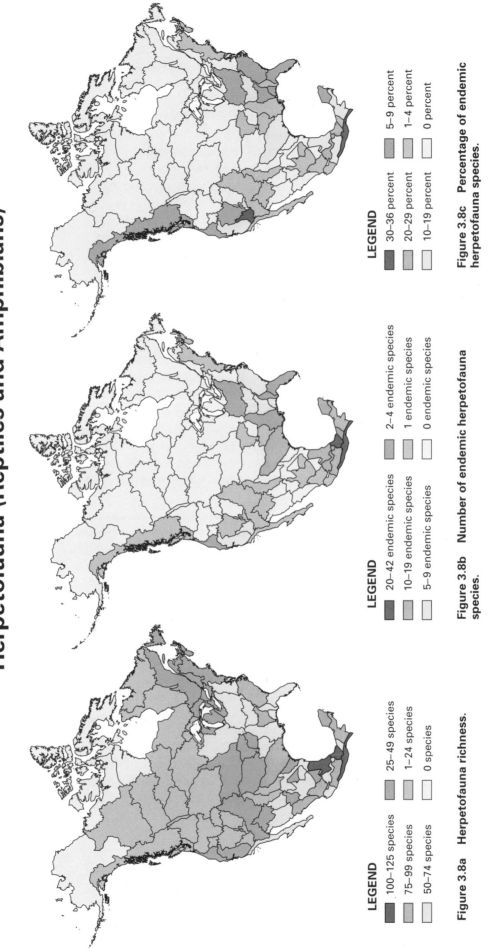

LEGEND

- 100–125 species
- 75–99 species
- 50–74 species
- 25–49 species
- 1–24 species
- 0 species

Figure 3.8a   Herpetofauna richness.

LEGEND

- 20–42 endemic species
- 10–19 endemic species
- 5–9 endemic species
- 2–4 endemic species
- 1 endemic species
- 0 endemic species

Figure 3.8b   Number of endemic herpetofauna species.

LEGEND

- 30–36 percent
- 20–29 percent
- 10–19 percent
- 5–9 percent
- 1–4 percent
- 0 percent

Figure 3.8c   Percentage of endemic herpetofauna species.

in ecoregions with the greatest richness for a given taxa (figure 3.2). Fish species endemism is highest in the Tennessee-Cumberland [35] ecoregion (N = 67), followed by the South Atlantic [40] (N = 48) and Mobile Bay [36] (N = 47) ecoregions (figure 3.4b). These three ecoregions also support the highest numbers of endemic crayfish (figure 3.6b) and mussels (figure 3.7b; with the addition of the Teays-Old Ohio [34] ecoregion for mussels). High numbers of endemic crayfish tend to be found in ecoregions that ring the Gulf of Mexico, with low numbers of endemics in the West, Midwest, North, and most of Mexico. Of the fifty-five ecoregions with mussel data, thirty-five had no endemic mussel species, and five had only a single endemic; this highlights the extraordinary endemism of the Mississippi and Atlantic complexes.

In contrast, ecoregions from the Rockies westward, and many in Mexico, support a moderate number of endemic fish. In Mexico especially, the Tamaulipas-Veracruz [68], Grijalva-Usumacinta [75], and Lerma [69] ecoregions stand out for their twenty-nine, twenty-nine, and thirty endemic fish species, respectively. The endorheic Lerma basin has been the site of high speciation of atherinid fishes (mainly *Menidia*), and Grijalva-Usumacinta and Tamaulipas-Veracruz both have high numbers of poeciliids and cichlids; these three groups are all poorly represented in temperate ecoregions.

For herpetofauna, the highest number of endemics by far is found in the Tehuantepec [74] ecoregion, where there are forty-two endemic species (figure 3.8b). Five additional ecoregions in southern Mexico also support high numbers of endemics compared with the rest of the continent. In North America north of Mexico, the East Texas Gulf [32] ecoregion stands out with thirteen endemic species; several of these are restricted to caves and springs in central Texas (see essay 5). All other ecoregions have fewer than ten endemic aquatic reptile and amphibian species.

E S S A Y 5

Subterranean Freshwater Biodiversity in Northeastern Mexico and Texas

Dean A. Hendrickson and Jean Krejca

The Gulf of Mexico is broadly surrounded by deep Cretaceous limestones, which, at least in Texas and northeastern Mexico, dip toward the gulf from higher inland terrains (Carillo-Bravo 1971; Smith and Veni 1994). Where exposed by Laramide orogenic events of the late Cretaceous to early Tertiary periods, and fractured by Miocene-epoch faulting, the solution of highly soluble formations has produced vast karst terrains. These are complex and often very deep systems of caves and channels containing water in cavities ranging from interstitial spaces and narrow cracks to massive caverns with underground streams, rivers, and lakes (Fish 1977; Sharp 1990; Veni 1994). Where bounded top and bottom by less soluble formations, water moving down the dip of the strata may develop artesian pressure, flowing to the surface through natural fractures to produce springs, or more recently, through drilled wells (Brune 1981).

It is generally believed that as these karst aquifers developed, surface ancestors of today's subterranean aquatic fauna invaded cavities and began to adapt, many to the point of becoming obligate residents of subterranean habitats. Through convergent adaptation, stygobites, as obligate aquatic cave organisms are known, have become characterized by depigmentation and lack of eyes. Compared with their surface relatives, they also typically have more highly developed sensory systems, lower metabolic rates, larger fat or lipid accumulations, smaller body sizes, attenuated appendages, lower fecundity (but larger eggs), and longer life cycles (Culver 1982; Barr and Holsinger 1985). Study of this fauna is difficult. Caves provide severely limited access, usually only to peripheral aquifer habitats, and their exploration often requires technical climbing and cave SCUBA diving. Despite limitations, a diverse fauna has been documented in our region, although studies go little beyond species descriptions, and many questions remain about biology and life histories.

Texas's eighty-one stygobitic species are unparalleled in nearly every other region of the world (Longley 1981; Reddell 1994). Much of the diversity is from the narrow Edwards Aquifer, which arcs from the southwest to the center of the state, and the Edwards Plateau, which extends westward from the aquifer and recharges it. Though small in surface area, the Edwards Aquifer has forty-seven stygobites, thirty-five (75 percent) of them endemic. Within the aquifer, the deep habitat of the San Antonio Pool provides a relatively stable environment and a large share of endemics, such as the catfishes *Satan eurystomus* and *Trogloglanis pattersoni*, known only from deep artesian wells (300–600 m) in the city of San Antonio (Longley and Karnei, Jr. 1979a; Longley and Karnei, Jr. 1979b; Longley and Karnei, Jr. 1993). Apparently restricted to a narrow zone west of the interface of a saline "bad-water" mass to the east and a "good-water" zone to the west, catfishes and other species likely rely on the unique ecological conditions of the interface area. Many species may feed directly or indirectly on bacteria and fungi that utilize the organic matter (Longley 1981) or perhaps the unique chemical conditions of the interface. Shallower areas of the aquifer support highly adapted stygobitic salamanders of the genus *Typhlomolge* and as many as eight species of *Eurycea* (David M. Hillis, pers. comm.), many yet to be described. However four nonvertebrate phyla, Platyhelminthes, Mollusca, Annelida, and Arthropoda, contribute the majority of the aquifer's stygobites, with most from the class Crustacea.

Karst aquifers and stygobitic faunas continue south through much of eastern Mexico, where documentation of regional hydrologic relationships among aquifers, and of their biotas, is even more limited. Though most of the literature consists of species descriptions (Reddell 1982, 1994), a few papers address relationships and biogeography of the regional fauna (Longley 1986; Botosaneanu and Notenboom 1986; Holsinger 1986, 1993). Distributions of genera indicate that at least some share common biogeographic histories spanning large areas, possibly pointing to extensive present or past interconnections among regional aquifers—including U.S.–Mexico subterranean connections. Five genera of amphipod and isopod crustaceans extend from the Edwards Aquifer of Texas throughout northeastern Mexico, including two isopod species, *Cirolanides texensis* (Cirolanidae) and *Lirceolus* n. sp. (Aselidae), which range from San Marcos, Texas, to northern Coahuila (unpublished data). Also now known to be widely distributed is the Mexican blind catfish genus, *Prietella* (Ictaluridae). Thought for forty years to be a single species endemic to a single spring in central Coahuila (Carranza 1954), it consists of two species which together range over 750 km, from northern Coahuila (unpubl. data), just south of the southern tip of the Edwards Aquifer, to southernmost Tamaulipas (Walsh and Gilbert 1995). Recent DNA sequence data (unpubl. data) for all ictalurid genera except the San Antonio blind catfishes contradict earlier hypotheses of relationships (Lundberg 1982, 1992) and leave open the possibility that *Prietella* may be more closely related to *Trogloglanis* than was previously believed. The same DNA data indicate a relative lack of genetic differentiation among populations of the northern species of *Prietella*, thus indicating substantial gene flow and subterranean hydrologic connections, throughout a large portion of northern Coahuila (García de León et al. 1998).

Almost completely unexplored are faunas of the deep artesian aquifers of Coahuila, Nuevo León, and Tamaulipas. Accessible only by wells, the water in these aquifers ranges in quality from fresh to saline and sulfurous, as does the water of the *cenotes* of the Tamaulipan coastal plain.

These *cenotes* extend well beyond the reach of divers who have explored to greater depths than 250 m (James Bowden, pers. comm.). It thus seems likely that a "good-water"/"bad-water" interface like that underlying San Antonio extends through Tamaulipas as well, harboring a similar, and perhaps related, fauna (Henderson and Krejca 1997).

Threats to karst aquifers are severe in our region. A large number of Texas springs have been dried in this century by aquifer pumping (Brune 1981). The "good-water" side of the San Antonio Pool of the Edwards Aquifer is the sole water source for the two million inhabitants of the San Antonio metropolitan area (Crowe and Sharp 1997), whose rapid growth poses many threats to the aquifer. Pumping continues at high rates, resulting in endangerment of a number of spring endemics (Edwards et al. 1984; Crowe and Sharp 1997). In contrast, the diverse deep San Antonio Pool fauna has gone unsampled for twenty years. Recent efforts to list some of this fauna as endangered failed because of lack of knowledge (U.S. Fish and Wildlife Service, Division of Endangered Species 1998), yet efforts to study it have been impeded by well owners concerned that, should explorations uncover new endangered species, the political system may further complicate water management. In Mexico, irrigated agriculture and industry are relying increasingly on aquifer pumping. Many springs have already dried as a result, and the large city of Monterrey searches more and more widely to satisfy the thirst of its rapidly growing population (Contreras-Balderas and Lozano-Vilano 1994; Vázquez-Gutiérrez 1997). Farther west, near Torreón, and in Sonora, aquifer overdrafts have reversed groundwater flow, resulting in major and detrimental shifts in well-water quality. Mexican officials have recently contested plans to install toxic waste dumps near the border in Texas, contending that they may contaminate Mexican aquifers, and rapid industrial development all along the border also poses threats of contamination. The future holds much uncertainty for the little-studied stygobitic fauna of the region.

Although our approach emphasizes the importance of conserving the largest number of endemic species possible, a strict evaluation of the number of endemics can obscure patterns of endemism in species-poor ecoregions. When we examine the percentage of each ecoregion's species that are endemic, different patterns emerge (figure 3.3). For freshwater fish, eleven ecoregions have values of 50 percent or greater, and all of these ecoregions belong to either the endorheic, xeric, or coastal MHTs. One ecoregion, Lake Catemaco [65], has an astounding 100 percent endemism in its twenty-six fish species (Several species are shared with neighboring ecoregions, but according to endemism decision rules employed in this assessment, species with limited ranges may be considered endemic to multiple ecoregions. See appendix A for additional details.) A map of the percentage of endemism in fish shows the heightened importance of western North America, particularly the United States and central Mexico, when compared with a map of the total number of endemic species (figure 3.4c).

For crayfish, the percentage of endemism statistic reveals high rates of endemism in a substantial number of ecoregions. Five ecoregions, all within Mexico and belonging to either endorheic or xeric MHTs, have 100 percent crayfish endemism, although all have only one or two species. Twenty ecoregions have endemism of 50 percent or more, and an additional eight have between 20 and 50 percent. The main difference in geographic pattern between percentage of endemism and the number of endemics is that several Mexican ecoregions are highlighted when percentage of endemism is mapped (figure 3.6c).

Percentage of endemism in mussels does not reach the high levels found in crayfish and fish, except in the Yukon [55] ecoregion, where the single mussel species is endemic. Percentage of endemism generally follows a geographic pattern similar to that for the number of endemics, with the highest values (except the Yukon) in ecoregions of the Mississippi and Atlantic complexes (figure 3.7c). The Pacific Central Valley [6] ecoregion stands out in the West with 20 percent endemism, though this percentage is derived from a single endemic mussel species. Ongoing research suggests that many mussel species may remain to be classified.

As with mussels, percentage of endemism values for herpetofauna do not reach the same high levels as for fish and crayfish, perhaps because ecoregion boundaries were not derived from herpetofauna distributions. The highest values for percentage of endemism in herpetofauna are found in the endorheic Death Valley [11] ecoregion, where 36 percent of the small herpetofauna assemblage (N = 11) is endemic, and in the Tehuantepec [74] ecoregion, where 34 percentage of the ecoregion's 125 species are found nowhere else. Values greater than 20 percent also occur in the species-rich Papaloapan [71] ecoregion, and in the endorheic Lerma basin [69]. The percentage of endemism map looks similar to that for the number of endemics, but with a greater emphasis on the western United States and southern Mexico (figure 3.8c).

The value of examining both endemism and percentage of endemism is illustrated by the cases of the Tennessee-Cumberland [35] and Lerma [69] ecoregions. The Tennessee-Cumberland ecoregion has sixty-seven endemic fish, far more than any other ecoregion, and these sixty-seven constitute 29 percent of all native fish species in the ecoregion. The Lerma ecoregion, in contrast, has thirty endemic fish, but they constitute 63 percent of its fish fauna. In conservation terms, if the fish fauna of the Tennessee-Cumberland ecoregion were decimated, sixty-seven species would be lost from the earth, but 71 percent of the ecoregion's extensive fauna would still exist elsewhere. If the Lerma ichthyofauna disappeared, less than half as many fish would be lost, but only 27 percent of the Lerma ecoregion's species would be found in other ecoregions. Clearly, both faunas deserve protection.

Patterns of endemism among MHTs are in many cases quite different from richness (table 3.2). A ranking of MHTs by the average number of fish species highlights temperate headwaters and lakes, large temperate lakes, and large temperate rivers. A ranking by the average number of endemic fish species, on the other hand, shows generally high endemism in headwater, coastal, and endorheic MHTs, but not in large rivers or lakes. Finally, endorheic, xeric, and subtropical coastal MHTs exhibit a far greater degree (percentage) of endemism among fish than any of the other MHTs.

As with richness, within MHTs there is often substantial variation among endemism values (table 3.2). For example, 29 percent of the fish species in the Tennessee-Cumberland [35] ecoregion are endemic, compared with the second-highest value in that MHT of 12 percent endemism in the Teays-Old Ohio [34] ecoregion. Comparison of ecoregions within the same MHT, then, often reveals ecoregions that have clearly distinct biodiversity components.

TABLE 3.2   Mean Biodiversity Indicators of MHTs, and Associated Ranks.

MHT	Avg. Fish Richness	Rank, Fish Richness	Avg. Fish Endemism	Rank, Fish Endemism	Avg. Fish % Endemism	Rank, Fish % Endemism
ARL	31	6	1	7	1	7
LTL	103	2	0	8	0	8
THL	115	1	12	1	6	6
LTR	94	3	5	6	8	5
ERLS	19	8	11	3	55	1
XRLS	29	7	9	5	42	2
TCRL	83	4	11	3	11	4
SCRL	51	5	12	1	30	3

MHT	Avg. Crayfish Richness	Rank, Crayfish Richness	Avg. Crayfish Endemism	Rank, Crayfish Endemism	Avg. Crayfish % Endemism	Rank, Crayfish % Endemism
ARL	1	6	0	5	0	7
LTL	9	4	0	5	0	7
THL	21	1	11	1	31	2
LTR	12	3	5	3	17	6
ERLS	1	6	1	4	42	1
XRLS	0	8	0	5	18	5
TCRL	17	2	8	2	23	4
SCRL	7	5	4	4	31	2

MHT	Avg. Herp Richness	Rank, Herp Richness	Avg. Herp Endemism	Rank, Herp Endemism	Avg. Herp % Endemism	Rank, Herp % Endemism
ARL	3	8	0	6	0	7
LTL	27	7	0	6	0	7
THL	34	5	1	4	2	4
LTR	41	2	1	4	2	4
ERLS	32	6	3	2	9	2
XRLS	36	4	0	6	2	4
TCRL	41	2	3	2	5	3
SCRL	76	1	10	1	10	1

MHT	Avg. Mussel Richness	Rank, Mussel Richness	Avg. Mussel Endemism	Rank, Mussel Endemism	Avg. Mussel % Endemism	Rank, Mussel % Endemism
ARL	1	5	0	4	13	1
LTL	29	2	0	4	0	5
THL	45	1	4	1	4	3
LTR	26	3	1	3	2	4
ERLS						
XRLS						
TCRL	21	4	4	1	9	2
SCRL						

Summary Table (mussel ranks excluded)			
MHT	Average Rank	Mode of Ranks	Range of Ranks
ARL	6.7	7	3
LTL	6.1	7	6
THL	2.8	1	5
LTR	4.0	3	4
ERLS	3.7	1	7
XRLS	5.1	5	6
TCRL	3.5	4	2
SCRL	2.6	1	4

Rare Ecological or Evolutionary Phenomena

Unlike richness or endemism, rare ecological and evolutionary phenomena cannot be counted, ranked, or comprehensively mapped. However, these phenomena represent one of the essential components of biological distinctiveness that we seek to conserve (see essay 6). Our evaluation of the biological distinctiveness of these phenomena is based primarily on expert assessment. Additionally, we included in this category the presence of endemic

---

ESSAY 6

Connecting the Land to the Sea: Anadromous Fish

Peter B. Moyle

Along the Pacific Coast of North America flows a powerful, cold, ocean current. It starts in the Gulf of Alaska and flows southward, gradually weakening until its remnants are spun offshore by Point Conception in Southern California. The current generates upwelling of nutrient-rich bottom waters along its length, making the coastal ocean extraordinarily productive of life, from whales to bacteria. The coast along which this current flows consists of abrupt mountains of rough and twisted rock, pierced by volcanoes and riven with faults from the interactions of the giant crustal plates that come together here. The unstable mountains intercept the rain and fog rising from the ocean, growing giant forests and creating turbulent rivers that carry the water back to its origins. The mountains and the ocean, however, are connected by more than just water; they are connected by all the life forms that move between them, most visibly the fish. The fish carry nutrients from the rich ocean into the rocky interior, fertilizing streams and estuaries, supporting large numbers of bears, eagles, and other symbolic creatures, and transferring the fertility of the ocean to the land and forests.

The fish most obviously involved in this transfer are salmon—pink (*Oncorhynchus gorbuscha*), chum (*O. keta*), sockeye (*O. nerka*), chinook (*O. tshawytscha*), coho (*O. kisutch*)—because they are so universal in the region, so large, and so many. Yet there are many other fish involved as well: Pacific lamprey (*Lampetra tridentata*), river lamprey (*L. ayresi*), white sturgeon (*Acipenser transmontanus*), green sturgeon (*A. medirostris*), dolly varden (*Salvelinus malma*), cutthroat trout (*O. clarki*), steelhead trout (*O. mykiss*), inconnu (*Stenodus leucichthys*), longfin smelt (*Spirinchus thaleichthys*), eulachon (*Thaleichthys pacificus*), prickly sculpin (*Cottus asper*), coastrange sculpin (*C. aleuticus*), threespine stickleback (*Gasterosteus aculeatus*), and others. Each of these species has given rise to many, sometimes hundreds, of distinct local forms adapted for a particular combination of stream, coastal, and ocean conditions. River lamprey, cutthroat trout, and threespine stickleback, for example, have given rise to forms that do not return to sea in practically any of the watersheds they inhabit. These forms typically coexist with sea-run (anadromous) forms, with or without interbreeding. Taxonomists have thrown up their hands in despair of ever naming all the forms, while evolutionary biologists have delighted in the complex and dynamic interactions of the species and their environments.

Dynamic species are required to live in the dynamic environment of the Pacific Coast, an environment where ocean productivity waxes and wanes; where volcanoes erupt and sterilize river systems with hot volcanic sludge; where landslides block rivers; where glaciers form and melt; where new streams appear when earthquakes move mountains. Thus it is not surprising that the total number of anadromous species is small, although the number of variations on the theme of each species is large.

Conservation of such dynamic species is difficult because they require large amounts of habitat to persist, especially in streams and rivers. If the coastal current weakens (as it does periodically), ocean productivity declines followed by a decline in the survival of fish that depend on that productivity. Thus survival of anadromous fishes in fresh water has to be high in order to survive the periods of low

---

Migratory fish and intact habitat. Once-abundant Pacific salmon (*Oncorhynchus* sp.) migrating up coastal streams to spawn are important ecological phenomena (left photo) (photo by G. Haknel, courtesy of US Fish and Wildlife Service). Some of the best remaining intact habitat for salmon is located in the Arctic, such as in the Kaguyak River, Kodiak Island, Alaska (right photo) (photo by Dominick Dellasala).

survival in salt water. Yet if there is any one thing that lies at the bottom of the widespread and calamitous decline of Pacific Coast anadromous fishes, it is the failure of modern humans to recognize the importance of the ocean-stream connection. We build dams that deny salmon and lampreys access to historic spawning and rearing areas. We overharvest fish so too few make it back to replenish the supply or to maintain the genetic diversity needed to respond to sudden change. We log, road, graze, farm, urbanize, and otherwise use delicate watersheds in ways that make the streams inimical to fish. In short, we diminish the resiliency of our ecological systems. No wonder that even anadromous fishes that are not harvested by humans are disappearing from many streams.

Nowhere is the decline of anadromous fish more evident than in California, where most of these fish reach their natural southern limits. Lampreys are now scarce in the Eel River, named for the wriggling masses of migrants. Green sturgeon now spawn in just three rivers on the entire coast, two in California. Eulachon, once an important Indian fishery, have virtually disappeared from the Klamath and neighboring rivers. Longfin smelt in the Sacramento-San Joaquin estuary and delta are in danger of extinction. Coho salmon, once abundant in all coastal streams from Santa Cruz to the Oregon border, are at less than 5 percent of their historic numbers. The four runs of chinook salmon in the Central Valley, once numbering in the millions, are either listed as endangered or are proposed for listing. Steelhead, perhaps the most resilient of all the anadromous fishes, are listed as threatened in all parts of the state except the North Coast (where they are in decline). Pink salmon are extinct.

To restore the diversity of anadromous fishes requires renewing the ocean-freshwater connection, both literally (by restoring flows to dewatered streams) and figuratively (by restoring watersheds in ways that allow them to accept the gifts of the salmon and other fish). There seems to be growing recognition of this need, reflected in the growth of citizens' watershed organizations and in large studies on the possibilities of ecosystem-based restoration. One example is that of Putah Creek, a tributary to the Sacramento River that flows through orchards and tomato fields. It has been systematically degraded for more than 130 years. Yet the Putah Creek Council was formed by citizens who recognized the possibilities and delights of having a healthy stream nearby once again. Their efforts and wet years have conspired to bring chinook salmon and perhaps steelhead back to the creek, joining a struggling lamprey population. Suddenly the council's dreams have an unexpected lining of reality and a new dream emerges: black bears following the stream's course down from the hills, to feed on the salmon and reconnecting the land to the sea.

The efforts of the Putah Creek Council are connected indirectly to a much larger, official multiagency Ecosystem Recovery Program for the Bay-Delta region. The Bay-Delta program exists because it has finally become obvious that the health of California's economy rests on healthy ecological systems. In neither case is anyone expecting miracles, only slow change, but even reversing present downward trends in the anadromous fish populations will be a sign that the renewal of the ocean-freshwater connection in Central California is at least possible. It should be elsewhere as well.

higher taxonomic groups (genera, families) and unique species assemblages. Eight ecoregions were considered globally outstanding (Cuatro Ciénegas [22], Florida [39], Manantlan-Ameca [64], Chapala [65], Río Verde Headwaters [67], Tamaulipas-Veracruz [68], Lerma [69], and Catemaco [72]), and an additional forty ecoregions were judged continentally outstanding (table 3.3, figure 3.9).

Rare Habitat Type

Of the eight MHTs identified for North America, the large temperate lake MHT had the smallest representation, with four ecoregions covering the Great Lakes area. Because there were more than three examples of each MHT within North America, none was considered continentally outstanding. Likewise, as none of the MHTs was represented by fewer than eight examples worldwide (Lakes Baikal, Balkhash, Ladoga, and the Aral Sea increase the total number of large temperate lake ecoregions to at least eight), none of the MHTs was considered globally rare. No ecoregions, therefore, had their biological distinctiveness rankings increased because of MHT.

A number of ecoregions contained rare habitat types not found on a scale large enough to warrant their own MHT. Where these habitat types were judged to be continentally or globally rare, the biological distinctiveness scores of the ecoregions containing them were elevated accordingly (table 3.3). Examples include extensive spring systems and their associated above- and below-ground habitats (see essay 7), and more ephemeral habitats such as vernal pools (essay 8). Other freshwater habitats not particularly rare but of important biological value are described in essay 9 and appendix G.

Synthesis of Biological Distinctiveness Data

When the richness and endemism scores were summed for each ecoregion and biological distinctiveness categories were assigned, ten ecoregions were considered globally outstanding, sixteen continentally outstanding, twenty-five bioregionally outstanding, and twenty-five nationally important (table 3.1, figure 3.10). When ecoregion status was elevated following an evaluation of phenomena and rare habitat types, the resulting tally was fifteen globally outstanding ecoregions, forty-one continentally outstanding, four bioregionally outstanding, and sixteen nationally important (figure 3.11).

Both before and after consideration of phenomena and rare habitat types, all MHTs contained at least one globally outstanding ecoregion, with the exception of arctic rivers and lakes and large temperate lakes. With consideration of phenomena, five xeric ecoregions and four subtropical coastal ecoregions were assessed as globally outstanding. Although neither the arctic nor large lakes MHTs had globally outstanding ecoregions, each had a large proportion of its ecoregions assessed as continentally outstanding once phenomena were considered (three of eight for arctic, two of four for large lakes).

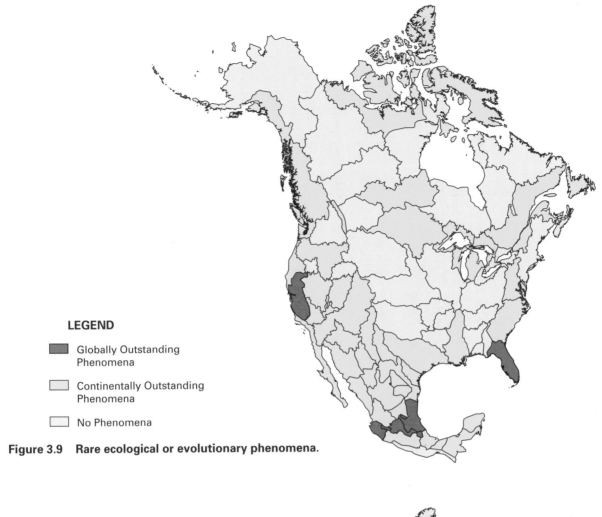

LEGEND

■ Globally Outstanding
  Phenomena

□ Continentally Outstanding
  Phenomena

□ No Phenomena

Figure 3.9   Rare ecological or evolutionary phenomena.

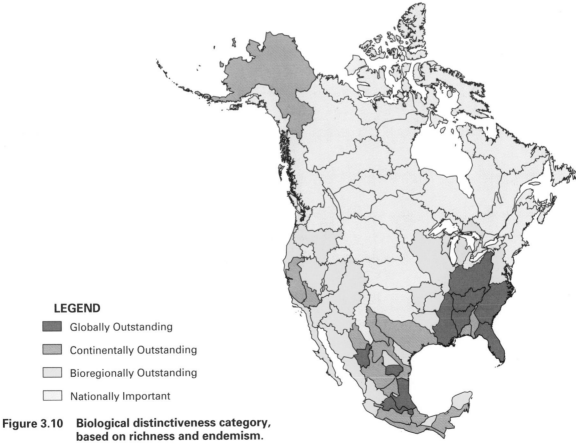

LEGEND

■ Globally Outstanding

▦ Continentally Outstanding

□ Bioregionally Outstanding

□ Nationally Important

**Figure 3.10   Biological distinctiveness category,
based on richness and endemism.**

TABLE 3.3 Rare Ecological and Evolutionary Phenomena.

Ecoregion	Phenomenon Category	Description of Phenomenon
North Pacific Coastal [1]	Continentally Outstanding	one endemic fish genus (*Novumbra*); major populations of anadromous fish (seven *Oncorhynchus*, three *Lampetra*); karst-formed caves with unique invertebrate spring fauna
Columbia Unglaciated [3]	Continentally Outstanding	numerous runs of anadromous fish (chinook, coho, steelhead, sockeye); near-endemic genus of fish (*Oregonichthys*)
Pacific Mid-Coastal [5]	Continentally Outstanding	southernmost populations of most Pacific Coast anadromous species and their derivatives; one endemic fish genus (*Deltistes*) and two near-endemics (*Oregonichthys* and *Eucyclogobius*); and near-endemic salamander genus (*Hydromantes*); vernal pools fauna
Pacific Central Valley [6]	Globally Outstanding	four runs of chinook; southernmost populations of five anadromous fish; near-endemic salamander genus (*Hydromantes*); five endemic fish genera (*Archoplites, Pogonichthys, Orthodon, Lavinia, Mylopharodon*) and only freshwater representative of *Hysterocarpus*; vernal pools fauna
South Pacific Coastal [7]	Continentally Outstanding	southernmost populations of anadromous Pacific lamprey and steelhead; near-endemic fish genus (*Eucyclogobius*), found in coastal lagoons
Bonneville [8]	Continentally Outstanding	has roughly 100 internally drained basins, centers of which include the Great Salt Lake and Bear Lake (with four endemic fish species); high beta-diversity; endemic fish genus (*Iotichthys*)
Lahontan [9]	Continentally Outstanding	three endemic or near-endemic fish genera (*Eremichthys, Relictus,* and *Crenichthys*)
Death Valley [11]	Continentally Outstanding	complete desert spring and river system; among the most extreme conditions inhabited by aquatic life (Devil's Hole, Cottonball Marsh); high beta-diversity; endemic fish genus (*Empetrichthys*)
Colorado [12]	Continentally Outstanding	endemic fishes specially adapted to high-flow large-river system; three endemic or near-endemic fish genera (*Xyrauchen, Lepidomeda, Plagopterus; see also 13 and 14 below*)
Vegas-Virgin [13]	Continentally Outstanding	four endemic or near-endemic fish genera (*Moapa, Lepidomeda, Plagopterus, Crenichthys*)
Gila [14]	Continentally Outstanding	three endemic or near-endemic fish genera (*Meda, Plagopterus, Tiaroga*)
Upper Rio Grande [15]	Continentally Outstanding	one endemic fish and invertebrate endemism in thermal springs (*Fontelicella, Tryonia, Thermosphaeroma*)
Guzmán [16]	Continentally Outstanding	endorheic habitats; high levels of endemism among fish; spring and subterranean habitats
Rio Conchos [17]	Globally Outstanding	only free-flowing large-river habitat left in Rio Grande catchment; high levels of endemism in cave and spring habitats
Pecos [18]	Continentally Outstanding	high fish richness and endemism for the Rio Grande system, including three pupfish (*Cyprinodon*), *Etheostoma, Cichlasoma, Cyprinella,* and *Gambusia*
Mapimí [19]	Continentally Outstanding	extinct, monotypic fish genus (*Stypodon*) and extant endemic fish genus (*Megupsilon*)
Rio Salado [21]	Continentally Outstanding	outstanding fish endemism, including minnows, darters, mosquitofish, and platyfish
Cuatro Ciénegas [22]	Globally Outstanding	diverse complex of hundreds of geothermal springs, lakes and streams; high beta-diversity
Rio San Juan [23]	Globally Outstanding	high levels of fish endemism and spring fauna
Mississippi Embayment [25]	Continentally Outstanding	large-river fish that exhibit many adaptations for constantly turbid environments
East Texas Gulf [32]	Continentally Outstanding	contains Edwards Aquifer and associated spring systems, with unique species assemblages; one endemic salamander genus (*Typhlomolge*); hypogean and spring fauna
West Texas Gulf [33]	Continentally Outstanding	two endemic fish genera (*Satan, Trogloglanis*); Edwards Aquifer hypogean and spring fauna
Teays-Old Ohio [34]	Globally Outstanding	very large assemblage of species, especially in fish (208) and mussels (122)
Tennessee-Cumberland [35]	Continentally Outstanding	one endemic fish genus (*Speoplatyrhinus*) and one near-endemic salamander genus (*Leurognathus*)

Area	Status	Description
Mobile Bay [36]	Globally Outstanding	wide variety of habitats; highest level of aquatic diversity in the eastern Gulf; high levels of fish, mollusk, and amphibian endemism
Apalachicola [37]	Continentally Outstanding	wide variety of habitats; spring fauna; 45% endemism in unionid mussels
Florida Gulf [38]	Continentally Outstanding	high levels of unionid mussel endemism; rich fish fauna
Florida [39]	Globally Outstanding	diverse aquatic landscapes, including Order 1 springs, tremendous freshwater swamps and marshes, and the Everglades; high beta-diversity; one endemic crayfish genus (Troglocambarus)
South Atlantic [40]	Continentally Outstanding	one endemic salamander genus (Leurognathus); one endemic crayfish genus (Distocambarus)
Chesapeake Bay [41]	Continentally Outstanding	anadromous fish (striped bass, blueback herring, alewife, American and hickory shad, shortnose sturgeon, and Atlantic sturgeon), plus semianadromous fish (e.g. white perch) require fresh water to spawn
Superior [43]	Continentally Outstanding	largest temperate freshwater lake (in terms of surface area); contains recently evolved cisco complex and other "new" endemics
Michigan-Huron [44]	Continentally Outstanding	with Superior, contains recently evolved complex of deepwater fish (ciscos), which played dominant ecological role before overharvest
Lower St. Lawrence [47]	Continentally Outstanding	home to healthiest surviving stocks of Atlantic salmon in the world
Canadian Rockies [49]	Continentally Outstanding	glacial refugium, with endemism in gastropods, isopods, amphipods, and fish; also unusual subterranean habitats
Lower Saskatchewan [51]	Continentally Outstanding	rich in oligotrophic, glacial lakes and clear streams; streams provide food for summering polar bears
English-Winnipeg Lakes [52]	Continentally Outstanding	prairie rivers; three big, shallow, and productive lakes (Lakes Winnipeg, Winnepegosis, and Manitoba)
Yukon [55]	Continentally Outstanding	four endemic fish despite Arctic location; areas escaped glaciation.
North Arctic [58]	Continentally Outstanding	migration of Arctic char and other anadromous fish from Mackenzie Delta to tributary streams for feeding runs
Arctic Islands [60]	Continentally Outstanding	some lakes with postglacial marine relict fishes and invertebrates
Sonoran [61]	Continentally Outstanding	three endemic snails and eight endemic fish; adaptations to xeric freshwater conditions
Sinaloan Coastal [62]	Continentally Outstanding	endemism among fish and crayfish; subtropical high-gradient habitats
Santiago [63]	Continentally Outstanding	endemic fish (including cichlids and catfish); warm spring habitats for aquatic microfauna
Manantlan-Ameca [64]	Globally Outstanding	five endemic or near-endemic fish genera (Allodontichthys, Xenotaenia, Ameca, Ilyodon, Skiffia)
Chapala [65]	Continentally Outstanding	two near-endemic fish genera (Chapalichthys, Skiffia)
Llanos el Salado [66]	Continentally Outstanding	very high endemism in fish, mostly pupfish; endorheic
Rio Verde Headwaters [67]	Globally Outstanding	La Medialuna; monotypic genera in Cyprinodontidae and Goodeidae
Tamaulipas-Veracruz [68]	Globally Outstanding	two near-endemic fish genera (Prietella, Xenoophorus)
Lerma [69]	Globally Outstanding	four endemic or near-endemic fish genera (Hubbsina, Skiffia, Chapalichthys, and Evarra [all E. extinct]) and one near-endemic salamander genus (Rhyacosiredon)
Balsas [70]	Continentally Outstanding	three near-endemic fish genera (Ameca, Chapalichthys, Ilyodon) and one near-endemic salamander genus (Rhyacosiredon)
Papaloapan [71]	Continentally Outstanding	17% percent fish endemism, including Priapella, Gambusia, and Heterandria; high levels of endemism (26%) for aquatic herpetofauna
Catemaco [72]	Globally Outstanding	ancient, volcanic crater; 44% fish endemism
Coatzacoalcos [73]	Continentally Outstanding	very high fish endemism, especially among poeciliids
Tehuantepec [74]	Continentally Outstanding	outstanding fish endemism (29%) and extraordinary richness and endemism in aquatic herpetofauna
Grijalva-Usumacinta [75]	Continentally Outstanding	large variety of habitats, including large rivers, endorheic basins, and large wetlands; 41% fish endemism
Yucatán [76]	Continentally Outstanding	subtropical karst habitats, including springs and aquifers

Western Springs: Their Faunas and Threats to Their Existence

W. L. Minckley and Peter J. Unmack

Springs occur where subsurface water rises to the land surface. In many parts of western North America, they are more constant and more reliable than other aquatic habitats and often form sources for surface streams and lakes. With desertification, aquatic habitats shrink and springs soon become isolated archipelagos in seas of aridity, continuing to flow long after perennial lakes and streams are gone. Ultimately they may become the only natural refuges for whole biotas. A substantial proportion of aquatic life in deserts as well as of terrestrial organisms reliant on perennial water is intimately associated with springs and spring-fed systems.

A suite of organisms is typically associated with springs throughout the world. Best known are fishes, of diverse phylogeny but sharing many common morphological and physiological characteristics; most are small in size, with short generation times and broad physiochemical tolerances.

Springs as fish habitat present an unresolved dilemma. Despite a significant number of species relying on such habitats (Meffe 1989; Minckley et al. 1991), not many consistently occupy the immediate zones where water rises to the surface. We rarely know why one fish succeeds in headsprings and another fails (Courtenay and Meffe 1989; Hubbs 1995). A few springs are too hot or present an equally insurmountable challenge such as low levels of dissolved oxygen (Sumner and Sargent 1940; Hubbs and Hettler 1964; Hubbs et al. 1967), yet most seem quite hospitable to human senses. Springheads must present one or more chronic extremes too severe for some species to succeed (e.g., constancy may exclude those requiring seasonal change for gonad maturation). They are often occupied by a single fish species, sometimes locked into the habitat by some specialization such as stenothermy (narrow temperature tolerance).

In contrast, some spring-inhabiting fishes are equally capable of living elsewhere, often under harsh conditions. Examples are pupfishes (genus *Cyprinodon*), some of which thrive equally well in highly fluctuating habitats as in headsprings. Outflow gradients are often occupied by pupfishes and some colonize terminal lakes fed by springs where salinities exceed that of sea water by factors of 3.0 or more and annual temperature variation approaches 40°C (Soltz and Naiman 1978). Most of these fishes are apparently noncompetitive. Despite broad tolerance to physiochemical extremes, they tend to remain allotopic to other fishes in habitats marked by constancy (as in springs) or severity (e.g., high temperature, hypersalinity), excluding all but themselves. Collectively, springs and spring-fed places support a large percentage of the ichthyofauna in the arid, often endorheic basins of western North America. Single-spring (or spring-group) endemism is high in fishes and far higher in other groups.

Unlike fishes, many other spring-dwelling organisms tend to be restricted to headsprings and the uppermost outflows. Hydrobiid snails are abundant and display tremendous diversity. For example, fifty-eight new species of a single genus were described in 1998 from springs in the Great Basin (Hershler 1998). Numbers of hydrobiid taxa are often proportionate to size, number, and proximity of springs to one another. Individual springs may have only a single species, small groups of springs between one and four species of one or two genera. Large springfields like Ash Meadows, Nevada, and Cuatro Ciénegas, Coahuila, have eleven and thirteen species in two and nine genera, respectively (Hershler 1984, 1985; Hershler and Sada 1987). Amphipod and isopod crustaceans are also frequent (e.g., Cole 1984), but as with most other, smaller animals, have scarcely been studied.

Springs and their biotas are critically threatened by human intervention. When water quality is suitable for irrigation or domestic use, excessive pumping soon stops natural water flow, killing entire surface ecosystems. Overuse of subsurface water can be insidious and difficult to detect, since effects at an outflow may appear only after years of pumping 10 to more than 100 km away. Agricultural pumping destroyed some of the largest Texas springs (Brune 1975). Water extraction for residential development and agriculture in Ash Meadows ended only after a favorable judgment by the U.S. Supreme Court (Deacon and Williams 1991); and entire biotas were destroyed in less than a decade in Valle de Sandía, Nuevo León, where at least four endemic pupfishes and unknown numbers of invertebrates disappeared with surface water before they could be fully recorded (Lozano-Vilano and Contreras-Balderas 1993; Contreras-Balderas and Lozano-Vilano 1996). Mexican springs are especially endangered because of a new availability of electricity in remote areas and expanding irrigation to support burgeoning populations. Almost 100 springs in Mexico are known to have dried in the last few decades. To date, however, there seems to be no strong support for conservation and laws are not well enforced (Contreras-Balderas 1991; Contreras-Balderas and Lozano-Vilano 1994).

Springs not pumped dry are often capped or diverted, or, if large enough, converted for recreation, all detrimental to their biotas (Contreras-Balderas 1984, 1991; Sheppard 1993). Intensive livestock grazing degrades springs through vegetation removal and trampling, as well as through the deposit of excrement and dead carcasses. Once springs are disturbed, fencing to protect them may have an opposite effect, as vegetation invades to choke headsprings and outflows. They also suffer from the effects of introduced species: the stocking of sportfish and bait such as crayfish jeopardizes springs in Owens Valley, California (Minckley et al. 1991), Ash Meadows and the White River system, Nevada (Courtenay et al. 1985; Pister 1991), and Cuatro Ciénegas, Coahuila (original data). Introduced aquarium fishes and invertebrates, such as poeciliid and cichlid fishes and *Melanoides* snails, are now common (Williams et al. 1985; Contreras-Arquieta et al. 1995). Mosquitofish have also been stocked worldwide, supposedly to control mosquitoes (Courtenay and Meffe 1989).

Because springs are so small and isolated, their absolute numbers of species are small; nonetheless, they contain and support a disproportionate amount of biodiversity, as they often represent the only existing surface water. Thus, habitat loss and alteration are highly destructive of both local and regional biodiversity on a relative scale. This is especially true since relict and endemic species, and whole ecosystems, are lost in a single stroke. Only public demand for conservation, coupled with swift and definitive action, can reverse the imminent extinction of a large percentage of arid-land springs and their irreplaceable biotas.

ESSAY 8

California Vernal Pools

Barbara Vlamis

Vernal pools and swales (vernal wetlands) are unique freshwater beacons in a larger mosaic of grassland, oak woodland, and riparian woodland in the California landscape. They are ponds and drainages varying in size from a few square feet to many acres, perched over restrictive soil layers or bedrock. The impermeable layers permit these wetland wonders to support extensive biotic activity during the winter and spring wet seasons and then radically transform into hot, dry grasslands until the following rainy season.

The endemic biota in the vernal wetlands is what makes these ephemeral havens such biological wonders. Vernal wetlands provide habitation and foraging for many special-status species (species considered rare or imperiled). Four freshwater special-status shrimp are found intermittently in pools: conservancy fairy shrimp (*Branchinecta conservatio*), vernal pool fairy shrimp (*Branchinecta lynchi*), longhorn fairy shrimp (*Branchinecta longiantenna*), and vernal pool tadpole shrimp (*Lepidurus packardi*). The shrimp are an integral part of this mosaic, providing food-chain support for migratory waterfowl and other native animals (Krapu 1974; Swanson et al. 1974; J. King pers. comm. 1992). The western spadefoot toad (*Scaphiopus hammondii*) utilizes vernal pools for breeding and is a species of special concern on California's Endangered Species List.

Numerous listed birds rely on the grasslands surrounding the vernal wetlands for foraging, including the Swainson's hawk (*Buteo swainsoni*), Aleutian Canada goose (*Branta canadensis leucopareia*), ferruginous hawk (*Buteo regalis*), golden eagle (*Aquila chrysaetos*), American peregrine falcon (*Falco peregrinus anatum*), merlin (*Falco columbarius*), northern harrier (*Circus cyaneus*), prairie falcon (*Falco mexicanus*), sharp-shinned hawk (*Accipiter striatus*), white-tailed kite (*Elanus leucurus*), greater sandhill crane (*Grus canadensis tabida*), long-billed curlew (*Numenius americanus*), short-eared owl (*Asio flammeus*), western burrowing owl (*Athene cunicularia hypugea*), and loggerhead shrike (*Lanius ludovicianus*).

Special-status plant species found in many local pools and associated grasslands include the Bogg's Lake hedge-hyssop (*Gratiola heterosepala*), Red Bluff dwarf rush (*Juncus leiospermus* var. *leiospermus*), Ahart's rush (*Juncus leiospermus* var. *ahartii*), legenere (*Legenere limosa*), wooly meadowfoam (*Limnanthes floccosa* ssp. *floccosa*), vieny monardella (*Monardella douglasii*), hairy Orcutt grass (*Orcuttia pilosa*), slender Orcutt grass (*Orcuttia tenuis*), Ahart's paronychia (*Paronychia ahartii*), Bidwell's knotweed (*Polygonum bidwelliae*), and Greene's tuctoria (*Tuctoria greenei*). Butte County meadowfoam (*Limnanthes floccosa* ssp. *california*) is a federally and state-listed endangered species and is found only along a narrow 25-mile strip from central Butte County to its northern border. There are only eleven known populations of this species.

Vernal wetland savannas perform many valuable functions for humans as well as wildlife. They assist in stormwater detention, minimizing flooding. These open grasslands also provide passive recreation opportunities (e.g., hiking, photography, bird-watching, wildflower viewing). The wetlands filter pollutants from stormwater runoff, enhancing local water quality. These savannas also store archaeological artifacts left by the Indian residents who dug for wild onions and, it is imagined, also used them for protein enhancement and hunting.

California as a whole has lost 91 percent of the estimated 5 million wetland acres that were present in the 1780s, leaving less than 450,000 acres in the state (National Audubon Society 1992). Vernal pools have not fared much better, with less than 30 percent of California's original vernal pools intact today (Holland 1978) and the remainder in imminent danger from urban sprawl, vineyard creation, and inadequate agency standards and staffing. Approximately 200,000 acres of wetlands remain from the original 4 million acres in California's Central Valley (ecoregion [6]) (National Audubon Society 1992; Kempka and Kollash 1990), making the extant vernal wetland savanna acreage critical to the vitality of the wildlands in this region.

The Central Valley is growing at twice the rate of the rest of the state. Given this burgeoning human population and agricultural expansion into previously marginal land, the remaining isolated wetlands in the valley are severely threatened. Major conservation efforts are needed to protect the remaining wildlands through voluntary conservation easements, rewards for compatible agricultural stewardship, and fee title acquisition. In the northern Central Valley, the Butte Environmental Council, Altacal Audubon, the Sierra Club, the Native Plant Society, and others have worked closely with local government, the U.S. Army Corps of Engineers, the EPA, the USFWS, and the California Department of Fish and Game. Still, high-value areas of remaining vernal wetland savanna continue to be lost. For example:

- Ten thousand acres of a vernal pool complex were filled in Tehama County by the Simpson Paper Company without a permit.
- Illegal wetland filling continues to occur on parcels of all sizes. There is little wetland enforcement capability in the area because of the distance from agency headquarters in urban centers and a shortage of personnel.
- Properties have been split to avoid thresholds that would have required more protection of the vernal pool landscape.

National regulatory guidelines for the destruction or degradation of wetlands are extremely limited at best, and at worst, nonexistent. At the head of the list is the infamous Corps' Nationwide Permit (NWP) 26. Before January 1997, this permit allowed wetland destruction without notice on plots of less than 1 acre, as well as losses of between 1 and 10 acres, often with minimal review.

The Sacramento Corps office alone allowed the destruction of more than 1,000 acres of wetlands between 1988 and 1996 under NWP 26. In the past two years, thresholds were lowered to require no reporting for wetland losses of less than one-third of an acre, with NWP 26 covering permits for up to 3 acres. The Corps intended to end the use of NWP 26 in December 1998 but is considering maintaining it in a revised form.

Vernal wetlands are frequently under an acre in size. For example, 180 acres of rich biological vernal wetlands contain only 5.6 acres of wetlands under the Corps' jurisdiction. Though the supporting landscape is critical for ponding, ecosystem health, and foodchain survival, it is neither evaluated nor protected under the Clean Water Act. When the Corps approves either an individual permit or one under NWP 26 to destroy wetlands, hundreds of acres of supporting grassland and oak woodland habitat may be part of the bargain, and also destroyed.

The USFWS's guidelines to protect endangered species may heighten mitigation requirements included on a permit, but to date no projects have been stopped by the Corps or the USFWS in the northern Central Valley. Additionally, loopholes allow the degradation of wetlands and endangered species by off-road vehicles and agricultural activity as long as soil is not *filling* the wetland.

There is clearly an urgent need for more restrictive regulations to protect the remaining natural wetland acres in California. Vernal wetlands are a wondrous portion of the natural heritage of California, and this landscape deserves to be preserved for future generations of humans and wildlife.

ESSAY 9

Prairie Potholes in Decline

Mark H. Stolt

Landscapes of the northern Midwest and Great Plains regions of North America are dotted with innumerable marsh-like freshwater wetlands referred to as prairie potholes. At the end of the last glaciation, ten to fifteen thousand years ago, retreating ice and raging meltwaters left the flat-barren landscape with 25 million of these shallow depressional wetlands. Prairie potholes range in size from less than 1 acre to several hundred. Many are lake-like in the spring from rain and snowmelt, and completely dry during summer droughts. Prairie potholes are dynamic ecosystems that support a wide range of wetland-dependent species and are critical for the functioning of landscape-level ecological processes.

The prairie pothole region covers roughly 300,000 square miles and extends into parts of Iowa, North Dakota, South Dakota, Montana, and Minnesota in the United States, and Alberta, Manitoba, and Saskatchewan in Canada (ecoregions [24], [26], [27], and [52]). Humans and animals rely heavily on this critical resource. Some of the most important agricultural soils in the world relative to wheat and corn production are found in the prairie pothole region. In contrast, innumerable hydrophytes including more than 30 submerged and floating aquatic plants, 300 migratory birds, 60 reptiles and amphibians, and more than 60 mammals find refuge and habitat in these wetlands. Half of the North American waterfowl are produced in the prairie pothole region, with the result that the area is called "North America's duck factory."

Water level dynamics, salinity, and the degree of agricultural disturbance are the primary factors controlling the array of plants and animals finding suitable habitat in the prairie potholes. Water levels are governed by the season, annual precipitation, and local and regional groundwater flow patterns. Surface water entering potholes is governed by both seasonal and long-term precipitation cycles. Most water entering potholes at the surface comes from snow or rain trapped in the winter and spring. In summer months considerable water is lost through evapotranspiration and the wetlands begin to dry down. Water may also be released to the regional and local groundwater through seepage. These groundwater-recharge wetlands are normally high on the landscape and the soils may have nutrient deficits. Many potholes receive groundwater on one end and release the water on the other (i.e., they are flow-through wetlands). Prairie potholes low on the landscape often receive groundwater. These recharge wetlands tend to be nutrient rich, but also quite saline.

The dynamic surface and subsurface hydrology, driven by yearly wet and dry seasons and periodic long-term droughts or multiple years of excess wetness, results in five different zones of vegetation within the pothole. Vegetation zones follow a gradient from high to low elevation, and are called wetland meadow (low grasses); shallow marsh (coarse sedges); emergent marsh (zones of tall herbaceous plants such as cattails); shallow open water (submerged aquatic vegetation); and open water. Not all zones can be found in each pothole, as vegetation comes and goes depending on extended wet periods or drought. Biomass can change by as much as twenty times from a dry to a wet year. Seed banks remain in the wetland and vegetation returns when the environmental conditions are conducive for a particular plant's emergence. Potholes with elevated salinity often have a reduced diversity of wetland flora.

Birds are the most important prairie pothole users. Almost 30 percent of the 800 migratory birds in North America use the prairie pothole region for breeding grounds, or for feeding and resting stops during annual fall and spring migrations. Some of these, such as the Baird's sparrow (*Ammodramus bairdii*), black tern (*Chlidonias niger*), interior least tern (*Sterna antillarum*), loggerhead shrike (*Lanius ludovicianus*), northern goshawk (*Accipiter gentilis*), peregrine falcon (*Falco peregrinus*), piping plover (*Charadrius melodus*), western burrowing owl (*Athene cunicularia*), and whooping crane (*Grus americana*) are either threatened, endangered, or being considered for federal listing.

Although approximately 5 percent of the wetlands in the lower forty-eight states are found in the prairie pothole region, publications consistently report that more than 50 percent of the ducks produced in North America each year come from these wetland resources. Their productivity can be attributed to the fertile soils, the rich supply of aquatic invertebrates for breeding females, and the diversity of vegetation. The five vegetation zones provide numerous islands of habitat for breeding pairs of ducks. Fifteen species of ducks find habitat in prairie potholes, with mallards (*Anas platyrhynchos*), blue-winged teals (*Anas discors*), and northern pintails (*Anas acuta*) constituting more than 60 percent of the population. Duck numbers in the region vary from year to year depending on the habitat each species seeks out, climatic conditions, and the degree of anthropogenic effects.

Land-use conflicts between agricultural production and wetland habitat have fragmented the landscape and severely reduced the number of prairie potholes functioning as habitat for plants and animals. Before settlement, an estimated 20 million acres of prairie potholes existed. Losses, however, occurred at rates as high as 33,000 acres per year, leaving less than half of the original 7 million acres of prairie potholes in North and South Dakota. In Minnesota, an estimated 9 million acres of potholes have been drained. Iowa has lost 98 percent of its pothole wetlands. Most of the losses are related to draining and alteration of the pothole hydrology for agricultural production. Burning, mowing, grazing, and the introduction of sediment, pesticides, and nutrients from adjacent agriculture have also affected the plant and animal population within prairie potholes.

Ducks have shown the greatest declines related to drainage of potholes. More than 15 million ducks were inventoried in the prairie pothole region in the mid-1950s. Today, only about 5 million ducks are produced each year. Mallards and northern pintails nest early and consequently have shown the most drastic declines as a result of draining. For example, the decline in the number of northern pintails for the decades of the 1970s and 1980s was 60 percent. Blue-winged teal numbers over this same period have declined by more than 30 percent.

Efforts to maintain the number of prairie potholes have met with minimal success. The USFWS has made a strong effort to acquire many prairie potholes. However, about 90 percent of the prairie wetlands are still owned by private parties. Because wetland use for agriculture is not restricted under the Clean Water Act, these prairie potholes continue to be drained for production. Provisions under national legislation (swamp-buster provisions in farm bills) have provided incentives to reduce the drainage of wetlands. About 18 percent of the prairie potholes in private hands fall under these or similar provisions and are somewhat protected, leaving 70 percent of these wetland resources without protection. Recent joint efforts involving the USFWS, a number of conservation organizations, and state and local government agencies have amplified the efforts to protect the prairie pothole region until the next glaciation. Much more still needs to be done to protect these diverse and critical prairie wetlands.

Figure 3.11   Final biological distinctiveness.

While biological distinctiveness categories were fairly well distributed within MHTs, the xeric and endorheic ecoregions stood out for the generally higher levels of biological distinctiveness within their MHTs. Of the eight endorheic ecoregions, seven were assessed as either continentally or globally outstanding, after consideration of phenomena. Likewise, all eleven xeric ecoregions were continentally or globally outstanding. In these MHTs, then, there are potentially more opportunities for conservation of important biodiversity features, as well as more possibilities that such important biodiversity features could be lost.

CHAPTER 4

Conservation Status of North American Ecoregions

"We are not faced with a catastrophic loss of freshwater biodiversity; the catastrophe has already happened."
W.L. Minckley (pers. comm.)

In this chapter we present broad trends detected in two components of the conservation status analysis. The first part presents the snapshot conservation status—the current situation as of 1997. The second part presents the final assessment modified by the threat analysis (projected threats over the next twenty years). More detailed information on the threats to individual ecoregions is located in appendix G.

Snapshot Conservation Status

Clear patterns of ecoregion-scale degradation and intactness, as of 1997, are observable both by looking at a map of North America (figure 4.1) and by examining trends among MHTs (table 4.1). The continent-wide map shows that all critical ecoregions are found in the western United States and Mexico, and that endangered ecoregions are concentrated in these regions as well as in the Great Lakes and the southeastern United States (table 4.2). The most intact ecoregions, as one might expect, are found in the arctic region, with addition of the Ozark Highlands [29] ecoregion of Arkansas and Missouri. Ecoregions assessed as relatively stable and vulnerable are found throughout the continent.

Examination of the snapshot assessment results by MHT shows that ecoregions of certain habitat types are far more degraded than others (table 4.1). Arctic ecoregions are all relatively intact or stable because of low human population and impact. In contrast, no large temperate lake, large temperate river, or endorheic ecoregion is relatively intact or stable; in fact, the majority of both large lakes and large rivers are endangered or critical. Xeric, temperate coastal, and subtropical coastal ecoregions are nearly as degraded. Only the temperate headwaters and lakes MHT appears to occupy the middle ground, with nearly all ecoregions either vulnerable or endangered.

Overall, the results of the snapshot assessment suggest that few ecoregions remain intact. A relatively small number are currently so degraded as to be potentially beyond restoration. The large number of ecoregions assessed as endangered and vulnerable

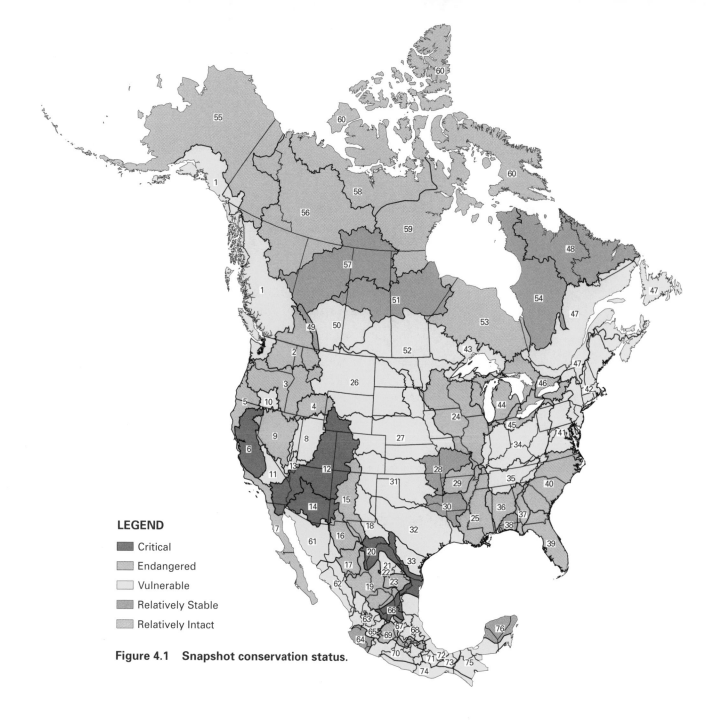

Figure 4.1 Snapshot conservation status.

LEGEND
- Critical
- Endangered
- Vulnerable
- Relatively Stable
- Relatively Intact

TABLE 4.1 Distribution of Ecoregions by MHT for the Snapshot (1997) Conservation Status Index.

Major Habitat Type	Snapshot Conservation Status Index (CSI)					
	Critical	Endangered	Vulnerable	Relatively Stable	Relatively Intact	Total
Arctic Rivers/Lakes	0	0	0	2	6	8
Large Temperate Lakes	0	3	1	0	0	4
Temperate Headwaters/Lakes	0	0	5	4	1	10
Large Temperate Rivers	2	3	2	0	0	7
Endorheic Rivers/Lakes/Springs	1	4	3	0	0	8
Xeric Rivers/Lakes/Springs	1	3	5	2	0	11
Temperate Coastal Rivers/Lakes	1	7	5	2	0	15
Subtropical Coastal Rivers/Lakes	0	1	9	3	0	13
Total	5	21	30	13	7	76

TABLE 4.2 Conservation Status Rankings.

Ecoregion	Major Habitat Type MHT	Snapshot Total (% possible)*	Snapshot Conservation Status Ranking (1–5)	Snapshot Conservation Status Category	Likelihood of Future Threats (H-M-L)	Final Conservation Status, Modified by Future Threat (1–5)	Final Conservation Status Category
South Hudson [53]	ARL	26.6	1	Rel. Intact	M	1	Rel. Intact
East Hudson [54]	ARL	42.2	2	Rel. Stable	M	2	Rel. Stable
Yukon [55]	ARL	29.7	1	Rel. Intact	M	1	Rel. Intact
Lower Mackenzie [56]	ARL	15.6	1	Rel. Intact	L	1	Rel. Intact
Upper Mackenzie [57]	ARL	37.5	2	Rel. Stable	H	3	Vulnerable
North Arctic [58]	ARL	15.6	1	Rel. Intact	L	1	Rel. Intact
East Arctic [59]	ARL	12.5	1	Rel. Intact	L	1	Rel. Intact
Arctic Islands [60]	ARL	12.5	1	Rel. Intact	L	1	Rel. Intact
Superior [43]	LTL	55.6	3	Vulnerable	M	3	Vulnerable
Michigan-Huron [44]	LTL	80.6	4	Endangered	H	5	Critical
Erie [45]	LTL	81.9	4	Endangered	H	5	Critical
Ontario [46]	LTL	84.7	4	Endangered	H	5	Critical
Central Prairie [28]	THL	34.1	2	Rel. Stable	M	2	Rel. Stable
Ozark Highlands [29]	THL	27.3	1	Rel. Intact	L	1	Rel. Intact
Ouachita Highlands [30]	THL	43.2	2	Rel. Stable	M	2	Rel. Stable
Southern Plains [31]	THL	68.2	3	Vulnerable	L	3	Vulnerable
Teays-Old Ohio [34]	THL	54.5	3	Vulnerable	L	3	Vulnerable
Tennessee-Cumberland [35]	THL	58	3	Vulnerable	H	4	Endangered
Canadian Rockies [49]	THL	47.7	2	Rel. Stable	H	3	Vulnerable
Upper Saskatchewan [50]	THL	67	3	Vulnerable	H	4	Endangered
Lower Saskatchewan [51]	THL	48.9	2	Rel. Stable	L	2	Rel. Stable
English-Winnipeg Lakes [52]	THL	61.4	3	Vulnerable	M	3	Vulnerable
Colorado [12]	LTR	87.5	5	Critical	H	5	Critical
Upper Rio Grande [15]	LTR	85	4	Endangered	H	5	Critical
Lower Rio Bravo [20]	LTR	87.5	5	Critical	H	5	Critical
Mississippi [24]	LTR	72.5	4	Endangered	M	4	Endangered
Mississippi Embayment [25]	LTR	77.5	4	Endangered	H	5	Critical
Upper Missouri [26]	LTR	67.5	3	Vulnerable	L	3	Vulnerable
Middle Missouri [27]	LTR	62.5	3	Vulnerable	L	3	Vulnerable
Bonneville [8]	ERLS	70	3	Vulnerable	H	4	Endangered
Lahontan [9]	ERLS	80	4	Endangered	H	5	Critical
Oregon Lakes [10]	ERLS	65	3	Vulnerable	M	3	Vulnerable
Death Valley [11]	ERLS	58.3	3	Vulnerable	M	3	Vulnerable
Guzmán [16]	ERLS	80	4	Endangered	H	5	Critical
Mapimí [19]	ERLS	75	4	Endangered	H	5	Critical
Llanos el Salado [66]	ERLS	86.7	5	Critical	H	5	Critical
Lerma [69]	ERLS	71.7	4	Endangered	H	5	Critical
South Pacific Coastal [7]	XRLS	84.4	4	Endangered	H	5	Critical
Vegas-Virgin [13]	XRLS	65.6	3	Vulnerable	H	4	Endangered
Gila [14]	XRLS	87.5	5	Critical	H	5	Critical
Rio Conchos [17]	XRLS	75	4	Endangered	H	5	Critical
Pecos [18]	XRLS	68.8	3	Vulnerable	H	4	Endangered
Cuatro Ciénegas [22]	XRLS	37.5	2	Rel. Stable	H	4	Endangered
Rio Salado [21]	XRLS	56.3	3	Vulnerable	H	4	Endangered
Rio San Juan [23]	XRLS	75	4	Endangered	M	4	Endangered
Sonoran [61]	XRLS	62.5	3	Vulnerable	H	4	Endangered
Chapala [65]	XRLS	60.9	3	Vulnerable	H	4	Endangered
Rio Verde Headwaters [67]	XRLS	34.4	2	Rel. Stable	M	2	Rel. Stable
North Pacific Coastal [1]	TCRL	56.8	3	Vulnerable	H	4	Endangered
Columbia Glaciated [2]	TCRL	77.3	4	Endangered	M	4	Endangered
Columbia Unglaciated [3]	TCRL	77.3	4	Endangered	M	4	Endangered
Upper Snake [4]	TCRL	75	4	Endangered	M	4	Endangered
Pacific Mid-Coastal [5]	TCRL	72.7	4	Endangered	H	5	Critical
Pacific Central Valley [6]	TCRL	93.2	5	Critical	H	5	Critical
East Texas Gulf [32]	TCRL	59.1	3	Vulnerable	M	3	Vulnerable
Mobile Bay [36]	TCRL	75	4	Endangered	H	5	Critical
Apalachicola [37]	TCRL	79.5	4	Endangered	H	5	Critical
Florida Gulf [38]	TCRL	50	2	Rel. Stable	M	2	Rel. Stable
South Atlantic [40]	TCRL	75	4	Endangered	H	5	Critical
Chesapeake Bay [41]	TCRL	59.1	3	Vulnerable	H	4	Endangered
North Atlantic [42]	TCRL	58	3	Vulnerable	H	4	Endangered
Lower St. Lawrence [47]	TCRL	64.8	3	Vulnerable	H	4	Endangered
North Atlantic-Ungava [48]	TCRL	47.7	2	Rel. Stable	H	3	Vulnerable
West Texas Gulf [33]	SCRL	61.4	3	Vulnerable	M	3	Vulnerable
Florida [39]	SCRL	80.7	4	Endangered	H	5	Critical
Sinaloan Coastal [62]	SCRL	54.5	3	Vulnerable	H	4	Endangered
Santiago [63] †	SCRL			Vulnerable	H	4	Endangered
Manantlan-Ameca [64]	SCRL	40.9	2	Rel. Stable	M	2	Rel. Stable
Tamaulipas-Veracruz [68]	SCRL	54.5	3	Vulnerable	H	4	Endangered
Balsas [70]	SCRL	65.9	3	Vulnerable	H	4	Endangered
Papaloapan [71]	SCRL	56.8	3	Vulnerable	H	4	Endangered
Catemaco [72]	SCRL	38.6	2	Rel. Stable	L	2	Rel. Stable
Coatzacoalcos [73]	SCRL	63.6	3	Vulnerable	H	4	Endangered
Tehuantepec [74]	SCRL	52.3	3	Vulnerable	M	3	Vulnerable
Grijalva-Usumacinta [75]	SCRL	59.1	3	Vulnerable	M	3	Vulnerable
Yucatán [76]	SCRL	45.5	2	Rel. Stable	M	2	Rel. Stable

†No conservation status data available; rank taken from Olson et al. (1997).   *"% possible" refers to proportion of total possible Conservation Status points for each MHT.

signals that the majority of North America's freshwater species assemblages may already be seriously disrupted.

Conservation Snapshot Criteria

Degree of Land-Cover (Catchment) Alteration

Land-cover alteration is strongly correlated with human population density in areas of urban growth, but it may also be extensive in largely agricultural areas where population is low. For this reason, the only discernible geographic pattern exhibited by this indicator was generally low values (signifying low levels of alteration) in arctic ecoregions, where human presence is minimal (figure 4.2). Otherwise, ecoregions in every region and every habitat type received high scores for this indicator, with few scores below 3 out of a possible 4 points. Exceptions were in some of the more inhospitable ecoregions, such as the Atlantic-Ungava [48] and Bonneville [8] ecoregions, and in several of the more rural headwater ecoregions of the Mississippi complex. This indicator had the highest average score: twenty-one ecoregions (28 percent) each received a score of 4 points. No ecoregions received the lowest score of zero, and only four arctic ecoregions were given a score of 1 point. Even in largely unsettled ecoregions, activities such as mining have left their mark on freshwater ecosystems.

Water Quality Degradation

Water quality is a function of all upstream inputs, and as a result more downstream ecoregions and large lakes with slow turnover rates tend to exhibit greater water quality problems (figure 4.3). All four Great Lakes received the most severe score for water quality degradation, a result of well-known pollution stemming from heavy industry. All large-river ecoregions, with the exception of the Upper Rio Grande [15], received scores of either 3 or 4 out of a possible 4 points. Additionally, xeric and endorheic ecoregions had poor water quality, due to a combination of irrigation for agriculture and concentration of pollutants. Other ecoregions receiving high scores (e.g., the Apalachicola [37], Lower St. Lawrence [47], Southern Plains [31]) tended to be located in or downstream of areas of high agricultural or industrial activity.

In total, eighteen ecoregions (24 percent) received the most severe score for this indicator. No ecoregions received the lowest score, suggesting that, at an ecoregion scale, no waters in North America remain untouched by human activity. In even the most remote arctic ecoregions, for instance, distant human activities are manifested in the form of acid deposition. However, despite wide public familiarity with and concern about issues of water quality, this threat was on average less serious across all ecoregions than either of the other habitat indicators, altered hydrographic integrity and habitat fragmentation.

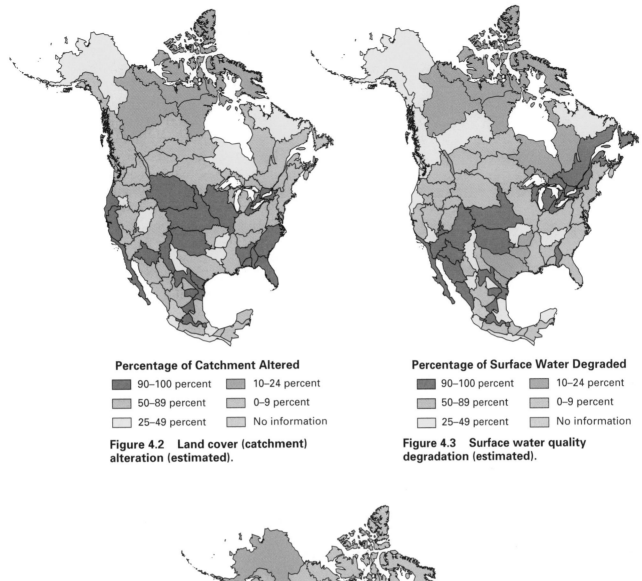

Percentage of Catchment Altered

- ☐ 90–100 percent
- ☐ 50–89 percent
- ☐ 25–49 percent
- ☐ 10–24 percent
- ☐ 0–9 percent
- ☐ No information

Figure 4.2   Land cover (catchment) alteration (estimated).

Percentage of Surface Water Degraded

- ☐ 90–100 percent
- ☐ 50–89 percent
- ☐ 25–49 percent
- ☐ 10–24 percent
- ☐ 0–9 percent
- ☐ No information

Figure 4.3   Surface water quality degradation (estimated).

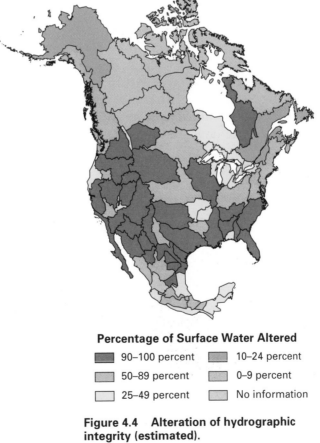

Percentage of Surface Water Altered

- ☐ 90–100 percent
- ☐ 50–89 percent
- ☐ 25–49 percent
- ☐ 10–24 percent
- ☐ 0–9 percent
- ☐ No information

Figure 4.4   Alteration of hydrographic integrity (estimated).

Alteration of Hydrographic Integrity

The high average value for this indicator underscores the very serious threat it poses in the majority of ecoregions (figure 4.4). Uniformly high scores were given to large rivers ecoregions, which have been extensively impounded, and to endorheic ecoregions, where groundwater drawdown has affected surface water levels. Other MHTs had mixed though generally high values; in particular, most temperate coastal ecoregions draining to the Pacific and Gulf Coasts were assessed as having seriously altered hydrographic integrity. For this indicator, the greatest number of ecoregions were given the highest possible score, stressing the seriousness of the threat.

Arctic ecoregions had very low scores for this indicator, with the exception of the East Hudson [54], where large-scale hydropower projects have drastically changed water flow. The large lakes ecoregions had moderate scores, as water level is manipulated to a certain extent in the Great Lakes to accommodate the lakes' multiple uses. Several temperate headwater ecoregions also had moderate scores; the relatively small size of headwater streams does not lend them to large flow-harnessing structures, though small dams may be abundant, and channelization in agricultural areas is widespread. Among the headwater ecoregions, only the Southern Plains [31] received the highest score for this indicator.

Degree of Habitat Fragmentation

Habitat fragmentation tends to follow patterns similar to those of alteration of hydrographic integrity (figure 4.5), because dams tend to cause both impacts. (Exceptions include where dams are operated to mimic the natural hydrograph, where dams for navigation do not substantially alter flow, or where dams may be designed to allow passage of fish.) Ecoregions with markedly different scores for the two indicators are the Upper Mackenzie [57] and the Lower St. Lawrence [47]. In nearly all other cases, small differences in scores for the two ecoregions were reflected in slightly higher scores for altered hydrographic integrity, suggesting that habitat fragmentation is a lesser problem in most ecoregions. Seventeen ecoregions

Two widespread threats to freshwater biodiversity. On the left, habitat loss in a channelized (straightened) stream in Parchman, Mississippi (photo by Mike Dawson). On the right, a crayfish encrusted with exotic zebra mussels (see Essay 10) (photo by Don Schloesser). Photos courtesy of U.S. Fish and Wildlife Service.

Freshwater Ecoregions of North America

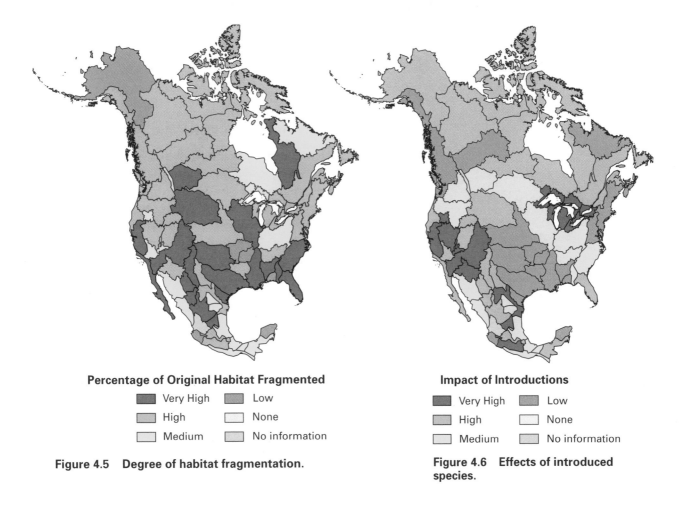

Percentage of Original Habitat Fragmented

- ■ Very High
- ■ High
- ■ Medium
- ■ Low
- □ None
- ▨ No information

Figure 4.5   Degree of habitat fragmentation.

Impact of Introductions

- ■ Very High
- ■ High
- ■ Medium
- ■ Low
- □ None
- ▨ No information

Figure 4.6   Effects of introduced species.

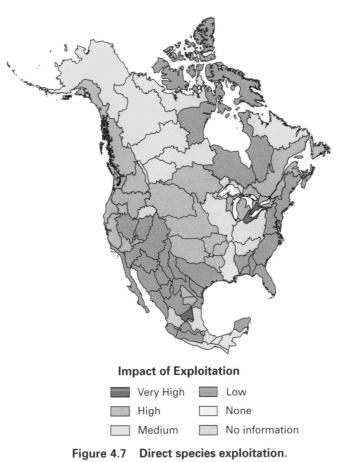

Impact of Exploitation

- ■ Very High
- ■ High
- ■ Medium
- ■ Low
- □ None
- ▨ No information

Figure 4.7   Direct species exploitation.

received the highest score for this indicator, compared with four ecoregions (all arctic) receiving the lowest score.

Additional Losses of Intact Original Habitat

This indicator was used at the discretion of experts to account for habitat loss not captured by the other, more specific, indicators. Experts evaluating ecoregions in the West and Mexico tended to give higher scores for this indicator, suggesting that important habitat loss was caused by factors other than water quality degradation, alteration of hydrographic integrity, and fragmentation. An example would be the multiple effects associated with the loss of riparian cover, such as reduced organic matter inputs and increased light levels.

Effects of Introduced Species

Introduced species present some of the most pervasive and persistent threats to freshwater systems (figure 4.6; see essay 10). Only in arctic ecoregions, and in a few other more isolated or upstream areas, were introduced species assessed as constituting a minor or nonexistent impact. On the other hand, in large temperate lakes, endorheic habitats, and xeric habitats, introduced species have had severe impacts on biodiversity in nearly all ecoregions. All four Great Lakes ecoregions were ranked as highly degraded, and all endorheic ecoregions received scores of either 3 or 4 out of a possible 4 points. These ecoregions, because of their insularity, tend to support endemic species that are particularly susceptible to competition with and predation by more hardy invasives. In total, fifteen ecoregions received the highest possible score for this indicator, a relatively small number compared with the first four indicators discussed.

It is often through disruption and degradation of habitat that such natives are rendered vulnerable and exotics are able to gain a foothold. In all but one case (the Lake Superior [43] ecoregion), ecoregions receiving the highest score for the introduced species indicator also received the highest score for one or more other conservation status indicators. In many cases the effects of recently introduced species have yet to be measured, but such species are considered a substantial future threat. A prime example is the zebra mussel, introduced less than two decades ago but rapidly spreading outward from its point of introduction in the Great Lakes.

Direct Species Exploitation

The impacts of direct species exploitation on freshwater biodiversity have been greatest in the Great Lakes and in ecoregions along the Pacific Coast (figure 4.7). Heavy fishing pressure in the Great Lakes has led to the extirpation of many game fish species, particularly ciscos (*Coregonus* spp.). In the ecoregions comprising the coastal complex of the Pacific bioregion (ecoregions [1]–[7]), overexploitation of anadromous salmonids has generated international concern and discussion but remains a serious threat to the distinctiveness of these ecoregions.

Invasive Nonindigenous Species: A Major Threat to Freshwater Biodiversity

James D. Williams and Gary K. Meffe

Introduction and establishment of nonindigenous species constitutes one of the most important issues confronting biodiversity conservation and natural resource management. The resulting homogenization of the world's flora and fauna, which represents millions of years of separate evolutionary history, has created an ecological holocaust of major proportions. Introduction of nonindigenous species into the continental United States began with the arrival of the first European settlers and has continued at an increasing rate. Many early introductions were intentional and generally viewed as positive enrichments of the native biota. As the numbers increased and effects on native biota became apparent, perception about many nonindigenous species shifted— the once welcome additions become unwanted pests. While some introductions have provided economic and recreational benefits, others are economically and ecologically expensive. The magnitude of the nonindigenous species problem has reached a critical level, and development of a coherent national policy is now required.

What are nonindigenous species? There has been considerable confusion over the definition of nonindigenous species in both the popular and scientific literature. It is important to distinguish between natural biological invasions, such as range expansions, and introductions associated with human activity. One widely accepted definition of nonindigenous species is "the condition of a species being moved beyond its natural range or natural zone of potential dispersal, including all domesticated and feral species and hybrids." This definition, taken from the Nonindigenous Aquatic Nuisance Prevention and Control Act of 1990, embodies the most critical biological aspect of nonindigenous species—the introduction of a species into an area beyond its native range, caused by humans. Ecosystems receiving nonindigenous introductions respond based on a suite of biological and ecological interactions, irrespective of the political jurisdiction from which the species originated.

The effects of nonindigenous species on the structure and function of populations, communities, and ecosystems is well documented (Elton 1958; Mooney and Drake 1986; Vitousek et al. 1987; Drake et al. 1989). Alteration of water, nutrient, and energy cycles, and of the productivity and biomass of ecosystems, directly affects human society. Ecosystem-level consequences of invasions by nonindigenous species have major ecological and economic implications and directly affect human health. Some progress has been made in addressing habitat degradation resulting from chemical contaminants and physical alterations. However, little attention has been given to and almost no progress made toward addressing the problem of invasive nonindigenous species. Biologists estimate there are more than 6,500 species of established, self-sustaining populations of nonindigenous animals, plants, and microbes in the United States (Office of Technology Assessment 1993; USGS, Gainesville, FL, unpublished plant and fish data). Of these species, approximately 6,271 are native to areas outside the United States. These numbers point to a serious problem and become alarming when one considers evidence that only 5–10 percent of introduced species become established and 2–3 percent are able to expand their ranges (di Castri 1989). Equally alarming is the growing number of introductions that result primarily from increases in the movement of people and the transportation of products, and reductions in travel time between destinations.

Nonindigenous fishes have originated from a variety of sources and geographic areas. Intentional game- or forage-fish stockings have been made by management agencies. However, many game fishes have been illegally stocked by misguided anglers. Several species, such as the grass carp (*Ctenopharyngodon idella*), were widely introduced for biological control of aquatic plants, many of which had also been introduced (Schmitz 1994). Aquarium fishes represent another group of nonindigenous fishes that have been widely introduced. Many escaped from fish culture facilities, but more were released by aquarists. Most aquarium fishes are from tropical areas and their distributions within the United States are limited to the southern tier of states or to thermal springs in colder areas (Courtenay et al. 1984).

The effects of nonindigenous fishes on aquatic biodiversity will probably increase significantly during the next twenty-five years. This prediction is based on the tremendous increase in introduced fishes during the past forty-eight years. Analysis of more than fifteen thousand records of introduced fishes revealed that between 1831, the date of the first known introduction of nonindigenous fishes to the United States, and 1950, a period of 120 years, fewer than 120 fishes were introduced. Between 1950 and 1995, more than 458 fish species were introduced. Introduction of fishes occurred in all states and watersheds, but more occured in California, Florida, Colorado, and Texas than in other states (figure 4.8).

Freshwater fishes and other aquatic species, like island species, are especially vulnerable to the effects of nonindigenous species. A review of worldwide animal extinctions since 1600 revealed that 75 percent were island species (Groombridge 1992). Freshwater ecosystems are similar to islands because they are aquatic habitats surrounded by land. The effects of nonindigenous species in freshwater ecosystems have been magnified by widespread habitat disturbance.

Native fishes of the desert Southwest are seriously threatened by nonindigenous fishes (Minckley and Deacon 1991). This area, characterized by low native fish diversity and high endemism, has received the most fish introductions and suffered the greatest loss of native fishes. Species such as the bonytail chub (*Gila elegans*) and razorback sucker (*Xyrauchen texanus*), which inhabit large rivers in the Colorado Basin, and the Sonoran topminnow (*Poeciliopsis occidentalis*) and several pupfishes and springfishes, inhabitants of small desert springs, are directly threatened by the presence of numerous predaceous nonindigenous fishes.

Currently more than 100 fishes are on the USFWS's list of endangered and threatened species, and the effects of nonindigenous species were cited as a contributing factor in the decline of more than half. Nonindigenous species were at least partially responsible for twenty-four of the thirty fish species extinctions in the United States. Involved in the extinctions of fishes were such nonindigenous species as the parasitic sea lamprey, mosquitofish, trouts, sunfishes, basses, and bullfrogs. Causes of extinctions included predation, competition for food and space, and genetic swamping through hybridization (Miller et al. 1989).

Two freshwater nonindigenous bivalve mollusks, the Asian clam (*Corbicula fluminea*) and the zebra mussel (*Dreissena polymorpha*), have contributed to the decline of native mussels (Ricciardi et al. 1995). Of the approximately 300 freshwater mussels found in the United States, about 73 percent are considered imperiled (Williams et al. 1993). The Asian clam, currently the most widespread nonindigenous mollusk in the United States, arrived on the West Coast in the 1930s and invaded the southeastern United States in the 1950s (McMahon 1983). In some areas, Asian clams carpet the stream bottom, reaching densities of several thousand individuals per square meter (J. Williams, pers. obs.).

One of the most recent molluscan introductions, the zebra mussel, arrived in the Great Lakes in 1988 via the ballast water of cargo ships. Since its introduction, the zebra mussel has caused serious economic and ecological damage to the Great Lakes. The cost to the power industry alone over the next ten years is projected to exceed $3 billion (Office of Technology Assessment 1993). Assigning a dollar cost to the alteration of freshwater ecosystem structure and function and the local extirpation of native mussels and other aquatic species is difficult. The economic cost of species extinction and ecosystem alteration is generally not included in the cost of nonindigenous species unless a commodity is lost. To meet the challenge of established nonindigenous species and future introductions of them requires policy development, enforcement, education, and research. Existing legislation on nonindigenous species is fragmented and not comprehensive. The most critical need is in the area of policy development and enforcement, supported by an aggressive public awareness and education campaign.

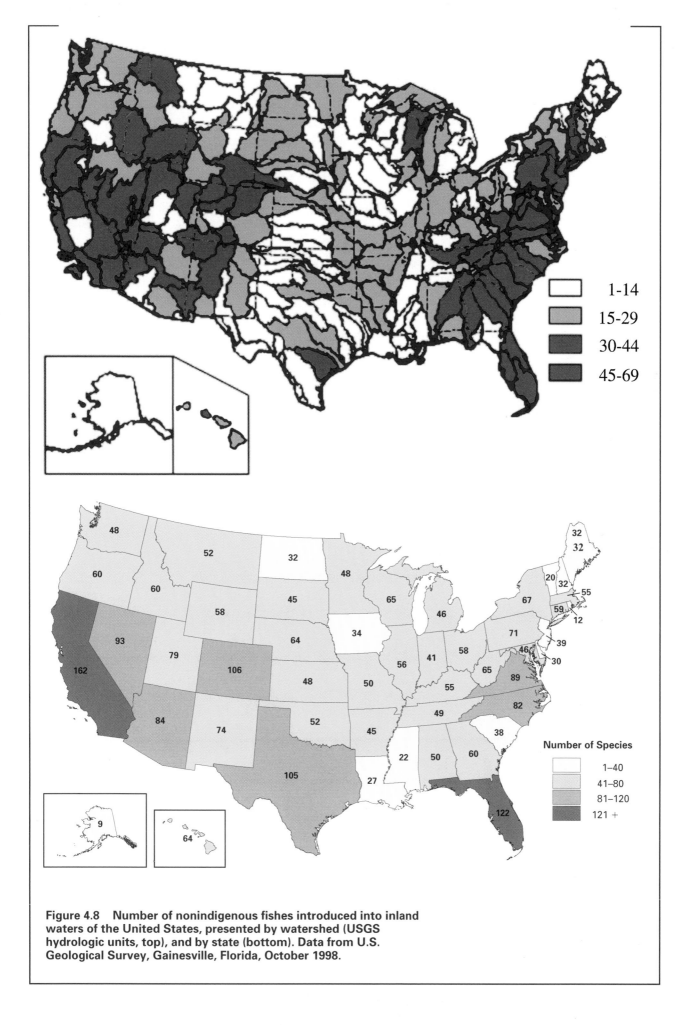

Figure 4.8   Number of nonindigenous fishes introduced into inland
waters of the United States, presented by watershed (USGS
hydrologic units, top), and by state (bottom). Data from U.S.
Geological Survey, Gainesville, Florida, October 1998.

Despite the heavy exploitation of fish for food, and of certain mussels for the pearl industry, no ecoregion received the highest score for the species exploitation indicator (Lake Erie [45] came the closest, with 3.5 out of 4 points assigned). This indicator had the lowest average score of the seven indicators, with twelve ecoregions assessed as having no impact from species exploitation, and only twelve ecoregions given a score of 3 or 3.5. Species exploitation is perhaps the most easily regulated and monitored threat to freshwater biodiversity, and its targeting of individual species makes its direct impact measurable to a large degree. However, declines or losses of target species can have wide repercussions; target fish species tend to be top predators, and mussels may play key roles in habitat maintenance.

Threat Assessment

For the threat assessment, we ranked the ecoregions as facing high, medium, or low threats to biodiversity over the next twenty years. Forty-three (57 percent) ecoregions were considered to face high threats; this number includes all ecoregions assessed as critical in the snapshot assessment, but none assessed as relatively intact (figure 4.9). Twenty-two (29 percent) ecoregions face medium threats, and only eleven (14 percent) face low threats. A map of the threat assessment results shows that the West and Southeast, as well as much of Mexico, tend to face high threats, whereas the Midwest and Canada are more mixed in degree of threat (figure 4.9).

Final Conservation Status

Modification of the snapshot assessment results with the threat assessment resulted in a change, toward greater imperilment, of thirty-eight ecoregions (table 4.3, figure 4.10). In general, the biodiversity of more degraded ecoregions is predicted to become increasingly imperiled, and more intact ecoregions are predicted to be fairly stable. After modification for threat, twenty-one (28 percent) ecoregions were assessed as critical, twenty-four (32 percent) were endangered, fifteen (20 percent) were vulnerable, nine (12 percent) were relatively stable, and seven (9 percent) were relatively intact. In other words, according to expert judgment, over one-fourth of the ecoregions are expected to be critically degraded within the next twenty years unless immediate action is taken to reverse current trends.

Additional Conservation Status Data

Imperiled Fauna

The degree to which an ecosystem's biota is imperiled is a good indicator of conservation status, and relatively comprehensive data are available for several of the faunal groups considered in this

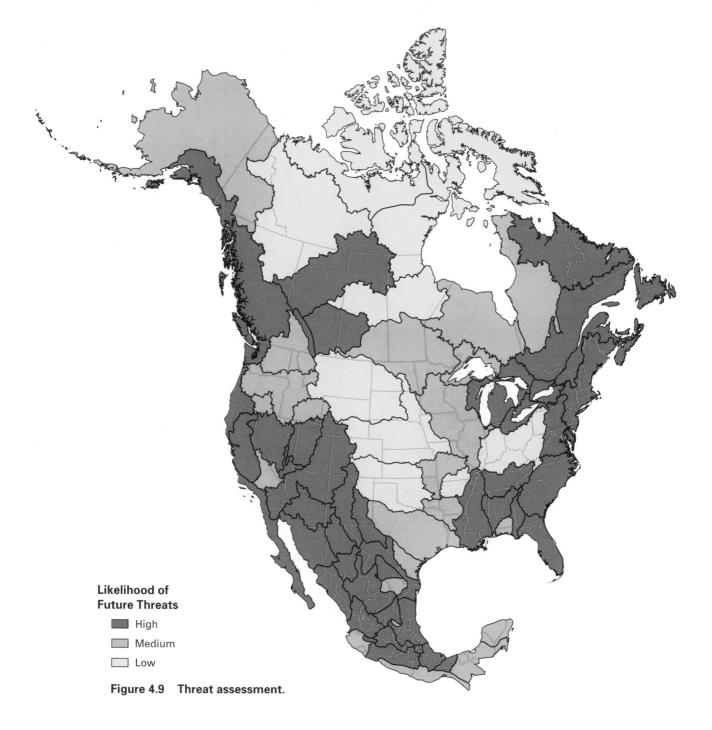

**Likelihood of
Future Threats**

■ High

■ Medium

□ Low

Figure 4.9   Threat assessment.

study. However, these data were not incorporated into the formal
conservation status assessment for three reasons. First, species
imperilment is generally a result of habitat degradation, exploitation, exotic species, and other threats, and a consideration of
imperilment in addition to these threats would, in effect, lead to
double-counting. Second, species imperilment would be inadequate as the only measure of conservation status, because it does
not address causes of imperilment. Furthermore, very rare species
are normally considered imperiled because of their vulnerability to
extinction, but their rarity is not always human induced. Finally,
declines in fish, herpetofauna, mussel, and crayfish populations
may not parallel trends in other, less-studied taxa, whereas an evaluation of habitat parameters would likely encompass broad threats
to most or all groups.

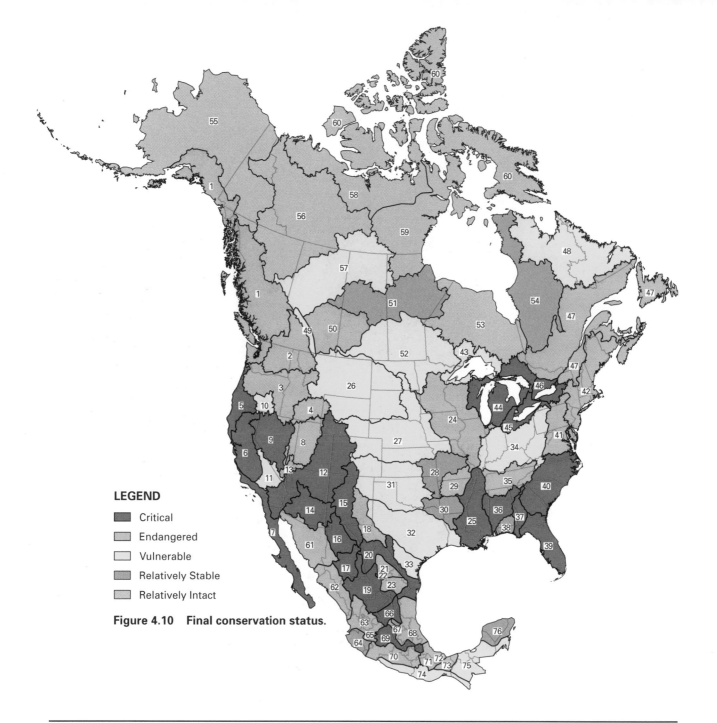

Figure 4.10 Final conservation status.

LEGEND

- Critical
- Endangered
- Vulnerable
- Relatively Stable
- Relatively Intact

TABLE 4.3 Distribution of Ecoregions by MHT for the Final Conservation Status Index (Snapshot Modified by Threat).

Major Habitat Type	Final Conservation Status Index (CSI)					
	Critical	Endangered	Vulnerable	Relatively Stable	Relatively Intact	Total
Arctic Rivers/Lakes	0	0	1	1	6	8
Large Temperate Lakes	3	0	1	0	0	4
Temperate Headwaters/Lakes	0	2	4	3	1	10
Large Temperate Rivers	4	1	2	0	0	7
Endorheic Rivers/Lakes/Springs	5	1	2	0	0	8
Xeric Rivers/Lakes/Springs	3	7	0	1	0	11
Temperate Coastal Rivers/Lakes	5	7	2	1	0	15
Subtropical Coastal Rivers/Lakes	1	6	3	3	0	13
Total	21	24	15	9	7	76

Despite these difficulties in using species imperilment data to judge conservation status, the data can be useful as supplementary information, especially to identify ecoregions where the conservation status score does not seem appropriate to the degree of imperilment. Species imperilment data came from several sources. Data from The Nature Conservancy's Natural Heritage database were used to identify species in the United States and Canada, and some in Mexico, that are *critically* imperiled globally (G1), or imperiled globally (G2; see essay 11) (Natural Heritage Central Databases 1997). For fish, additional data came from Williams et al. (1989), and from CONABIO's ranking of Mexican species. For herpetofauna, CONABIO'S extensive list provided most of the information for Mexico, with additional rankings from Gonzales et al. (1995); Natural Heritage data were again used for the United States and Canada.

ESSAY 11

Categorizing Imperiled and Vulnerable Species

Lawrence L. Master

Assessing relative imperilment is critical in directing conservation efforts toward those species in greatest need. Which species and ecological communities are thriving and which are at the brink of extinction? To answer these questions, the Network of Natural Heritage Programs and Conservation Data Centers and TNC have developed a consistent method for evaluating the status of both species and ecological communities. This assessment leads to the designation of conservation status ranks, which are an estimation of the relative imperilment of each species or community. For species this provides an approximation of their risk of extinction (Master 1991b; Stein et al. 1995).

In determining a conservation status rank, natural heritage biologists and their collaborators evaluate a number of different criteria that contribute to a species' risk of extinction. These criteria include considerations related to rarity, viability, trends, threats, and fragility. Four distinct components of rarity for each species are assessed: the number of different populations of a species; the extent of its area of occupancy; the breadth of its geographic range; and its total population size. The viability of existing populations is also an important factor, especially for species whose occurrences are reduced in number or extent. Viability is a function of a population's size, condition (e.g., reproductive output, intactness of ecological processes), and landscape context (e.g., genetic connectivity). Short- and long-term trends in overall population size, condition of populations, area of occupancy, and geographic range extent are also key ranking considerations. Threats to a species, human and natural, also must be considered because these are important predictors of future decline. Related to threat is consideration of the number of populations that are legally "protected" from threats through land use and tenure. Other considerations include the environmental or habitat specificity of a species, and the degree to which a species is sensitive to nondestructive human intrusion.

Natural heritage biologists and collaborators assign a conservation status rank to each species by categorizing and documenting the species' status related to each of the factors described above. Global conservation status ranks are based on a one-to-five scale (See the accompanying table) ranging from critically imperiled (G1) to demonstrably widespread, abundant, and secure (G5). Species and communities known to be extinct, or missing and possibly extinct, also are recorded. In general, globally vulnerable or at-risk species are those that have a conservation status rank of G1, G2, or G3. A numeric "range rank" (e.g., G2G3) is used to denote the range of uncertainty about the exact status of a species or community.

Definitions of Global Conservation Status Ranks.

Global Rank	Description
GX	*Presumed Extinct* Believed to be extinct throughout its range. Not located despite intensive searches of historic sites and other appropriate habitat, and virtually no likelihood that it will be rediscovered.
GH	*Possibly Extinct* Of historical occurrence; still some hope of rediscovery.
G1	*Critically Imperiled Globally* Extreme rarity or other factor(s) make it especially vulnerable to extinction; typically five or fewer occurrences, or very few remaining individuals, or very small range or area of occupancy.
G2	*Imperiled Globally* Rarity or other factor(s) make it very vulnerable to extinction; typically six to twenty occurrences, or few remaining individuals, or small range or area of occupancy.
G3	*Vulnerable* Often very rare and local throughout its range, found only in a restricted range (even if abundant at some locations), or other factors (e.g., significant ongoing decline over a majority of range) make it vulnerable to extinction; typically 21 to 100 occurrences.
G4	*Apparently Secure* Uncommon but not rare, and usually widespread; possibly cause for long-term concern; typically more than 100 occurrences.
G5	*Secure* Common, typically widespread and abundant.

Using a "T" rank as part of the global rank indicates the global conservation status of an infraspecific taxon (subspecies or variety). Rules for assigning T ranks follow the same principles outlined above. For example, the global rank of a critically imperiled subspecies of an otherwise widespread and common species would be G5T1.

A similar numeric scale is also used to assign national (N1 through N5) and subnational (S1 through S5) ranks. For these ranks, the status of a species or community is evaluated within specific national or subnational (e.g., state, province) jurisdictions rather than on a global or rangewide basis. Species and communities that are at risk locally, but relatively secure in at least some other portions of their ranges, have G4 or G5 global ranks but local ranks of N1, N2, or N3 (or S1, S2, or S3). In this way the three levels in the conservation status ranking system allow independent distinction of global, national, and more local (subnational) conservation status.

Conservation status assessments are continually reviewed, refined, and documented in an Internet-displayed database. Each year natural heritage and TNC scientists and their collaborators appraise and update the status of several thousand species and natural communities in the Western Hemisphere. These conservation status determinations are based on the best available information, including such sources as natural history museum collections, scientific literature, and other surveys and reports by knowledgeable biologists. Most changes in status assessment tend to reflect improved scientific understanding of the actual condition of a species or community, often based on targeted field surveys.

Other systems have also been developed to assess extinction risk for species. For example, at an international level the World Conservation Union has developed threat categories, which are quantitatively defined for taxa at high risk of extinction and qualitatively defined for taxa at lower risk of extinction (IUCN 1994). For North American species, the American Fisheries Society and "Partners in Flight" have assessed the degree of imperilment of fishes, mussels, crayfishes, and birds. In the United States, the USFWS has listed species as endangered or threatened under the terms of endangerment defined by the U.S. Endangered Species Act. For a comparison of rankings of U.S. species under these different systems, see Master et al. (in prep).

Data from the various sources were combined by assuming that G1 ranks were equivalent to "endangered" rankings, and G2 ranks corresponded with "threatened" and "vulnerable" categories. A rank of G3, assigned to species with restricted ranges, was assumed to be equivalent to the categories of "rare" as well as "special concern." These assumptions were tested where given species were ranked by more than one source, and categories could be compared; in all but a handful of cases, ranks were equivalent among sources. Where there were discrepancies, the category representing the higher degree of imperilment was chosen. Analysis of the data focused on the two highest categories. The third category of "rare/restricted range/special concern" was not used because it does not distinguish between natural and human-induced rarity, and an accounting of naturally rare species is similar to an accounting of endemics. For the sake of clarity, we refer to the categories of imperilment as "endangered," "threatened," and "rare," although these should not be confused with legal definitions used in the United States or elsewhere.

As an indicator of conservation status, the most descriptive measure of species imperilment is the percentage of an ecoregion's species that are at risk. When calculated for North American ecoregions, this figure shows unequivocally that many ecoregions' faunal assemblages are in a precarious state. It also suggests that the fauna in ecoregions from the Rockies westward, and throughout Mexico, are particularly imperiled.

Twenty-five percent or more of the fish species in eleven ecoregions are endangered or threatened, and all of these ecoregions are found in the United States and Mexico, from the Rio Grande to the west and the south (figure 4.11). In three ecoregions 50 percent or more of the fish species are endangered or threatened: Death Valley [11], Vegas-Virgin [13], and the Rio Verde Headwaters [67]. The highest degree of imperilment of fish is found in the Vegas-Virgin [13] ecosystem, where 64 percent of the ecoregion's eleven fish species are at risk.

While none of these ecoregions has many fish species, they do tend to exhibit high endemism, and this translates to a highly endangered endemic fauna. For instance, 100 percent of the endemic fish species in the Vegas-Virgin [13] ecoregion are at risk. In seven additional ecoregions 100 percent of the endemic fish are imperiled, though most of these ecoregions have only one (Upper Snake [4], Upper Rio Grande [15], Mississippi [24], Michigan-Huron [44]) or two (Lower St. Lawrence [47]) endemic species. The exception is the Colorado [12] ecoregion, with all of its seven species at risk. However, other ecoregions with more endemics have rates of endangerment nearly as high; tiny Cuatro Ciénegas [22], for example, has eight endemic fish species, 63 percent of which are imperiled. In twenty-two ecoregions, at least 50 percent of the endemic fish were ranked as endangered or threatened.

Geographic patterns of imperilment for herpetofauna were similar to those for fish, though values were substantially lower (figure 4.12). The highest degree of imperilment was in the Death Valley [11] ecoregion, where 36 percent of the eleven aquatic herpetofauna were ranked as endangered or threatened. More striking, perhaps, is

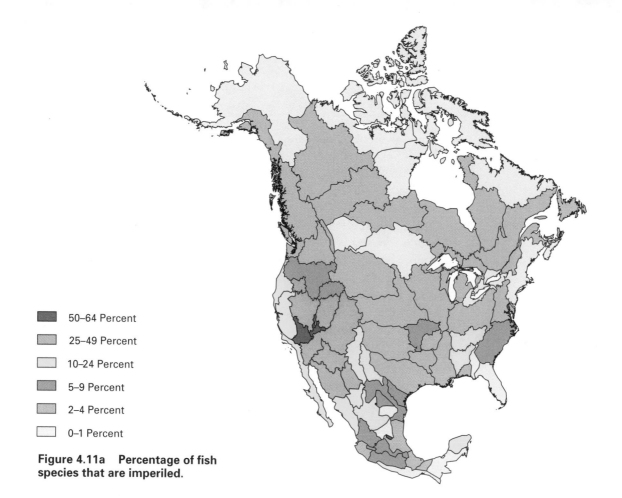

50–64 Percent

25–49 Percent

10–24 Percent

5–9 Percent

2–4 Percent

0–1 Percent

Figure 4.11a   Percentage of fish species that are imperiled.

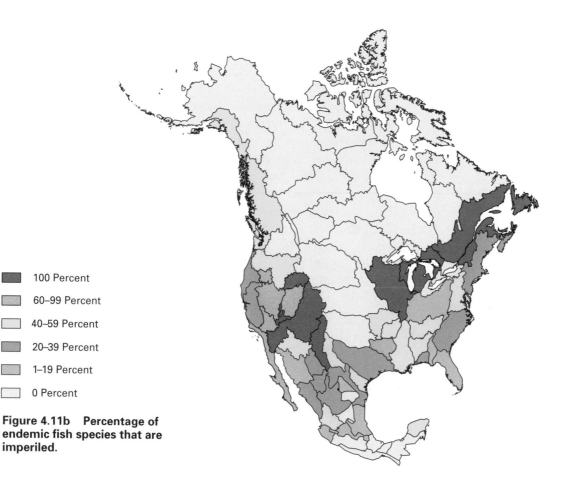

100 Percent

60–99 Percent

40–59 Percent

20–39 Percent

1–19 Percent

0 Percent

Figure 4.11b   Percentage of endemic fish species that are imperiled.

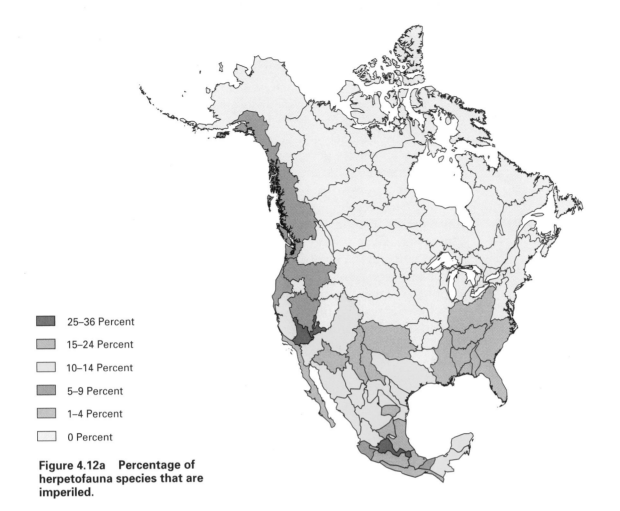

25–36 Percent
15–24 Percent
10–14 Percent
5–9 Percent
1–4 Percent
0 Percent

Figure 4.12a   Percentage of herpetofauna species that are imperiled.

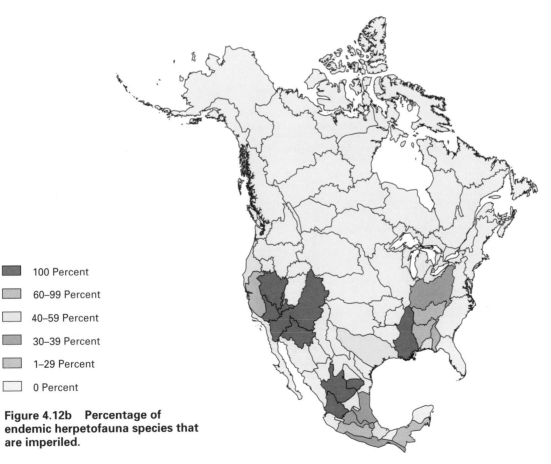

100 Percent
60–99 Percent
40–59 Percent
30–39 Percent
1–29 Percent
0 Percent

Figure 4.12b   Percentage of endemic herpetofauna species that are imperiled.

the case of the more speciose Rio Lerma [69], where 31 percent of sixty-two species are at risk. As with fish, ecoregions with the most imperiled herpetofauna were all located in the West and in Mexico.

Imperilment differed substantially among MHTs (see appendix E for statistical analysis). In xeric and endorheic ecoregions, on average, 23 and 24 percent of the fish faunas were at risk, respectively. In contrast, an average of 1 percent of the fauna was imperiled in large temperate lake and arctic ecoregions (see table 4.4). The differences among ecoregions are even more striking for the percentage of endemic fish that were imperiled: 56 percent in large temperate river ecoregions, versus 0 percent in arctic ecoregions.

For herpetofauna, in both large temperate lake and arctic ecoregions an average of 0 percent of the total species were imperiled, though it should be noted that ecoregions in both MHTs are species poor for reptiles and amphibians. Temperate coastal, large temperate river, and temperate headwater and lake ecoregions all had relatively low values for percentage of imperiled herpetofauna

TABLE 4.4   Average Faunal Imperilment of MHTs.

MHT	All Fish Endangered/Threatened		Endemic Fish Endangered/Threatened	
	Avg. Percent	Rank	Avg. Percent	Rank
TCRL	4	3	39	6
XRLS	23	7	37	5
ERLS	25	8	45	7
LTR	9	6	56	8
THL	4	3	21	3
SCRL	5	5	12	2
LTL	1	1	25	4
ARL	1	1	0	1

MHT	All Herpetofauna Endangered/Threatened		Endemic Herpetofauna Endangered/Threatened	
	Avg. Percent	Rank	Avg. Percent	Rank
TCRL	4	4	20	5
XRLS	14	7	27	6
ERLS	15	8	42	8
LTR	5	5	29	7
THL	1	3	6	3
SCRL	13	6	12	4
LTL	0	1	0	1
ARL	0	1	0	1

Summary Table

MHT	Average Rank
TCRL	4.5
XRLS	6.25
ERLS	7.75
LTR	6.5
THL	3
SCRL	4.25
LTR	1.75
ARL	1

(4, 5, and 1 percent, respectively), whereas xeric, endorheic, and subtropical coastal ecoregions had high values (14, 15, and 13 percent). Because endorheic ecoregions have high herpetofaunal endemism, there was a high degree of imperilment among their endemic species as well (42 percent). Interestingly, there was a relatively high degree of imperilment among endemic species in large temperate river ecoregions, 29 percent. When MHTs were ranked according to values for percentage of imperilment of various taxa, xeric, endorheic, and large temperate rivers MHTs were uniformly high (high imperilment), while large temperate lakes and arctic MHTs were low.

These data should not obscure the fact that there are large numbers of imperiled fauna in more speciose ecoregions. In the Tennessee-Cumberland [35] ecoregion, thirty-one fish species are endangered or threatened, and twenty-one fish species are imperiled in the Mobile Bay [36] ecoregion. In the Tamaulipas-Veracruz [68], Papaloapan [71], and Tehuantepec [70] ecoregions, there are twenty-five, twenty-four, and twenty-three imperiled herpetofauna species, respectively. However, neither the percentage nor the total number of imperiled species provides information about why species are at risk. An evaluation of habitat modification, exotic species, overexploitation, and other threats can indicate which threats are most prevalent in each ecoregion, and whether these threats correspond with patterns of imperilment.

National Watershed Characterization Data

As part of its Index of Watershed Indicators (IWI) project (formerly the National Watershed Assessment Project), the EPA has aggregated data describing the condition and vulnerability of all U.S. watersheds, as delineated by the United States Geological Survey (USGS). "Condition" is characterized by the following indicators: (1) assessed rivers meeting all designated uses set in state or tribal water quality standards; (2) fish and wildlife consumption advisories; (3) indicators of source water condition for drinking water systems; (4) contaminated sediments; (5) ambient water quality data (four toxic pollutants); (6) ambient water quality data (four conventional pollutants); and (7) wetlands loss index. "Vulnerability" is characterized by: (1) aquatic or wetlands species at risk; (2) discharge loads discharged above permitted discharge limits (toxic pollutants); (3) discharge loads discharged above permitted discharge limits (conventional pollutants); (4) urban runoff potential; (5) index of agricultural runoff potential; (6) population change; (7) hydrologic modification caused by dams; and (8) estuarine pollution susceptibility index. A combination of all indicators resulted in a National Watershed Characterization, which was in draft form when this book was being written (figure 4.13; U.S. EPA 1997).

Figure 4.13   U.S. National
Watershed Characterization—
water quality (from EPA 1997).

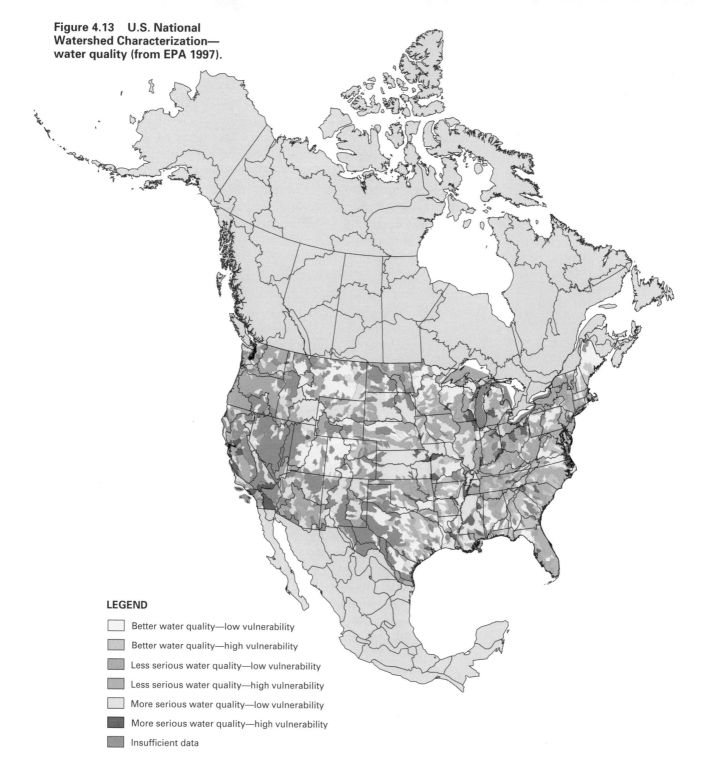

LEGEND

Better water quality—low vulnerability

Better water quality—high vulnerability

Less serious water quality—low vulnerability

Less serious water quality—high vulnerability

More serious water quality—low vulnerability

More serious water quality—high vulnerability

Insufficient data

The National Watershed Characterization maps only the United States, but it is informative as a supplement to the conservation status assessment. We aggregated the watershed data of the IWI into our ecoregion data, then calculated weighted average values for each ecoregion, using the area of the component watersheds (figure 4.14). The final score determined to which of six categories the ecoregion would be assigned. The categories used in the National Watershed Characterization were as follows: more serious water quality–high vulnerability; more serious water quality–low vulnerability; less serious water quality–high vulnerability; less serious water quality–low vulnerability; better water quality–high vulnerability; and better water quality–low vulnerability. Vulnerability refers to the potential for future declines in aquatic health, based on the presence of pollutants or other stressors, and in a sense is akin to the threat assessment in this study (U.S. EPA 1997). One of the main weaknesses of aggregating the IWI data into our ecoregion data is that many watersheds did not have data values, and watersheds with data consequently influenced the final ecoregion score disproportionately.

A map of the aggregated National Watershed Characterization data shows a geographical pattern similar to that for imperiled species (figure 4.14). Ecoregions with the most serious water quality and both high and low vulnerability are exclusively in the West and Southwest. The remainder of the West, and nearly all of the Midwest, is characterized by less serious water quality and high vulnerability. The East Coast, parts of the South, and additional ecoregions within the Mississippi and St. Lawrence complexes were rated as having less serious water quality and low vulnerability. Ecoregions with better water quality, though still high vulnerability, are the Ouachita Highlands [30], the Apalachicola [37], and the Chesapeake Bay [38]. No ecoregions were rated as having better water quality and low vulnerability.

Unlike our conservation status assessment indicators, the indicators used to generate the original National Watershed Characterization data were not weighted according to habitat type or any other factor. The data are useful because they are based on more quantitative information than this study's conservation status indicators, but they address very few of the threats we consider the most important to freshwater systems. With the exception of the wetlands loss index, the condition indicators all relate to water quality, and as such are perhaps most comparable to this study's water quality indicator. The vulnerability indicators, on the other hand, address broader threats, many of which are considered major categories in our conservation status assessment. These vulnerability indicators are intended to quantify future threats, yet they are based on current trends. For this reason, the vulnerability score assigned to each watershed, and aggregated into ecoregions, can be compared with both this study's snapshot conservation status score and the threat assessment.

Experts were asked to rate water quality by considering factors including, but not limited to, changes in parameters such as pH, turbidity, dissolved oxygen, nutrients, pesticides, heavy metals, suspended solids, hydrocarbons, and temperature. This is a somewhat broader interpretation of water quality than that used by the EPA, which is skewed more toward human than aquatic health. It is possible that the water quality parameters considered by the experts to be most important to aquatic biota would not be captured entirely by the EPA data.

Statistical analyses of the relationship between the EPA data and this study's conservation status assessment scores for U.S. ecoregions suggests close agreement of the two data sets (see appendix E for a full discussion of statistical analyses). Ecoregions that we rated as having good water quality tended to be ranked

Figure 4.14 U.S. National Watershed Characterization— compiled by ecoregion.

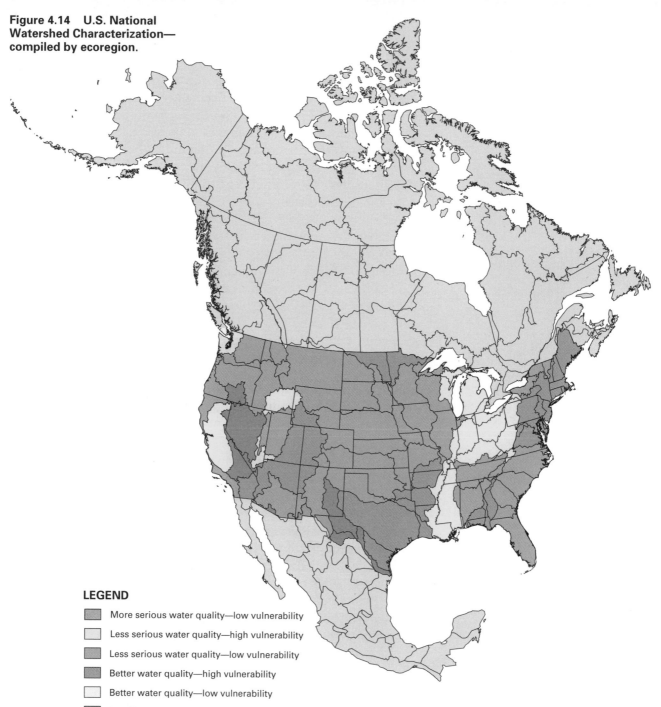

LEGEND

- More serious water quality—low vulnerability
- Less serious water quality—high vulnerability
- Less serious water quality—low vulnerability
- Better water quality—high vulnerability
- Better water quality—low vulnerability
- Insufficient data

similarly by the EPA data. The Ozark Highlands [29], Ouachita Highlands [30], East Texas Gulf [32], and West Texas Gulf [33] ecoregions, in particular, received the best water quality score in both data sets. The South Coastal [7] ecoregion, which includes Los Angeles, was given the worst water quality score in both.

The relationship between the EPA water quality score and this analysis' snapshot total score was also moderately strong, suggesting that poor water quality is associated with various categories of degradation. The relationship between the vulnerability score derived from EPA data and the snapshot total score from the conservation status assessment was not as strong as that for water quality; however, it still suggested that expert assessment was an appropriate method for judging degradation and future threats.

Comparable data for Mexico and Canada were not available for a similar analysis, but each country has developed important databases of its own, particularly through the work of CONABIO and Environment Canada (see essay 12 for an overview of freshwater conservation issues in Canada).

Canadian Freshwater Biodiversity

Don McAllister

Introduction

Canadian rivers drain about 9 percent of the world's renewable water supply. Canada's two million lakes, countless wetlands, groundwaters, and 100,000 glaciers store immense amounts of fresh water. Canada's land area (excluding rivers and lakes), which is essentially its drainage basin, occupies 9,215,430 km². Wetlands, including marshes, wooded swamps, bogs, seasonally flooded forests, and sloughs, cover about 14 percent of Canada's land surface (Linton 1997), while lakes and rivers cover 8 percent (SOE 1996).

Useful reviews of the status of fresh waters and their biodiversity include McAllister et al. (1997) on the global scale and Linton (1997) on the national scale. Canada's State of the Environment reports (SOE 1996) and Mosquin et al. (1995) provide general reviews of the state of Canada's environment and biodiversity. A review of Canadian aquatic biodiversity is being prepared by the author as a chapter in a forthcoming book on Canada's biodiversity.

Ecosystems

Canada possesses a variety of aquatic ecosystems including lakes, ponds, temporary pools, streams, rivers, springs, marshes, swamps, fens, bogs, groundwater, and estuaries. Areas of ice and snow, tree cavities, hot springs, and cave waters are some of the smaller but biotically most interesting aquatic ecosystems.

Some of Canada's river basins are biotically distinctive. The North Pacific Coastal Basin [1], for instance, contains a number of distinctive species, such as the redside shiner, *Richardsonius balteatus*, which occurs almost entirely in Pacific drainages.

Number of Taxa

Twenty-three percent of the known species of animals, plants, and microorganisms in Canada occur in fresh water, while 72 percent of the major phyla occur in that environment (Mosquin et al. 1995). The species include 177 freshwater, 26 anadromous, and 1 catadromous fishes (McAllister 1990); 179 freshwater mollusks (Clarke 1981); and 11 freshwater crayfishes (Hamr 1998). Of Canada's 15,200 fungi, 7 percent occur in fresh water, and of the 2,980 native flowering plants, 15 percent occur in fresh water (Mosquin et al. 1995). About 32 percent of the bacteria are freshwater species. Of the 745 species of mayflies, dragonflies, and stoneflies, 50 percent may be said to be freshwater species by virtue of their freshwater eggs and larvae.

Endemism

Canada was almost completely glaciated during the last, or Wisconsin, glaciation. Most of today's species reinvaded from glacial refugia, mainly in the United States. Some 1 to 5 percent of Canada's species are endemic and survived or developed in Canadian glacial refugia (Mosquin et al. 1995). Canadian endemics include a blind subterranean amphipod, *Stygobromus canadensis*, from Castleguard Cave, Banff National Park, Alberta (ecoregion [49]), and the copper redhorse, *Moxostoma hubbsi*, from the St. Lawrence Basin near Montreal (ecoregion [47]).

Alien Species

The number of alien species in Canada's fresh waters is increasing. Twenty-three percent of Canada's terrestrial and aquatic flowering plant species are alien (Mosquin et al. 1995), as are 18 percent of its crayfishes (Hamr 1998), 6 percent of its freshwater fishes (McAllister 1990), and 2 percent of its amphibian species (Cook 1984).

Some alien species like the zebra mussel, *Dreissena polymorpha*, and the purple loosestrife, *Lythrum salicaria*, have spread widely and are displacing native species and transforming ecosystems. The growth of the aquaculture industry poses an additional threat. Atlantic salmon (*Salmo salar*), escaped from aquaculture pens, are reproducing in Pacific Coast streams.

Geographic Distribution of Biodiversity

Freshwater species tend to be fewer towards northern Canada. For example, only three of Canada's fifty freshwater mussels and none of its eleven crayfishes are found north of 60 degrees north latitude (Clarke 1981; Hamr 1998).

Coexistent with the north-south cline in number of species are regional hot spots—areas with high numbers of species. The two prominent ones are in southern Ontario (ecoregions [43] through [46]) and southern British Columbia (ecoregion [1]). Southern Ontario has the most species of fishes, mollusks, and crayfishes across Canada.

Status of Biodiversity

By 1997, the Committee on the Status of Endangered Wildlife in Canada had classified sixty-two species, subspecies, or distinct populations of freshwater fishes as being extinct, endangered, threatened or vulnerable; this corresponds to almost one-third of all Canadian fish species. From the point of view of fisheries, ten of thirty basin stocks (Pacific, Arctic, and Atlantic) of important species are in poor or very poor condition (Linton 1997). According to Dr. David Green (pers. comm.), 25 percent of Canada's sixty-six amphibians are on the decline.

Nationally, 14 percent of all wetlands have been lost (SOE 1996). The wetlands are drained or filled for farmland, municipal development, or transportation, or are deprived of seasonal water supplies by dams. Loss of wetlands has contributed to a decline in duck populations from 200 million to about 30 million in this century.

Proximate Threats: Dams, Agriculture, and Forestry

A portion of land the size of Lake Ontario (about 20,000 km²), is now covered by hydroelectric dam reservoirs (SOE 1996). Dam impoundments convert running waters to still-water ecosystems, diminishing habitats for species adapted to currents. Dams block fish migrations and affect seasonality of flow, temperature, and turbidity. If we assume that downstream effects of Canada's 650 large dams (SOE 1996) extend 200 km, then these dams influence 130,000 km of rivers. The Bennett Dam in British Columbia disrupted annual flooding in the Athabaska-Peace Delta 1,200 km downstream, causing conversion of 25 percent of its wet grasslands to forest habitats.

In Canada 680,000 km² are devoted to agriculture on improved farmland (cropland, improved pasture, summer fallow) and unimproved farmland. That represents 7 percent of the land surface of Canada, or almost 14 percent of the land area of the southern half of Canada where most agricultural activity occurs.

Agricultural impacts on aquatic biodiversity include drainage of wetlands; turbidity and sedimentation resulting from runoff of bared, tilled soil; eutrophication from fertilizer or manure runoff; pollution from toxic pesticides; withdrawal of surface water or groundwater for irrigation; and removal of riparian vegetation by stock or for cropland.

Space permits us to analyze only one of the impacts, agricultural chemicals. Dr. Martin Ouellet and coworkers of the Redpath Museum, McGill University, studied deformities in thousands of specimens of amphibians from hundreds of localities in Quebec. Grotesque frogs were not uncommon along a 250 km stretch on both sides of the St. Lawrence River. On land that has not been sprayed with pesticides for many decades, an average of 1 percent of the frogs are deformed, while on working farms that used insecticides, fungicides, herbicides, and chemical fertilizers, 12 percent of the frogs are deformed. Such deformities can be expected to lower rates of survival.

That toxaphene, applied to crops thousands of kilometers away, is found in fishes in Lac Laberge, Yukon Territory, shows the extent to which atmospheric transport, followed by distillation in cold climates, may convey pesticides. Similar atmospheric distillation into high lakes in the Rockies has been recently demonstrated in studies by Dr. David Schindler (pers. comm., 1998).

Crop diversity within and between fields, permaculture, organic farming, biological controls, and integrated pest management are all techniques useful in eliminating or reducing use of pesticides. Such practices are modestly developed in Canada.

About 2,440,000 km^2, representing 24 percent of Canada's land area, are occupied by active forest harvesting areas or those with potential for future harvesting (SOE 1996). Of Canada's original forest cover, 42 percent has been lost (Bryant et al. 1997). Among the remaining forested areas, 21 percent of the frontier forests—large, relatively undisturbed tracts—are threatened by logging and other activities.

Forests provide a number of ecological services for aquatic ecosystems. They are a key element in hydrological cycles. Forests retard rapid runoff into streams, evening out the flow, diminishing freshets and low-water periods, and diminishing soil erosion. Clear-cutting, the primary harvesting method in Canada, short-circuits this cycle; more of the water flows rapidly to streams (encouraging erosion) and thence to the sea; freshets are higher and the lows are longer and shallower. Low-water periods reduce living space and increase temperature fluctuations.

Leaf, twig, branch, and tree trunk input into streams is a primary source of food. Invertebrates consume these items and in turn serve as food for species higher on the food chain. Fallen branches and trunks are vital elements in forming stream, lake, and even marine habitats and supply unique food chains (Maser and Sedell 1994).

In the hands of large forest-product companies with a focus on short-term profits, there is a push to cut more and harvest sooner. But MacMillan-Bloedel, one of the large West Coast forest-product companies, has just announced a phase-in of more sustainable practices over the next five years. May (1998) reviews the problems in Canada's forests. Hammond (1991) and Drengson and Taylor (1997) show that ecoforestry provides environmentally, socially, and economically viable alternatives to industrial forestry and maintains forest services to aquatic ecosystems.

Root Causes

Root causes are the forces that drive anthropogenic impacts on the environment. They include population growth, per capita consumption, poverty, market forces, and militarism. For a fuller discussion of root causes and freshwater biodiversity, see McAllister et al. (1997). Conserving Canadian fresh waters and their biota means dealing with both proximate impacts and root causes.

CHAPTER 5

Setting the Conservation Agenda: Integrating Biological Distinctiveness and Conservation Status

Efforts to conserve biodiversity must be undertaken in every ecoregion. However, some ecoregions support such outstanding biological diversity that they require immediate and proportionally greater attention from conservationists if that diversity is to be preserved. Some ecoregions are so degraded, and the causes so entrenched and irreversible, that no conservation investment could protect more than pockets of biodiversity. In contrast, some ecoregions remain relatively undegraded, and in these areas less intensive conservation intervention may preserve a proportionately larger degree of the native biodiversity.

Integrating the results from the biological distinctiveness and conservation status indices helps to categorize ecoregions with different conservation trajectories. Using the classification for the two indices, each ecoregion is placed in one of the twenty matrix cells. With this matrix, conservation planners and organizations can decide which scenarios best describe the situation in their ecoregions of concern and develop appropriate conservation strategies.

We propose the following method of grouping these twenty cells into final conservation classes (figure 2.5). This assignment of priorities differs substantially from that proposed for North American terrestrial ecoregions (Ricketts et al. 1999a) but is similar to that offered for freshwater ecoregions of Latin America and the Caribbean (Olson et al. 1997). Freshwater ecoregions differ from their terrestrial counterparts in two important and related ways. First, because of the connectedness of freshwater habitats, spatial and functional linkages across large distances are strong, with upstream activities manifested in downstream effects. Second, conservation of a given freshwater site must nearly always occur at the watershed scale. Considering that entire ecoregions must be the focus of any ambitious conservation action, North American freshwater experts agreed that critically imperiled ecoregions are likely beyond repair, and that the greatest biodiversity conservation may be achieved by focusing on endangered and vulnerable ecoregions with globally outstanding biodiversity. The specific nature of conservation efforts will vary among ecoregions in the same class because of differences in habitat type, levels of beta diversity, and resiliency.

This priority-setting matrix does not preclude or discourage conservation actions in any ecoregion; instead it suggests ecoregions where conservation dollars may do the most work for global- and continental-scale biodiversity conservation. Local-scale activities may be appropriate in critically imperiled ecoregions to save source pools of species, or in less imperiled ecoregions where opportunities exist for protecting intact areas without intensive restoration (see chapter 7).

However, in the face of severely limited resources for conservation, organizations may choose to focus their efforts on ecoregions containing globally unique species assemblages, or on ecoregions containing the last examples of important ecological processes. This assessment assists the process of allocating resources most strategically.

We offer two sets of results, one using the snapshot conservation status (table 5.1a and figure 5.1a) and the other using the final conservation status (table 5.1b and figure 5.1b). The snapshot results consider ecoregions in their current state of degradation; presumably, immediate conservation action in certain ecoregions facing high future threats could forestall further degradation. Eleven globally outstanding ecoregions are endangered or vulnerable, prior to consideration of threats, and would therefore receive first priority for conservation. However, when threats are considered, six of these ecoregions become critically imperiled, and they are downgraded to the second-priority class. On the other hand, one ecoregion (Cuatro Ciénegas [22]) moves up to the first-priority class with consideration of threats, as it is currently relatively stable but faces high risk of future degradation. We provide both the "before" and "after" sets of results here, and leave the choice of which to use up to individual conservation agencies and

TABLE 5.1A   Distribution of Ecoregions within Integration Matrix, Using Snapshot Assessment.

| Biological Distinctiveness | Snapshot Conservation Status | | | | | |
	Critical	Endangered	Vulnerable	Relatively Stable	Relatively Intact	Total
Globally Outstanding	0	7	4	4	0	15
Continentally Outstanding	5	9	20	4	3	41
Bioregionally Outstanding	0	1	0	2	1	4
Nationally Important	0	4	6	3	3	16
Total	5	21	30	13	7	76

TABLE 5.1B   Distribution of Ecoregions within Integration Matrix, Using Final Conservation Status.

| Biological Distinctiveness | Final Conservation Status | | | | | |
	Critical	Endangered	Vulnerable	Relatively Stable	Relatively Intact	Total
Globally Outstanding	6	3	3	3	0	15
Continentally Outstanding	13	12	10	3	3	41
Bioregionally Outstanding	0	1	0	2	1	4
Nationally Important	2	4	6	1	3	16
Total	21	20	19	9	7	76

organizations. The identities of the ecoregions in each cell can be found in appendix F; the matrix is reproduced separately for each MHT and individual ecoregions are listed in the cells they occupy.

There is an uneven distribution of ecoregions within integration cells, with more than one-fourth of the ecoregions assessed as continentally outstanding and vulnerable as of 1997 (table 5.1a). Overall, the majority of ecoregions are either endangered or vulnerable, and more than half are continentally outstanding.

Figure 5.1a   Priority classes of ecoregions using snapshot conservation status.

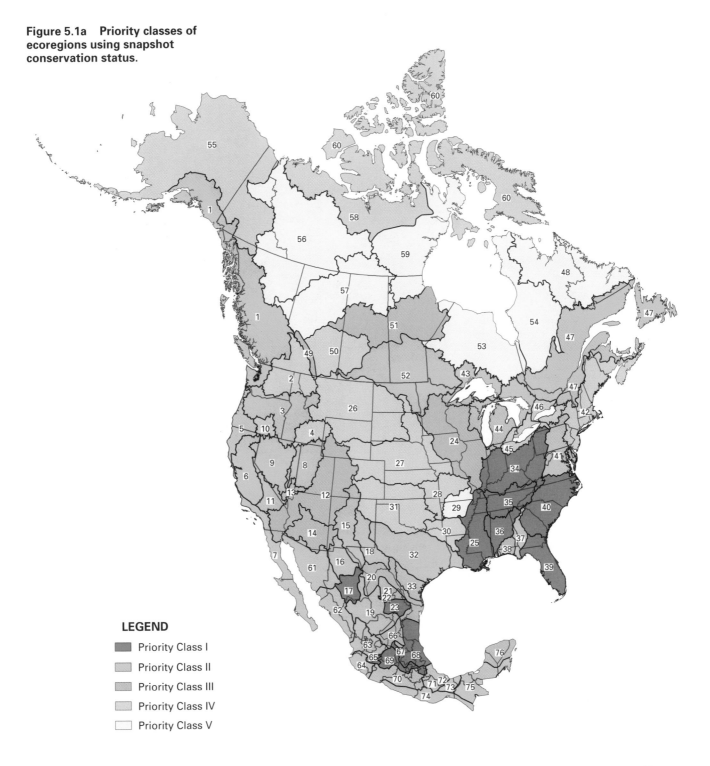

LEGEND

■ Priority Class I
■ Priority Class II
■ Priority Class III
□ Priority Class IV
□ Priority Class V

Figure 5.1b Priority classes of ecoregions using final conservation status.

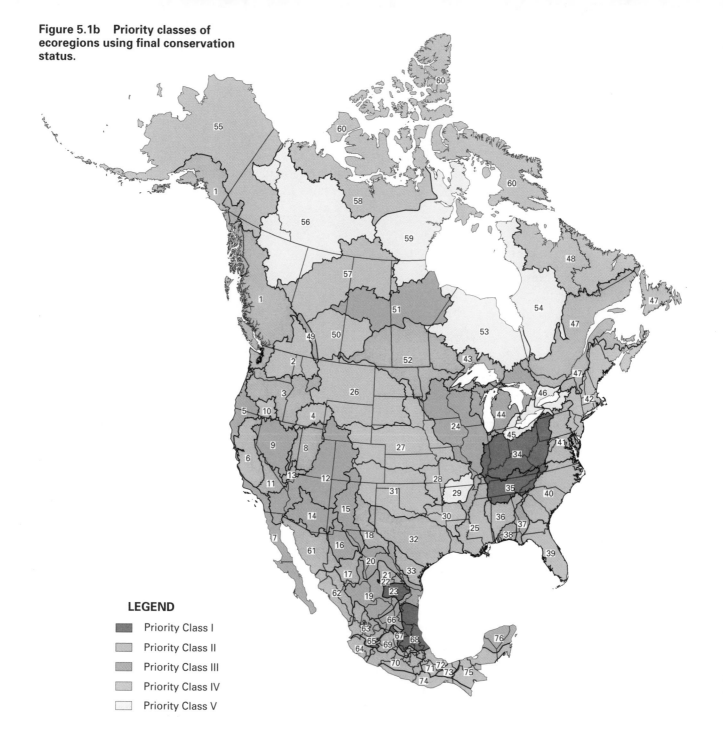

LEGEND

- ■ Priority Class I
- ■ Priority Class II
- ■ Priority Class III
- ■ Priority Class IV
- □ Priority Class V

A summary of the integration matrices by priority classes illustrates how ecoregions shift with consideration of future threats (tables 5.2a and b). The integration results are plotted onto the ecoregion map, showing the geographic distribution of the ecoregions in each class (figures 5.1a and b).

Priority I ecoregions have the highest biological distinctiveness values and are either endangered or vulnerable (table 5.3). These ecoregions are located in the eastern United States and in Mexico. Priority II ecoregions include the remaining globally outstanding ecoregions, as well as continentally outstanding units assessed as endangered or vulnerable. Continentally outstanding ecoregions in the Priority II class may be equally important for conservation at the continental or bioregional scale, and the large number of ecoregions

TABLE 5.2A   Distribution of Ecoregions by Priority Class, Using Snapshot Assessment.

	Priority I	Priority II	Priority III	Priority IV	Priority V	Total
Number of ecoregions	11	34	9	15	7	76
Percentage of total	14	45	12	20	9	100

TABLE 5.2B   Distribution of Ecoregions by Priority Class, Using Final Conservation Status.

	Priority I	Priority II	Priority III	Priority IV	Priority V	Total
Number of ecoregions	6	32	16	15	7	76
Percentage of total	8	42	21	20	9	100

TABLE 5.3   Distribution of Ecoregions by Priority Class.

Snapshot		Final Conservation Status	
Priority I		**Priority I**	
Teays-Old Ohio [34]	THL	Teays-Old Ohio [34]	THL
Tennessee-Cumberland [35]	THL	Tennessee-Cumberland [35]	THL
Mississippi Embayment [25]*	LTR	Tamaulipas-Veracruz [68]	SCRL
Lerma [69]*	ERLS	Cuatro Ciénegas [22]*	XRLS
Rio Conchos [17]*	XRLS	Rio San Juan [23]	XRLS
Rio San Juan [23]	XRLS	Chapala [65]	XRLS
Chapala [65]	XRLS		
Mobile Bay [36]*	TCRL		
South Atlantic [40]*	TCRL		
Florida [39]*	SCRL		
Tamaulipas-Veracruz [68]	SCRL		
Priority II		**Priority II**	
Superior [43]	LTL	Superior [43]	LTL
Michigan-Huron [44]*	LTL	Canadian Rockies [49]*	THL
English-Winnipeg Lakes [52]	THL	English-Winnipeg Lakes [52]	THL
Upper Rio Grande/Bravo [15]*	LTR	Mississippi Embayment [25]*	LTR
Bonneville [8]	ERLS	Bonneville [8]	ERLS
Lahontan [9]*	ERLS	Death Valley [11]	ERLS
Death Valley [11]	ERLS	Lerma [69]*	ERLS
Guzmán [16]*	ERLS	Vegas-Virgin [13]	XRLS
Mapimí [19]*	ERLS	Rio Conchos [17]*	XRLS
South Pacific Coastal [7]*	XRLS	Pecos [18]	XRLS
Vegas-Virgin [13]	XRLS	Rio Salado [21]	XRLS
Pecos [18]	XRLS	Rio Verde Headwaters [67]	XRLS
Rio Salado [21]	XRLS	Mobile Bay [36]*	TCRL
Cuatro Ciénegas [22]*	XRLS	South Atlantic [40]*	TCRL
Rio Verde Headwaters [67]*	XRLS	North Pacific Coastal [1]	TCRL
North Pacific Coastal [1]	TCRL	Columbia Unglaciated [3]	TCRL
Columbia Unglaciated [3]	TCRL	Pacific Central Valley [6]	TCRL
Pacific Mid-Coastal [5]*	TCRL	East Texas Gulf [32]	TCRL
Pacific Central Valley [6]	TCRL	Chesapeake Bay [41]	TCRL
East Texas Gulf [32]	TCRL	Lower St. Lawrence [47]	TCRL
Apalachicola [37]*	TCRL	Sonoran [61]	TCRL
Chesapeake Bay [41]	TCRL	West Texas Gulf [33]	SCRL
Lower St. Lawrence [47]	TCRL	Florida [39]*	SCRL
Sonoran [61]	TCRL	Sinaloan Coastal [62]	SCRL
West Texas Gulf [33]	SCRL	Santiago [63]	SCRL
Sinaloan Coastal [62]	SCRL	Manantlan-Ameca [64]	SCRL
Santiago [63]	SCRL	Balsas [70]	SCRL
Manantlan-Ameca [64]	SCRL	Papaloapan [71]	SCRL
Balsas [70]	SCRL	Catemaco [72]	SCRL
Papaloapan [71]	SCRL	Coatzacoalcos [73]	SCRL
Catemaco [72]*	SCRL	Tehuantepec [74]	SCRL
Coatzacoalcos [73]	SCRL	Grijalva-Usumacinta [75]	SCRL
Tehuantepec [74]	SCRL		
Grijalva-Usumacinta [75]	SCRL		

(continues)

TABLE 5.3 *Continued.*

Snapshot			Final Conservation Status		
Priority III			**Priority III**		
Canadian Rockies [49]*	THL		Michigan-Huron [44]*	LTL	
Lower Saskatchewan [51]	THL		Lower Saskatchewan [51]	THL	
Colorado [12]	LTR		Colorado [12]	LTR	
Lower Rio Grande/Bravo [20]	LTR		Upper Rio Grande/Bravo [15]*	LTR	
Mississippi [24]	LTR		Lower Rio Grande/Bravo [20]	LTR	
Llanos el Salado [66]	ERLS		Mississippi [24]	LTR	
Gila [14]	XRLS		Lahontan [9]*	ERLS	
Florida Gulf [38]	TCRL		Guzmán [16]*	ERLS	
Yucatán [76]	SCRL		Mapimí [19]*	ERLS	
			Llanos el Salado [66]	ERLS	
			South Pacific Coastal [7]*	XRLS	
			Gila [14]	XRLS	
			Pacific Mid-Coastal [5]*	TCRL	
			Apalachicola [37]*	TCRL	
			Florida Gulf [38]	TCRL	
			Yucatán [76]	SCRL	
Priority IV			**Priority IV**		
Yukon [55]	ARL		Yukon [55]	ARL	
North Arctic [58]	ARL		Upper Mackenzie [57]*	ARL	
Arctic Islands [60]	ARL		North Arctic [58]	ARL	
Erie [45]*	LTL		Arctic Islands [60]	ARL	
Ontario [46]*	LTL		Central Prairie [28]	THL	
Central Prairie [28]	THL		Ouachita Highlands [30]	THL	
Ouachita Highlands [30]	THL		Southern Plains [31]	THL	
Southern Plains [31]	THL		Upper Saskatchewan [50]	THL	
Upper Saskatchewan [50]	THL		Upper Missouri [26]	LTR	
Upper Missouri [26]	LTR		Middle Missouri [27]	LTR	
Middle Missouri [27]	LTR		Oregon Lakes [10]	ERLS	
Oregon Lakes [10]	ERLS		Columbia Glaciated [2]	TCRL	
Columbia Glaciated [2]	TCRL		Upper Snake [4]	TCRL	
Upper Snake [4]	TCRL		North Atlantic [42]	TCRL	
North Atlantic [42]	TCRL		North Atlantic-Ungava [48]*	TCRL	
Priority V			**Priority V**		
South Hudson [53]	ARL		South Hudson [53]	ARL	
East Hudson [54]	ARL		East Hudson [54]	ARL	
Lower Mackenzie [56]	ARL		Lower Mackenzie [56]	ARL	
Upper Mackenzie [57]*	ARL		East Arctic [59]	ARL	
East Arctic [59]	ARL		Erie [45]*	LTL	
Ozark Highlands [29]	THL		Ontario [46]*	LTL	
North Atlantic-Ungava [48]*	TCRL		Ozark Highlands [29]	THL	

*Designates ecoregions whose priority rank changes with consideration of threats.

in this priority class represents many opportunities for important conservation. Priority I ecoregions are described briefly below and in greater detail in appendix G.

1. The Teays-Old Ohio [34], the most northern Priority I ecoregion, exhibits high species richness and endemism across taxa, particularly along its unglaciated southern periphery. Restoration of stream channel form and removal of dams, reduction of pollution, reforestation of riparian areas, and implementation of a policy to control exotic species may save important biodiversity components.

2. The Tennessee-Cumberland [35] ecoregion, arguably the most diverse temperate headwater ecoregion in the world, faces every form of disturbance to flowing-water systems. Restora-

tion and protection have had some success through public education and efforts focused on selected stream reaches. Collaborations between grassroots organizations and the private sector show promise.

3. The Mississippi Embayment [25] ecoregion, dominated by the lower Mississippi River, harbors distinct species adapted to the turbid environment of a large river. The Mississippi is a "chemical alley" with a tremendously altered floodplain and riparian zone. It may present the biggest conservation challenge among the Priority I ecoregions.

4. The Lerma [69] ecoregion in south-central Mexico is an endorheic basin with a globally outstanding degree of endemism in fish. Surrounding land uses, especially industry, have had major impacts on freshwater habitats. Increasing development of the region requires immediate attention if the unique fish fauna is to be saved.

5. The Rio Conchos [17] ecoregion contains the only free-flowing large river habitat in the greater Rio Grande catchment. It still maintains relatively intact species assemblages in both river and spring habitats, which together support a highly endemic fish fauna. The ecoregion is degraded by nonpoint source pollution, flow regulation, exotic species, and overgrazing, but retains the potential for recovery of locally extirpated species.

6. The Rio San Juan [23] ecoregion, a tributary basin to the lower Rio Grande/Rio Bravo, is distinguished by a high degree of endemism in fish and crayfish. Impacts to this xeric ecoregion include overexploitation of water, salinization from agriculture and industry, exotics, and widespread human population growth. It may be partially restorable through improved water use and water quality regulations.

7. The Chapala [65] ecoregion exhibits nearly complete endemism among its fishes and shares two endemic fish genera (*Chapalichthys* and *Skiffia*) with neighboring ecoregions. Current impacts, particularly from exotic species, are high.

8. The Mobile Bay [36] ecoregion has the highest level of aquatic diversity in the eastern Gulf of Mexico, including an exceptional degree of endemism in crayfish and mussels. Agriculture and timber harvest, as well as interbasin water transfer, have had heavy impacts on this ecoregion. Public education aimed especially at protecting riparian areas may lead to restoration and protection.

9. The South Atlantic [40] is a large ecoregion spanning much of the mid-Atlantic and southeastern coastline. It contains high species richness and endemism across all freshwater taxa, yet this fauna remains one of the least studied. Impoundments, pollution, sedimentation, exotic species, and human population growth paint a grim picture for this ecoregion. Intact pockets remain where protection and restoration may be possible.

10. The Florida [39] ecoregion is distinguished by its diverse aquatic habitats, from springs, to extensive freshwater swamps and marshes, to the relatively intact Suwannee River. Development, interbasin water transfer, staggering human population growth, agricultural and mining pollution, and impoundments have degraded this ecoregion. The massive campaign waged to

restore the Everglades suggests that public education and habitat protection may work in favor of broader restoration across the ecoregion.

11. The Tamaulipas-Veracruz [68] ecoregion, along the eastern coast of Mexico, shares two endemic fish genera (*Prietella* and *Xenoophorus*) with neighboring ecoregions and harbors a highly endemic crayfish fauna as well as a large number of amphibian species. This ecoregion is degraded by the impacts of human population growth, including invasion of marine fish into increasingly saline rivers. Much of the original biodiversity remains to be preserved.

12. The Cuatro Ciénegas [22] ecoregion is recognized worldwide for its globally outstanding spring-fed pools and streams situated within a forbidding desert. It contains in its tiny area a degree of endemism across all taxa that is virtually unrivaled elsewhere. Cuatro Ciénegas faces potentially grave threats from water overexploitation and diversion, mining of nearby gypsum dunes, exotics, and excessive tourism. Currently it maintains most of its native species, but the existing protected area must be expanded and threats mitigated. Cuatro Ciénegas requires immediate attention (see essay 13).

ESSAY 13

The Valley of Cuatro Ciénegas, Coahuila: Its Biota and Its Future

Salvador Contreras-Balderas

The valley of Cuatro Ciénegas has received much attention because of its abundant endemic biota relative to its size. Minckley (1969) commented that it contains as many endemics as in three states of the United States. It is located nearly at the geographical center of the state of Coahuila in northeastern México. It is a W-shaped, desertic intermontane valley, approximately 30 by 20 km, hydrologically semiclosed, draining toward the headwaters of the Río Salado of the Río Bravo Basin. Its climate is xeric, with less than 200 mm of annual rainfall and temperatures that range from subzero to 44°C (García 1988).

Exploration of the valley began with E.P. Marsh's first biological survey in 1938. This was followed by a published report of some of the area's novelties, such as the endemic semiaquatic tortoise *Terrapene coahuila,* a member of a terrestrial woodland group (Schmidt and Owens 1944). More papers entered the literature as the valley was explored, intensively after 1959 and throughout the 1960s and into the 1970s.

Despite the region's scarce rainfall, the most distinguishing characteristic of the valley is its abundance of water. It contains numerous large and isolated headsprings, small fast creeks, seeps, marshes, lagoons, and evaporation trays. The valley and its mountains are home to a wide variety of habitats and formations, including the peculiar gypsum dunes, saline flats, marshes, prairies, shrubs, riparian woods, rocky hillsides and sierras, and temperate woodlands on the high mountains.

Endemism is pronounced in aquatic and semiaquatic groups such as fishes (Minckley 1984), snails (Taylor 1966; Hershler 1985), turtles (Schmidt and Owens 1944), and crustaceans (Cole 1984); in gypsum or rocky terrestrials such as cacti and compositae (Pinkava 1984); in scorpions (Williams 1968); in innumerable little-known insects; and in a few apparently endemic birds and mammals of recent discovery (A.J. Contreras pers. comm.). Such diversity across such different

A pair of freshwater *pozas* (pools) from Cuatro Ciénegas, Coahuila, Mexico. Dozens of warm and cool pools, linked by underground spring systems, dot the Chihuahuan Desert landscape (above photo) (photo by Colby Loucks). Underwater, the pools resemble freshwater coral reefs, harboring numerous endemic species such as the cichlid, *Cichlasoma minckleyi,* shown here (right photo) (photo by David Olson).

taxa is unknown anywhere else in the world. Most endemics are associated with gypsum dunes or gypsum-loaded waters, water-rich soils, or rocky areas. This biodiversity has been summarized by Minckley (1969); Minckley and Taylor (1969); and Contreras (1978a). An excellent compendium was produced from a symposium organized by the Desert Fishes Council (Marsh 1984).

The geographical and ecological characteristics of the area have been reported by Minckley (1969) and Alcocer and Kato (1995). Water chemistry varies for each spring (Minckley and Cole 1968). It is usually very aggressive, as testified by the rapid corrosion of metal objects left in the springs. Water may be cool or thermal, moving or stagnant, clear or in rare cases turbid.

The varied quality of water and aquatic habitats, coupled with habitat isolation through the fragmentation of drainages, underlies the diversity of evolutionary mechanisms known to act on the valley biota. Hershler and Minckley (1986) have reported microgeographical variation in snails of the genus *Mexipyrgus,* such that every spring has a variant population of the species. Trophic polymorphism comprising three described forms has been reported for the cichlid fish *Cichlasoma minckleyi* (Kornfield and Taylor 1983). The fish-eating form may consist of two different morphs, with opposite teeth patterns (pers. obs.).

Disjunct distributions have been reported for several amphibians (Schmidt and Owens 1944); tidal marsh plants such as *Salicornia, Suaeda,* and *Ruppia;* the local form of the snake *Natrix* (Minckley 1969); and the small blue butterfly *Brephidium* sp. (Contreras 1978a) around local saline flats. Relict species apparently coming from the same single vicariance event have given rise to close species pairs such as *Lucania parva* and *L. interioris* (Hubbs and Miller 1965), or less similar pairs such as *Cyprinodon variegatus* and *C. bifasciatus* (Miller 1968), the latter two species members of the coastal *variegatus* complex. However, the originally allotopic headspring species *C. bifasciatus* and its neighbor the lagoonal species *C. atrorus,* belonging to the Plateau *eximius* complex, occasionally hybridize despite being the two most divergent species in the genus. Their contact may be due to habitat disruption, often caused by humans.

Several genera of cave isopods and amphipods such as *Speocirolana, Sphaerolana,* and *Mexistenasellus* are members of groups that are related to Gulf of Mexico and Mediterranean taxa, remnants of a Tethyan biota that took refuge in the water-rich karstic rocks underlying the springs, all of marine ancestry.

An excellent bibliography of the natural sciences of the valley has been compiled by Minckley (1994a). Because of the diversity of evolutionary examples, the valley has long been used as a living laboratory for courses in evolution, biogeography, field biology, and natural resource conservation at Universidad Autónoma de Nuevo León and Arizona State University (1965 to date).

Professional and social pressure for the protection of the valley began in 1996, when W.L. Minckley (at Arizona State University) and the writer (at Universidad de Nuevo León) became involved in making colleagues and government officials aware of the biological richness of the valley, and proposed a national park. The government was reluctant and the Mexican scientists not very interested. Besides, parks at that time hardly would have offered the valley real protection. One of the biggest problems was opposition from inhabitants of Cuatro Ciénegas, living in the north-central part of the valley, who had to make a living from their environment.

Talks with local people began early, with little success, as they assumed that once the government or the scientists took control of the place, they would not be allowed access to it. It took time for the inhabitants to understand that most of the abundant water and the surrounding land are inadequate for standard agricultural or industrial development because the high salt content is harmful to both crops and installations. The cost of water treatment may be prohibitive. Consequently, the valley's inhabitants will need to seek out alternate methods of supporting themselves, which, to be profitable, must be low-profile options with high sustainability.

Historically, land use in the valley was agricultural, with fruit and wine production dominant. Although agricultural development has increased, harvests have generally declined. The good waters from the north have been reduced by overextraction for irrigation of artificial prairies. The town's orchards have lost their water to higher priority urban uses. Meanwhile, some industries have been established in the valley, two for extracting prime gypsum from the dunes and one for extracting salts. The gypsum facility using the prime dunes has been closed to protect the remainder of the dunes. The other makes a modest profit. Salt extraction has not been as profitable as planned. Some of these impacts have been recorded by Contreras (1984) and Minckley (1994b).

In 1988, federal law for the first time allowed for multilevel definition of areas to be protected differently for various purposes, which opened the possibility for a protected area in Cuatro Ciénegas. In 1994 parts of the valley were declared a Protected Area for Flora and Fauna, with the objective of protecting local biodiversity. A board of directors composed of area directors, municipal authorities, nongovernmental organizations (NGOs), and academicians, responsible for policymaking and overall control, was established in August 1998, and a management plan is being developed. Both encourage participation by the local community. Within the terms of the protected area, the local community must decide whether as individuals they want to apply modern conservationist agriculture or ecotourism, and fare a bit better, or keep operating their businesses on a lower level of productivity without further destroying local ecosystems.

A balance between biotic protection and rational use must be attained to conserve both important biota and the equally important human community. Achievement of this objective depends on informed participation of social groups and individuals, political conviction and coordination from local authorities, compromise on the part of enterprise leaders, and input from academicians. Working together, community members can contribute fresh and innovative ideas, which, coupled with outside support, will benefit them all.

There is hope.

The exclusive distribution of Priority I ecoregions in the Southeast and Mexico suggests an uneven distribution of priority classes among MHTs. Indeed, MHTs show remarkably different patterns (tables 5.4a and b).

Arctic rivers and lakes (ARL) ecoregions typically have low biological distinctiveness but are also relatively undegraded; consequently, all arctic ecoregions are in either Priority class IV or Priority class V. At the bioregional or national scale, these ecoregions represent important opportunities for conserving intact biotic assemblages and processes. However, the urgency associated with

TABLE 5.4A   Distribution of Priority Classes by MHT, Using Snapshot Assessment.

Major Habitat Type	Priority I	Priority II	Priority III	Priority IV	Priority V
Arctic Rivers/Lakes	0	0	0	38	63
Large Temperate Lakes	0	50	0	50	0
Temperate Headwaters/Lakes	20	10	20	40	10
Large Temperate Rivers	14	14	43	29	0
Endorheic Rivers/Lakes/Springs	13	63	13	13	0
Xeric Rivers/Lakes/Springs	30	60	10	0	0
Temperate Coastal Rivers/Lakes	13	50	13	19	6
Subtropical Coastal Rivers/Lakes	15	77	8	0	0
All Major Habitat Types	14	45	12	20	9

TABLE 5.4B   Distribution of Priority Classes by MHT, Using Final Conservation Status.

Major Habitat Type	Priority I	Priority II	Priority III	Priority IV	Priority V
Arctic Rivers/Lakes	0	0	0	50	50
Large Temperate Lakes	0	25	25	0	50
Temperate Headwaters/Lakes	20	20	10	40	10
Large Temperate Rivers	0	14	57	29	0
Endorheic Rivers/Lakes/Springs	0	38	50	13	0
Xeric Rivers/Lakes/Springs	30	50	20	0	0
Temperate Coastal Rivers/Lakes	0	50	25	25	0
Subtropical Coastal Rivers/Lakes	8	85	8	0	0
All Major Habitat Types	8	42	21	20	9

these efforts is not nearly as great as for ecoregions within the other MHTs.

The large temperate lake MHT (LTL) is unlike the others in that it contains only four ecoregions, covering the Great Lakes. Because none of these ecoregions is globally outstanding, no ecoregions of this MHT were placed in the Priority I class. All four ecoregions are assessed as currently endangered and are likely to become critical in the future. Conservationists focusing on the Great Lakes could use the results of this assessment to set priorities among its four ecoregions; the Superior [43] and Michigan-Huron [44] ecoregion have greater biological distinctiveness and might be the best targets.

Ecoregions of the temperate headwaters and lakes MHT (THL) are the most evenly distributed, with 40 percent of the ecoregions in Priority class IV and 20 percent in Priority class I. Overall, these ecoregions are not highly degraded; none were assessed as critical or endangered as of 1997. Located from the Midwest to the Appalachians, they represent some of the best opportunities for conserving important biodiversity features that remain, in some areas, relatively intact.

Large temperate rivers ecoregions (LTR), which are by definition downstream of the temperate headwaters and lakes ecoregions, suffer from centuries of intensive development. More than half of these ecoregions are globally or continentally outstanding, yet overall they are highly degraded and restoration is unlikely without both major investment and a major paradigm shift in river and floodplain management. The Mississippi Embayment [25] ecoregion is given Priority I status using the snapshot assessment, but if anticipated threats are not blocked the ecoregion will likely reach a critical, and irreversible, state. The Mississippi River and Upper Rio Grande/Rio Bravo harbor distinctive large river-adapted species, and conservationists concerned with preserving these forms will need to act immediately to protect remaining pockets of habitat.

Endorheic rivers, lakes, and springs ecoregions (ERLS), a special case of xeric habitats, are, on average, as degraded and imperiled as large rivers, if not more so. As of 1997, all ecoregions are ranked as vulnerable or worse, and five of eight face high future threats. The Llanos el Salado [66], Lerma [69], Lahontan [9], Guzmán [16], and Mapimí [19] ecoregions require immediate attention if their globally and continentally outstanding biodiversity features are to be protected, though many species have already been lost. Of primary importance is protecting water sources (especially groundwater) from overexploitation and pollution, and controlling exotic species.

Xeric-region rivers, lakes, and springs ecoregions (XRLS) face similar threats as endorheic systems, and the urgency for action is equally great. Xeric ecoregions, by virtue of their connections with other systems, tend to have higher diversity than endorheic systems and therefore are given higher priority status, on average. Five of the xeric ecoregions are globally outstanding, and the remaining six are continentally outstanding. With their highly localized endemic species, the biotas of these ecoregions are virtually irreplaceable, and once the source pools of species are gone no degree of restoration will be possible.

The temperate coastal rivers and lakes MHT (TCRL) is a group of fifteen ecoregions that represents all priority classes. The two globally outstanding ecoregions, Mobile Bay [36] and South Atlantic [40], will reach critical status without immediate restoration. Five additional ecoregions face high future threats. Eight ecoregions are continentally outstanding, several of which historically supported massive runs of migratory fish and today retain those runs to various degrees. Protection and restoration of these phenomena may be achievable only through strict regulation and cooperation among countries, states, and indigenous tribes. The impacts of urbanization along the coasts are far more difficult to regulate and control and in some cases may be irreversible.

The subtropical coastal rivers and lakes ecoregions (SCRL) tend to differ from coastal ecoregions to the north in two major respects. First, in general the subtropical coasts are not yet as developed as the temperate coasts. Second, species richness is far greater in the south because of the addition of new subtropical faunas. All subtropical coastal ecoregions are either globally or continentally outstanding, and over 90 percent of them were given either Priority I or Priority II status, both before and after consideration of threats. Florida [39] is the only ecoregion whose status becomes critical with inclusion of future threats; large-scale restoration efforts in the Everglades may be only the first step required to save the biodiversity of the larger ecoregion.

Use of the matrices in appendix F will allow conservationists to begin to identify those ecoregions that require the most immediate attention, depending on the geographic scale of concern (global, continental, bioregional, or national). International organizations such as WWF can use all the matrices together to develop a strategy for conserving the broadest range of biodiversity features possible, given real-world constraints.

Within these priority ecoregions, it is possible to identify specific watersheds that warrant particular attention. For example, TNC has recently employed its G-rank data to map the number of freshwater fish and mussel species at risk in each of 2,111 watersheds delineated by the USGS for the lower forty-eight states (Master et al. 1998). The resulting map shows which watersheds within priority ecoregions harbor the greatest number of imperiled species. As we might expect, these watersheds tend to be inhabited by species with highly limited ranges. The level of resolution provided by these data provides a good tool for moving to the next step of priority-setting, especially considering that whole-watershed approaches to conservation are critical for meaningful, long-term success.

CHAPTER 6

Recommendations

North America's Global Responsibilities for Biodiversity Conservation

This study helps put the biological distinctiveness of North American freshwater ecoregions in a global context. We rank fifteen North American ecoregions as globally outstanding—their levels of biodiversity attributes equal or exceed levels found in the most distinct ecoregions sharing the same MHT on other continents. Another way of illustrating the extraordinary biodiversity value of North American ecoregions is to display them along with other outstanding examples of the world's freshwater MHTs. The Global 200 ecoregions map portrays these units, aggregated in certain cases for global-scale analysis (Olson and Dinerstein 1998; see figure 6.1). The map shows that North American freshwater ecoregions are well represented among Global 200 ecoregions within the subtropical, temperate, and arctic regions.

With the designation of these globally outstanding ecoregions in North America comes a global responsibility to conserve them. This assessment highlights eleven Priority I ecoregions using the snapshot assessment, and six Priority I ecoregions when future threats are added into the equation. These ecoregions are listed below, with asterisks distinguishing those ecoregions that remain Priority I ecoregions after consideration of future threats.

- Rio Conchos [17]
- Cuatro Ciénegas [22]★
  (Priority II before threats consideration)
- Rio San Juan [23]★
- Mississippi Embayment [25]
- Teays-Old Ohio [34]★
- Tennessee-Cumberland [35]★
- Mobile Bay [36]
- Florida [39]
- South Atlantic [40]
- Chapala [65]★
- Tamaulipas-Veracruz [68]★
- Lerma [69]

None of the ecoregions with globally outstanding biodiversity is stable enough to remain secure without conservation attention.

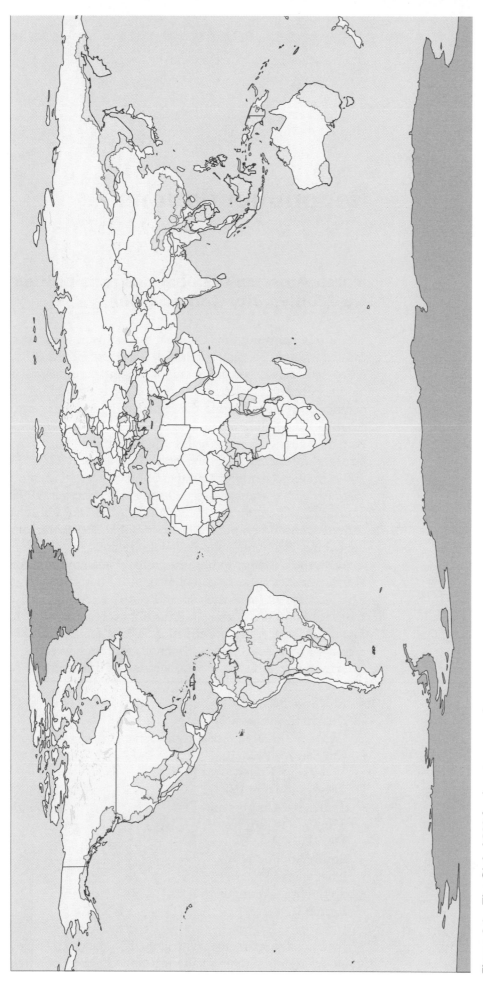

Figure 6.1   The Global 200 freshwater ecoregions.

This underscores the dire situation facing North American freshwater species—the majority of species are at great risk and will remain so over the next twenty years. Immediate action is crucial if remaining biodiversity is to be saved or restored for the future.

On a positive note, many ecoregions supporting continentally and bioregionally outstanding biodiversity are relatively undisturbed and face lower risks from future threats. These ecoregions tend to be less populated and developed by humans, but it would be foolish to leave the fate of their species uncertain. Now is our chance to protect this biodiversity, before development and other threats are entrenched. A prime example is that of exotic species—preventing their introduction takes vigilance, but removing them once they are established is often impossible.

WWF's Global 200 assessment, which identified ecoregions of global importance for biodiversity conservation, included seven freshwater ecoregions in North America (table 6.1; Olson et al. 1997). Because the Global 200 assessment evaluated all freshwater habitats (as well as terrestrial and marine habitats) on earth, it was conducted in ecoregions that were generally of a larger scale than that used for continental (e.g., North America, Latin America) assessments. For instance, the Global 200 assessment identifies Chihuahuan Desert rivers and springs as a single, globally outstanding ecoregion, whereas in this study the same area is covered by twelve ecoregions. Taken alone, each ecoregion is not globally outstanding, but when the twelve are considered together as a unit, they stand out for their high endemism in fish and other taxa. Similarly, only one of the Pacific Coast freshwater ecoregions in this study was considered globally outstanding, despite spectacular fish migrations, because the phenomenon occurs in every ecoregion and is therefore not globally rare. However, when these ecoregions are grouped together, their combined anadromous fish phenomena are indeed globally outstanding, hence the designation of the Pacific Northwest coastal rivers and streams as a Global 200 ecoregion.

TABLE 6.1 Global 200 Freshwater Ecoregions in North America (from Olson and Dinerstein 1998).

- Mississippi Piedmont rivers and streams
- Southeastern rivers and streams
- Pacific Northwest coastal rivers and streams
- Gulf of Alaska coastal rivers and streams
- Colorado River
- Chihuahuan rivers and springs
- Mexican Highland lakes

Targets Requiring Urgent Action

The expert assessment workshop, an integral part of this study, identified several important measures that must be taken to conserve—and in many cases restore—the biodiversity in North America's freshwater systems. Many of these measures require action in the terrestrial realm and should stand with companion recommendations proposed for terrestrial systems (Ricketts et al. 1999a). These recommendations offer no surprises, but their implementation will be challenging and will require the united efforts of local communities, resource agencies, and conservation groups. Most of these answers demand a political will by leaders in government and the private sector. Recommendations are grouped by the level (national/international, regional, local) at which implementation would primarily occur, though these levels are often overlapping. The measures include:

- enacting federal legislation that provides for the designation of freshwater systems themselves as natural protected areas
- educating the public and policy makers about the biodiversity hidden from view in freshwater systems, and the cumulative effects of land uses on downstream waters
- committing funds to systematic research on freshwater species, many of which may become extinct before they are described (see essay 14)
- establishing cooperation among nations for sustainable harvest of important migratory species

---

Essay 14

North America's Freshwater Invertebrates: A Research Priority

David L. Strayer

North America is home to well over ten thousand species of freshwater invertebrates, most of which are found nowhere else in the world. Many more species await discovery. Insects (e.g., flies, beetles, caddisflies, mayflies, stoneflies); crustaceans (e.g., copepods, scuds, crayfish, isopods); mollusks (snails and clams); and mites are especially rich in species. Invertebrates serve important roles in freshwater ecosystems as food for fish, grazers of algae, links in life cycles of parasites, and processors of materials, including autumn-shed leaves and biofilms in groundwaters.

Among habitats, North American streams and groundwaters contain more rare invertebrates than do lakes. For example, the Southeast's ancient rivers and groundwaters have supported the evolution of rich biological communities, but the region contains no ancient lakes equivalent to Russia's Baikal or Africa's Tanganyika, with their flocks of endemic lake-dwelling species.

Although most of the larger invertebrates of North American lakes and streams have been discovered, new species of insects, large crustaceans, and mollusks are still found regularly. Surveys of small invertebrates and groundwaters routinely turn up new species, sometimes literally by the dozen, and genera, families, and even orders of invertebrates new to North America or new to science altogether appear frequently. Clearly, hundreds of undiscovered invertebrate species still lie concealed even in the familiar and domesticated American countryside.

A vast amount of work on the physiology, behavior, life history, roles, and distribution of freshwater invertebrates remains to be done. Remarkable biological discoveries are still being made even regarding the larger and more familiar invertebrates. For example, some truly amazing lures used by pearly mussels to attract fish hosts were found only in the past decade, even though these animals were first seen 150 years ago.

With the exception of relatively well-studied mollusk, crayfish, and insect species, it is not known how many invertebrate species are declining, endangered, or extinct. It certainly seems likely that among poorly dispersing groups such as small crustaceans, species with small ranges have been endangered or extinguished by human activity. Given such a spotty knowledge of the distribution and ecology of many freshwater invertebrates, it is difficult to identify precisely the major threats to their existence. A concerted, immediate effort must be devoted to the investigation of North America's aquatic invertebrate taxonomy, ecology, and distribution. Otherwise, we will lack the information required to protect species that are described, while countless other species may go extinct before they are ever known.

- enforcing legislation to protect federally listed species, including the requirement of strict adherence by federal and state land and water management agencies
- increasing the effort to deter and punish illegal collection of nonlisted species such as pearly mussels

Regional

- promoting conservation at the watershed scale, which will require cooperation and communication among multiple agencies with varying jurisdictions
- restoring and protecting riparian habitats by limiting grazing, promoting buffer strips, and restricting development near stream and lake margins
- reducing sedimentation through minimization of logging, road-building, and damaging agricultural practices
- reducing water consumption in xeric areas through implementation of sustainable agriculture and restriction of nonessential water use
- minimizing water diversions and use to maintain water levels
- preventing the introduction and spread of exotics into freshwater systems through public education and vigilant monitoring and enforcement, and developing and implementing control programs for established exotics

Local

- immediately protecting the last relatively intact streams, springs, and lakes to act as source pools for future restoration
- increasing protection of cave and other underground freshwater habitats by restricting access and controlling pollution and extraction of source water (see essay 15)
- reducing groundwater pumping in sensitive areas
- restoring channelized streams to their original forms
- establishing natural flow regimes in rivers and streams by removing unneeded structures and modifying dam operations to resemble natural flow patterns
- reconnecting stream reaches and drainage networks by removing impoundments or creating structures to allow the passage of organisms and organic material and the flushing of nutrients
- removing flood-control structures such as levees to allow for reestablishment of floods
- restoring and protecting wetlands, which provide important filtering mechanisms for pollutants and contribute organic matter to freshwater systems
- removing point sources of pollution, and working to reduce nonpoint source pollution through education, monitoring, and enforcement at the local scale

Freshwater biodiversity, with the exception of the headline-grabbing controversy over the snail darter, has never been at the top of North America's conservation agenda. But the truly outstanding variety of life inhabiting this continent's streams, lakes,

ESSAY 15

Conservation of Aquatic Karst Biotas: Shedding Light on Troubled Waters

Stephen J. Walsh

The word *karst* originates from a Slavic word for bare, stony ground. In geology it denotes lands that are produced by the dissolutional action of water on soluble rocks, resulting in the development of internal, subsurface drainage conduits (Gines and Gines 1992). Most karst consists of limestone and dolomite. Water combines with carbon dioxide to form carbonic acid, which readily dissolves calcium (and magnesium) minerals to form bicarbonate plus calcium (or magnesium) ions, largely responsible for the development of most cavernous systems (Barr 1961).

Globally, about 20 percent of the earth's dry land surface is karst; in the midwestern and eastern United States, 40 percent of the land is composed of karst (White et al. 1995). These lands are sculpted by hydrological features that couple surface and subsurface drainages. Typically, karst environments are characterized by sinkholes, sinking streams, springs, and caves. Recharge of carbonate rock aquifers can occur by catchment through the porous rock itself (autogenic recharge) or by capture at point sources (allogenic recharge), such as sinkholes and sinking streams, where water passes over insoluble rocks at higher elevations. For instance, in the Appalachian Highlands, Interior Low Plateaus, and Ozark Plateaus, allogenic recharge is common where limestone outcrops form in narrow, folded bands on mountainsides or beneath valley floors (White et al. 1995). Sometimes, both types of recharge occur in one catchment basin, as in the dendritic and low-elevation aquifers of northern Florida that contain extensive sinkholes, subterranean caverns, and major artesian springs where water is discharged back to the surface (Florida Geological Survey 1977). In the United States the approximately 40,000 known caves (Holsinger 1988) range in length from just a few meters to the enormous Mammoth Cave system of Kentucky, with more than 500 km of charted passages. However, only caves with surface openings have been explored by humans; consequently the extent of subterranean habitats is vastly greater than what is currently known. Although karst areas of the eastern United States have been reasonably well explored, the distributions, systematics, and ecology of many species that occur in these regions are inadequately studied.

Because of their complex geological history, long isolation, and relatively stable physicochemical milieu, karst lands provide ecosystems in which unique biotas occur. The faunas of caves and karst habitats have long fascinated biologists because they are ideal subjects for the study of regressive evolutionary phenomena (Poulson and White 1969; Culver 1982). Many cave-dwelling animals exhibit similar morphological features resulting from convergence during exposure to common environmental conditions; these include absence of eyes, depigmentation, and sensory structures that are adaptive for life in total darkness (Holsinger 1988; Camacho et al. 1992; Jones et al. 1992). Moreover, cave animals exhibit unique physiological and ecological attributes as a result of the physical environment in which they have evolved.

A variety of terms are used to describe and categorize cave faunas. *Troglobitic* organisms are those found only in caves, *troglophilic* species occur in both cave and non-cave habitats, and *trogloxenes* are species that spend part of their life cycles in caves. Thus, troglobites are obligate cave dwellers, troglophiles are facultative, and trogloxenes may be either (Culver 1986).

Terrestrial organisms generally outnumber aquatic taxa in cave systems, and invertebrate troglobites vastly outnumber vertebrates. However, in some karst regions dominated by aquatic habitats, such as northern Florida, aquatic species may outnumber terrestrial ones (Franz et al. 1994). In addition to faunas associated with caves, there are many *epigean* (above-ground) aquatic and semiaquatic animals that are closely associated with spring or sinkhole habitats and that face similar conservation challenges.

Because karst lands are highly fragmented, cave and spring faunas exhibit high endemism and individual species often have small geographic distributions. Zoogeographically, karst faunas are

106   *Freshwater Ecoregions of North America*

analogous to those of islands (Culver 1970). Aquatic animals in karst areas generally have larger range sizes than do terrestrial species (Culver 1986; Holsinger 1988), but still much smaller than related epigean species. Most input of organic nutrients into subterranean habitats generally originates allochthonously, although a few systems are supported by chemautotrophic primary production (White et al. 1995). Thus, cave faunas are generally considered to be energy limited. Relatively stable physical conditions, limited food resources, and selection within cave environments generally result in small populations of animals that have reduced metabolic rates, low fecundity (reproductive capacity), slow development and growth, delayed maturation, and increased longevity. Geographic isolation, genetic drift, selection, and stochastic events in a myriad of karst habitats have resulted in speciation of many unique, diverse taxa in a wide range of unrelated groups.

In the United States, crustaceans, snails, fishes, and salamanders are the predominant groups that are well represented by a diverse number of species in aquatic karst habitats. An example of the faunistically rich eastern karst fauna was provided by Holsinger and Culver (1988), who documented approximately 335 invertebrate species representing 90 families and 173 genera from 500 caves in Virginia and eastern Tennessee. Included in this number were 140 troglobites, of which 42 were aquatic.

Increasingly, the biotas of karst terrains are seriously threatened by humans. Karst habitats are relatively fragile and highly susceptible to perturbation. Since many of the species they harbor have relict or limited distributions, have small population sizes, and are particularly sensitive to disturbance because of their ecological adaptations, these inconspicuous organisms are especially vulnerable. Single catastrophic events have the potential to seriously alter or eliminate entire communities of organisms in karst habitats, where some of these highly endemic species occupy single caves or springs. Minckley and Unmack (essay 7) have discussed the problems associated with arid spring habitats of western ecosystems; aquatic organisms in the rich karst lands of the eastern United States face similar jeopardy.

The most serious threats to karst areas are habitat loss and alteration, hydrological manipulation, environmental pollution, and overexploitation (Culver 1986; Proudlove 1997b). Physical habitat alteration results from many human activities, such as mining, quarrying, and deforestation, the latter causing sedimentation. Hydrological modifications include impoundment of fluvial systems and water removal or diversion. In the Mammoth Cave system, impounds on the Green River may have negatively affected the northern cavefish (*Amblyopsis spelaea*) by causing sedimentation or reducing levels of organic nutrients (Poulson 1992). Removal of water from aquifers and spring systems for human consumption, residential development, and irrigation is a pervasive threat that is not confined to arid regions. The impoundment, channelization, and pumping of water from a spring system near Huntsville, Alabama, resulted in the extinction of the whiteline topminnow, *Fundulus albolineatus* (Miller et al. 1979).

Environmental pollution is especially insidious and may involve eutrophication as well as agricultural and industrial runoff or seepage of fertilizers, pesticides, heavy metals, and sewage. Holsinger (1966) documented a decline in the diversity of aquatic cavernicoles associated with the overflow of a septic tank and pollution of a Virginia cave. A massive kill of an aquatic community, including the troglobitic pallid cave crayfish (*Procambarus pallidus*), may have been partially the result of groundwater pollution following a flood of the unconfined Florida Aquifer (Streever 1992). Rivers draining karst terrain are especially susceptible to eutrophication and pollution that can result from certain land-use practices. For example, the recent increase in nitrates and other nutrients in the central Suwannee River watershed is associated with agriculture, phosphate mining, and urbanization (Andrews 1994; Ham and Hatzell 1996). Thus, prevention of groundwater contamination from overlying land use is a major concern for the conservation of karst biodiversity (Mattson et al. 1995).

Overexploitation may result from zealous or unscrupulous collection of specimens for the pet trade or scientific study, as in the case of the Ozark cavefish (*Amblyopsis rosae*); indirect effects on communities may occur through excessive visitation to karst habitats by speleologists and others

(Culver 1986; Tercafs 1992). Ironically, human interest and attention in some cases may have a detrimental impact on karst-dwelling taxa.

Threats to the biotas of karst habitats are generally unfamiliar to the public and under-appreciated by many people within the scientific community. This is due in part to lack of awareness of the fauna and limited concern for noncharismatic organisms that fail to generate public support or sympathy. Yet these extraordinary animals merit urgent attention for assessment of their conservation status, given their vulnerability, our inadequate knowledge of their biology, and the management issues that will have to be addressed if we are to ensure their survival well into the next millennium. Some organizations, such as The Nature Conservancy and the Karst Waters Institute, have recently elevated awareness of the plight of karst ecosystems, but much more needs to be done to proactively ensure protection of this often mysterious and obscure component of global biodiversity.

and springs has rarely been broadcast. Sadly, the threats to these species and processes are as acute as the ecoregions are diverse.

This study puts North America's freshwater biodiversity in a global context, and maps areas of high distinctiveness using zoogeographically useful units. By utilizing freshwater ecoregions, we can see where hot spots occur at the ecoregion scale and then apply ecoregion-based conservation where it is needed most. As all conservation ultimately occurs at the site level, the use of freshwater ecoregions in conjunction with terrestrial ecoregions can illuminate places where interventions can potentially benefit both realms. An overlay of priority ecoregions derived from separate assessments can show areas where site-specific interventions will achieve both freshwater and terrestrial conservation.

This project hopes to offer a scientifically credible method for recommending areas for restoration, increased protection, and policy reform in ways that support biodiversity conservation first and foremost. North America's freshwater systems are truly imperiled across the continent, but there are still opportunities to save what remains, if we act swiftly and strategically.

Freshwater Ecoregions of North America

CHAPTER 7 **Site-Specific Conservation**

Continental-scale analyses can guide us to the most distinctive and threatened freshwater ecoregions, but conservation requires integrated actions at the scale of sites as well as whole ecoregions. For this we need to understand how biodiversity features are distributed within ecoregions and how individual sites, habitats, and assemblages fit into a broader conservation strategy. Ecoregion-based conservation (ERBC) approaches may be a useful way to begin to preserve or restore the distinct biological features highlighted in this study.

Within a given ecoregion, ERBC focuses on four general conservation targets: (1) distinct communities, habitats, and assemblages (e.g., foci of extraordinary endemism, rare habitats); (2) larger examples of intact habitats and intact biotas (e.g., unimpounded rivers, presence of top predators); (3) keystone ecosystems, habitats, or phenomena (e.g., intact floodplains); and (4) distinct large-scale ecological phenomena (e.g., anadromous fish migrations). Based on these targets, an ERBC approach identifies a portfolio of priority sites and activities. It results in a vision of success for the ecoregion—the extent and configuration of habitats to be conserved or restored across the landscape, and the conservation activities and guidelines that need to be implemented—in order to conserve the fullest range of biodiversity over the long term. The relationships between upstream and downstream conditions, and between the freshwater and terrestrial realms, are essential components of an ERBC strategy.

This continental-scale assessment identifies those North American ecoregions where an ERBC program should perhaps be initiated first, given the ecoregions' globally outstanding biodiversity and the immediacy of threats. However, all ecoregions contain important biodiversity features, and it is possible to identify readily apparent sites where conservation intervention would significantly contribute to achieving conservation goals, particularly where the sites are integrated into a larger-scale conservation plan.

Availability of resources limits the number of ecoregions in which comprehensive ERBC programs can be implemented. Recognizing this, we asked taxonomic and regional experts to undertake a preliminary identification of sites across North America where intervention—from dam removal to increased protection—would serve as a first step toward achieving one or more of the conservation targets listed above. Sites were selected based on the presence of important biodiversity targets. For example, some priority sites were selected because they are places where rare habitats remain intact or where important species assemblages could be restored. The resulting list of sites is not intended to be comprehensive, but instead offers examples of where conservation organizations and agencies working at the local scale could direct their efforts. The 136 sites that were identified are listed and mapped in table 7.1 and figures 7.1a–c.

TABLE 7.1 Important Sites for the Conservation of Freshwater Biodiversity in North America.

Ecoregion Number	Site Number	Site Name
42	1	Penobscot system, Maine
42	2	Connecticut River (below White River junction), Vermont/New Hampshire
42	3	Neversink River, New York
41	4	Sideling Hill Creek, Pennsylvania/Maryland
34	5	French Creek, Pennsylvania/New York
34	6	Big Darby Creek, Ohio
34	7	Wabash River, Indiana
34	8	Licking River, Kentucky
34	9	South Fork Kentucky River, Kentucky
34	10	Barren River (lock and dam to Gasper River), Kentucky
34	11	Green River, from Green River Lake Dam to Nolin River, Kentucky
35	12	Big South Fork Cumberland, Kentucky/Tennessee
35	13	Little South Fork Cumberland, Kentucky
35	14	Rockcastle River System, including Horse Lick Creek in Daniel Boone National Forest, Kentucky
34	15	Buck Creek, Kentucky
35	16	Bunches Creek and Criscillis Branch, Kentucky (Daniel Boone National Forest)
35	17	Sinking Creek, Addison Branch, Eagle Creek, and Cave Creek Cave in Daniel Boone National Forest, Kentucky
35	18	Powell River, Virginia/Tennessee
35	19	Clinch River, Virginia/Tennessee
35	20	North Fork River, Virginia
35	21	Holston River, Tennessee
35	22	Emory Creek, Tennessee
35	23	Tennessee River from Kentucky Dam to the Ohio River, Kentucky
35	24	Buffalo Creek, Tennessee
35	25	Shoal Creek, Tennessee/Alabama
35	26	Duck Creek, Tennessee
35	27	Tim's Ford Dam, Tennessee
35	28	Paint Rock, Tennessee/Alabama
35	29	Walden Ridge, Tennessee
35	30	Hiwassee River, Tennessee/Georgia
35	31	Little Tennessee, North Carolina
35	32	Marsh Creek, Kentucky/Tennessee (Daniel Boone National Forest)
40	33	Roanoke River in and above Roanoke, Virginia
40	34	Carolina Sandhills, North Carolina
40	35	Lake Waccamaw, North Carolina
40	36	Altamaha River basin, Georgia
39	37	Indian River Lagoon, Florida
39	38	Suwannee River–Okefenokee Swamp–St. Marys River, Georgia/Florida
37	39	Flint River tributaries, Georgia
37	40	Jim Woodruff Lock and Dam, Florida
37	41	Chipola River, Alabama/Florida
38	42	Choctawhatchee River, Alabama/Florida
38	43	Yellow River and feeder creeks on Eglin Air Force Base, Florida
38	44	Escambia/Conecuh River, Alabama/Florida
36	45	Cahaba River, Alabama
36	46	Coosa River basin, Georgia
36	47	Black Warrior tributaries, Alabama

(continues)

TABLE 7.1 *Continued.*

Ecoregion Number	Site Number	Site Name
36	48	Buttahatchie (Tombigbee tributary), Alabama/Mississippi
36	49	Sipsey (Tombigbee tributary), Alabama
36	50	Lower Alabama River, Alabama
25	51	Pascagoula/Leaf River, Mississippi
25	52	Pearl River, Mississippi
25	53	Big Sunflower River, Mississippi
25	54	Hatchie River, Tennessee/Mississippi
25	55	Cache River, southern Illinois
29	56	Upper Kings River, Carroll and Madison Counties, Arkansas
24	57	Fox River, Illinois
24	58	St. Croix River, Wisconsin/Minnesota
1	59	Mayer Lake on Queen Charlotte Island, British Columbia
1	60	Ecstall River Watershed, British Columbia
1	61	Great Bear Rainforest Watershed, British Columbia
1	62	Fraser Valley, British Columbia
1	63	Thanksgiving Cave, Vancouver Island
1	64	Cowichan Lake, Vancouver Island
1	65	Hourglass Cave, Vancouver Island
1	66	Elwha River, Washington
3	67	Sites within the Columbia Gorge and Lower Columbia River, Oregon/Washington
3	68	Lower Deschutes River, Oregon
4	69	Upper Snake River drainage, Idaho
3	70	Middle Snake River drainage, Idaho
5	71	Umpqua River, Oregon
5	72	Trinity River, California
5	73	Upper Klamath Basin, Oregon (including Klamath Lake)
10	74	Oregon Lakes, mostly Oregon
10	75	Goose Lake, Oregon/California
6	76	Mill, Deer, and Chico creeks (tributaries to Sacramento), California
6	77	Cosumnes River, California
6	78	Clavey River (tributary to Tuolumne), California
6	79	Upper Kings River—North and Middle Forks (above Pine Flat Reservoir), California
6	80	Upper Kern River (above Isabella Reservoir), California
9	81	Eagle Lake watershed, California
9	82	Soldier Meadows Valley, Nevada
9	83	Lake Tahoe–Truckee River–Pyramid Lake basin, Nevada/California
9	84	Steptoe Valley, Nevada
11	85	Owens Valley, California
11	86	Owens River and Lake, California
11	87	Ash Meadows/Death Valley, Nevada
7	88	Upper San Gabriel River, California
8	89	Bear Lake, Utah/Idaho
13	90	Pluvial White River, Nevada
13	91	Virgin and Moapa Rivers, Arizona/Nevada/Utah
12	92	Upper Green/Yampa Rivers, Utah/Colorado
12	93	Colorado River Delta wetlands, Baja and Sonora, Mexico
12	94	San Juan River, Utah/New Mexico/Colorado
12	95	Little Colorado River headwaters, Arizona/New Mexico
14	96	Upper Verde River, Arizona (above Camp Verde)
14	97	Upper reaches of East Fork Gila River, Catron County, New Mexico
14	98	Gila River in Cliff–Gila Valley, Grant County, New Mexico
14	99	Cajon Bonito, Sonora, Mexico
14	100	Mimbres River, including Moreno Spring, New Mexico
14	101	Upper Gila River, Arizona (above Safford, Arizona)
14	102	San Pedro River and Aravaipa Creek, Arizona
14	103	Headwaters of Santa Cruz, Santa Cruz, Arizona through N. Sonora & back into Santa Cruz County
14	104	Wilcox/Upper Yaqui, southeast Arizona/northeast Sonora, Mexico
61	105	Quitobaquito/Rio Sonoyta, Sonora, Mexico
61	106	Rio Bavispe, from Morelos to headwaters (above La Angostura Reservoir), Sonora, Mexico
61	107	Rio Sonora headwaters to south of Cananea, Sonora, Mexico
61	108	Rio Yaqui headwaters, Sonora, Mexico
61	109	Headwaters of Rios Papigochic/Aros/Sirupa, Mexico

(continues)

TABLE 7.1 *Continued.*

Ecoregion Number	Site Number	Site Name
16	110	Rio Casas Grandes, Chihuahua, Mexico
16	111	Laguna Guzman, northern Chihuahua, Mexico
16	112	Laguna Bavicora, Chihuahua, Mexico
16	113	Sauz Basin, Chihuahua, Mexico
15	114	Rio Grande, New Mexico
15	115	Willow Spring, central New Mexico
15	116	Tularosa Basin, south-central New Mexico
18	117	Pecos River, from Sumner Dam south, New Mexico/Texas
18	118	Leon Creek and Diamond Y Spring, Texas
20	119	Upper Conchos, west of Chihuahua City, Mexico
20	120	Devils River, Texas, northwest of Del Rio (includes Dolan Creek)
21	121	Chorro, southeast Saltillo, Mexico
21	122	Zona Carbonifera from Del Rio/Eagle Pass to Muzquiz/Sabinas, Coahuila, Mexico
22	123	Cuatro Cienegas, Coahuila, Mexico
17	124	San Diego, near San Diego de Alcola, Chihuahua, Mexico
17	125	Bustillos, central Chihuahua, Mexico
17	126	Upper Rio Conchos, including headwaters, Chihuahua, Mexico
19	127	Rio Cadena, southeast from Chihuahua City, Chihuahua, Mexico
19	128	Upper Nazas, Durango, Mexico
19	129	Mayran–Nazas complex, Durango, Mexico
19	130	Santiaguillo, 40–80 km north of Durango City, Durango, Mexico
19	131	La Concha spring and canyon, near Penon Blanco, Durango, Mexico
19	132	Parras Basin, Coahuila, Mexico
19	133	Upper Aguanaval, north Zacatecas, Mexico
19	134	Potosí, Ejido Catarino Rodriguez, Zacatecas, Mexico
23	135	Parque Cumbres de Monterrey, Nuevo León, Mexico
66	136	Iturbide, 100 km south of Monterrey, Nuevo León, Mexico
66	137	Sandía, Llanos de Salas, San Luis Potosí, Mexico
66	138	Venado, north San Luis Potosí, Mexico
66	139	Extorax, east San Luis Potosí, Mexico
67	140	Media Luna/Rio Verde, East San Luis Potosí, Mexico
68	141	Panuco, Querataro/Hidalgo, Mexico
63	142	Mezquital, around Durango City, Durango, Mexico
65	143	Chapala wetlands, Jalisco, Mexico
69	144	Cuitzeo wetlands, Michoacán, Mexico
69	145	Lerma River swamps, Mexico state, Mexico
73	146	Grijalva/Usumacinta delta swamps, Tabasco, Mexico

Dams

Dams represent one of the most serious threats to freshwater biodiversity, as they alter hydrologic integrity and water quality, fragment habitat, and cause additional habitat loss in the form of upstream and downstream changes in sediment and other parameters. The site-specific nature of dams makes them viable targets for conservation intervention, particularly where the dams no longer serve the purpose for which they were originally constructed or where their strategic decommissioning would greatly benefit important freshwater sites nearby. Experts were quick to recommend such nonessential or especially harmful dams for decommissioning or modification (figure 7.2), focusing on those that degrade or impair important elements of biodiversity. Each of these dams is listed below, with a brief description of the recommended activity. These dams will be considered by WWF-US for inclusion in its "Gifts to the Earth" campaign, in which WWF works with government or other authorities to secure commitments toward specific conservation action.

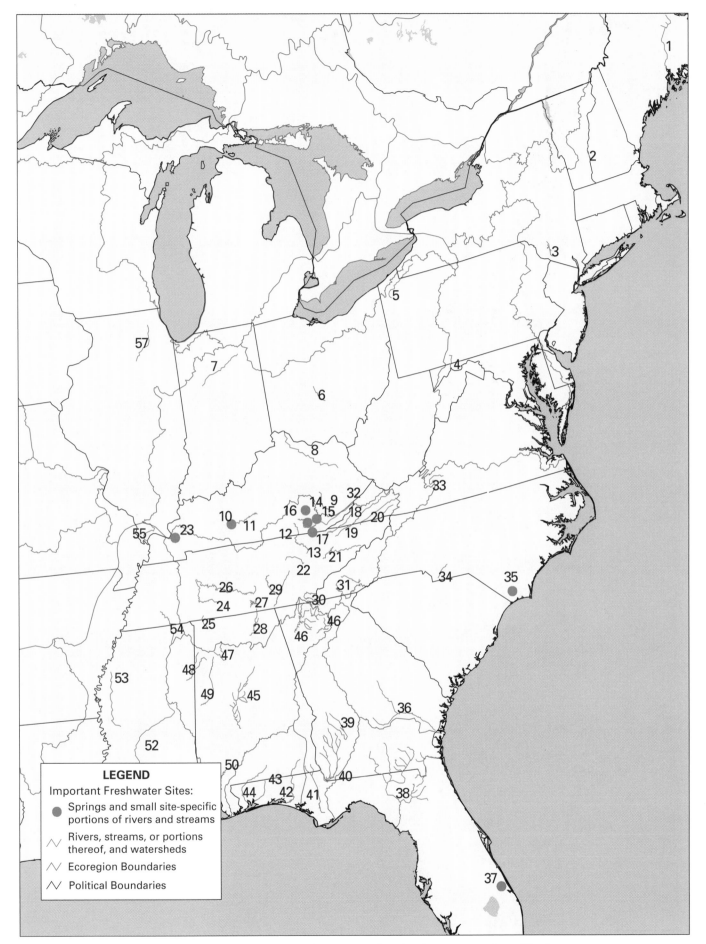

Figure 7.1a   Important sites for the conservation of freshwater biodiversity in North America—eastern United States.

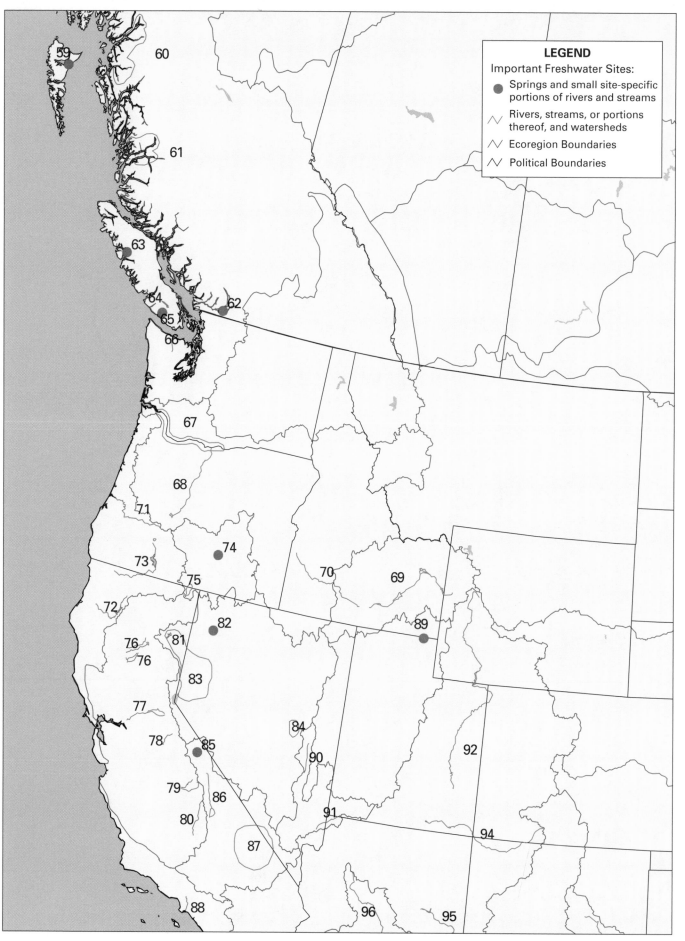

Figure 7.1b   Important sites for the conservation of freshwater biodiversity in North America—western United States and Canada.

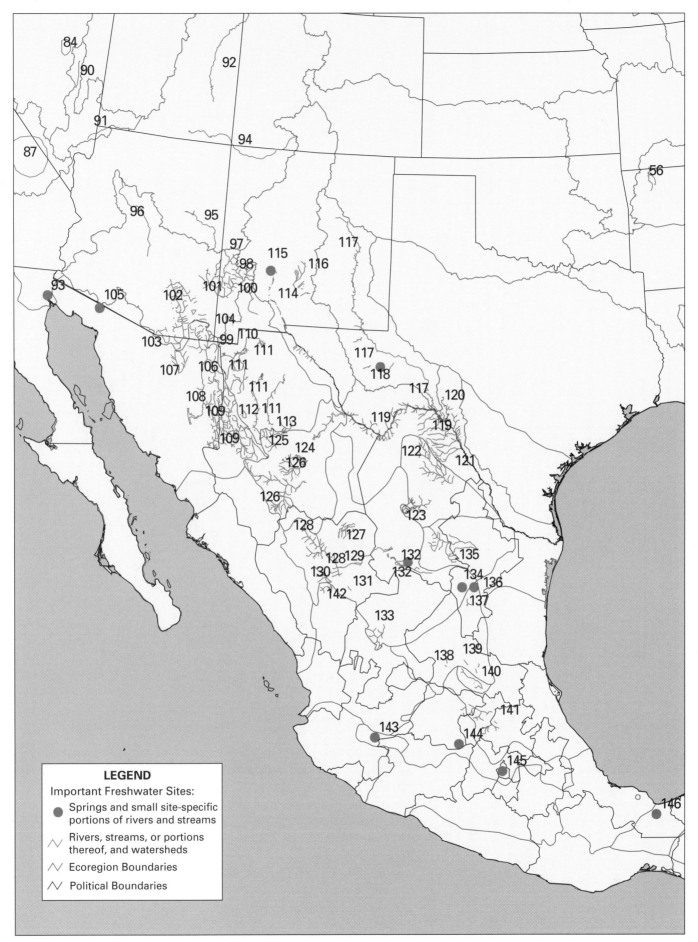

Figure 7.1c   Important sites for the conservation of freshwater biodiversity in North America—southwestern United States and Mexico.

Figure 7.2 Dams proposed for decommissioning or modification.

- remove dams at Dayton and Yorkville on Fox River between Aurora, Illinois, and confluence with Illinois River
- remove Green River Lake Dam, Kentucky
- remove dam on lower half of St. Croix River, Wisconsin/Minnesota
- install air injection on Allatoona Dam on Etowah River, Georgia
- remove Tim's Ford Dam on Elk River, Tennessee
- prevent impoundment construction on tributaries of Choctawhatchee River, Alabama
- remove dam at union of forks of Upper San Gabriel River, California
- remove dam that creates Crowley Lake Reservoir, Arkansas

- remove Trinity Dam, California
- remove Lewiston Dam, California
- remove barriers on Thomes Creek, Oregon
- remove diversion dams (and buy water rights) on Cosumnes River, California
- remove Jim Woodruff Lock and Dam on the Apalachicola River, Florida
- remove Rodman Dam on the Ocklawaha River, Florida
- remove Jackson Bluff Dam on the Ochlockonee River, Florida
- remove diversion dams on lower reaches of Deer, Mill, Pine, Dye, and Chico Creeks, California
- remove Matilija Dam on Ventura River, California
- remove Condit Dam on the White Salmon River, Washington

Dam removal will not occur without some opposition, yet recent large-scale efforts at restoring or approximating the natural flow of systems such as the Colorado River below Glen Canyon Dam and the Kissimmee River in Florida illustrate that the political will exists. Furthermore, the Federal Energy Regulatory Commission (FERC) has begun to consider seriously the needs of freshwater biota as it considers the relicensing applications for old dams (see essay 16). In a landmark action in November 1997, FERC ordered the removal of the Edwards Dam in Augusta, Maine, which for 160 years blocked spawning habitat for nine species of migratory freshwater fish.

Reoperation and Decommissioning of Hydropower Dams: An Opportunity for River Rehabilitation

Andrew Fahlund

For centuries, rivers have been harnessed for human use. Irrigation, navigation, flood control, water supply, and electricity generation have opened new lands, fueled economies, and provided for the livelihoods of individuals and entire civilizations. Not until the last twenty years have aquatic ecologists and resource managers recognized the significant harm caused by even small dams to fish and other wildlife, water quality, and entire riverine ecosystems.

According to the U.S. Army Corps of Engineers, approximately 75,000 dams more than 5 feet tall block the waterways of the United States. Only about 2,400 of these dams generate electricity; most of them are owned and operated by private interests. Because of the way they are sited and operated, hydropower dams are among the most harmful influences on riverine ecosystems and aquatic biodiversity. Of the 3.5 million miles of rivers in this country, 600,000 miles are impounded behind dams, and many more miles of downstream resources are adversely affected.

Hydropower Dams

Dams block free-flowing river systems, hindering the flow of nutrients and sediments and impeding the migration of fish and other wildlife. When water is diverted to hydropower dams for power production, healthy in-stream ecosystems are often completely dewatered. By withholding and then releasing water to generate power for peak periods, dams cause alternating flow in downstream stretches: periods of no water are followed by powerful surges that erode soil and vegetation and flood or strand fish and riparian-dependent species.

Stagnant reservoir pools and altered flow timing confound the reproductive cues of many fish species and significantly hinder both upstream and downstream passage. While following currents downstream, fish may be maimed or killed by power turbines. Reservoirs also flood valuable riffle and swift-water habitat, necessary for many species' life cycles. Water temperatures and oxygen levels, critical to species survival, are also altered by dams. Yet despite hydropower's harmful effects on rivers and fisheries, private interests have long harnessed the nation's rivers for their own profit without mitigating those impacts or providing compensation.

The cumulative results of these impacts, along with other stresses on our nation's watersheds, have been destruction of native plant and animal communities and overall reduction in species biodiversity. However, rivers are resilient systems which, given a chance to function under a specific range of baseline parameters, can bounce back from significant chemical and physical alteration. Short of dam removal, it is unlikely that affected rivers can be restored to conditions such as existed before dam construction. Nevertheless, steps can be taken to change how hydropower dams operate so that pre-project baseline conditions can be approximated and rivers rehabilitated. The widely reported 1996 simulated floods at the Glen Canyon Dam on the Colorado River were designed to restore natural function by affecting hydrology and geomorphology below the dam. For several years before the Glen Canyon experiment, resource managers, aquatic ecologists, and environmentalists had sought similar operational changes through FERC's relicensing process, with noted success.

Most nonfederal hydroelectric dams on rivers and streams in the United States come under the jurisdiction of FERC, which grants thirty- to fifty-year licenses for operation of hydropower dams on public waterways. FERC is required to give "equal consideration" to the power and non-power values of these projects. This public interest determination is intended to ensure that licensed dams are operated in a safe fashion, that their impacts on the environment are minimized, and that alternative uses of the affected rivers are considered and accommodated. When a hydropower dam license expires, the dam owner must renew it through a complex administrative process known as relicensing. Most of the dams currently under review were licensed before enactment of modern environmental laws such as the Clean Water Act and the Endangered Species Act, making this a once-in-a-lifetime opportunity to restore degraded freshwater systems through project reoperation.

Hydropower Dam Reoperation

Efforts undertaken over the last several years are beginning to show that dam reoperation can lead to successful river rehabilitation. By changing the flow regime of dams on several rivers in Michigan from a peaking flow to a more natural run of the river, thus improving both water quality and habitat structure, resource managers have begun to see a significant increase in natural fish recruitment below these dams. Along the Deerfield River in Vermont and Massachusetts, flows have been restored to more than 12 miles of previously dry riverbed. Because dam owners often manage large tracts of land in order to operate these facilities, relicensing has also presented opportunities for protecting wildlife such as the federally listed Indiana bat (*Myotis sodalis*) and the bald eagle (*Haliaeetus leucocephalus*).

In some cases, however, dam removal is the only alternative that will lead to ecosystem recovery. The Edwards Dam, on the Kennebec River in Maine, is scheduled for removal in 1999, opening 20 miles of prime spawning and rearing habitat to Atlantic salmon (*Salmo salar*), short-nosed sturgeon (*Acipenser brevirostrum*), and several other fish species that have suffered from habitat fragmentation and inability to utilize fish ladders. While this precedent has generated a great deal of discussion about removing dams in other watersheds, it will take time for many people to understand that dams are not monuments or permanent geological features on the landscape, structures that cannot or should not be altered.

Future Opportunities for Preserving Biodiversity

The next several years offer an unprecedented opportunity to rehabilitate rivers through hydropower dam reoperation. Between 1998 and 2010, licenses will expire for more than 500 dams, representing approximately half of the nation's nonfederal hydropower generating capacity. The vast majority of these projects are located along the West Coast where Pacific salmon and trout (*Oncorhynchus* spp.) continue their precipitous decline. A significant amount of relicensing activity will also occur in the Southeast, harboring perhaps the most diverse temperate-zone assemblage of aquatic species in the world, where we are only beginning to understand the threats to biodiversity. These regulatory proceedings offer opportunities for immediate improvement to flow and fish passage as well as creative solutions for off-site critical habitat protection and establishment of mitigation trust funds. It is critical that resource managers, dam owners, and the public work together to find solutions by which imperiled river systems can be restored.

The importance of ensuring that the multiple uses of the public's waterways reflect prevailing conditions and priorities was underscored by President Theodore Roosevelt in 1908:

> The public must retain control of the great waterways. It is essential that any permit to obstruct them for reasons and on conditions that seem good at the moment should be subject to revision when changed conditions demand. . . . Each right should be issued to expire on a specified day without further legislative, administrative, or judicial action. Provision should be made for the termination of the grant or privilege at a definite time, leaving to future generations the power or authority to renew or extend the concession in accordance with the conditions which may prevail at the time. (H.R. Report No. 507, 99th Congress, 2d Session. 1986. Legislative History of the Electric Consumers Protection Act of 1986.)

Additional Site-Selection Tools

Site-selection exercises apart from this one are valuable and their recommendations should be integrated, where appropriate, with those presented here. CONABIO, the biodiversity agency of Mexico, identified priority hydrologic basins across Mexico, using a broader range of criteria than used here (Arriaga et al. 1998). For example, CONABIO's analyses included terrestrial habitats associated with freshwater systems, such as riparian zones; water-use issues; and bird, mammal, rotifer, and aquatic plant species distributions. A total of 110 priority basins were identified, of which sixty-five were highlighted for their biodiversity value. The CONABIO results were not available in time for incorporation into the list of priority sites or watersheds (table 7.1), but they are noted here because they provide important additional information for conservation planners. For the United States, TNC has identified eighty-seven small watersheds as conservation priorities (Master et al. 1998).

We encourage planners to use the full range of available information to identify conservation priorities. Conservation organizations and agencies continue to share their findings and develop productive collaborations. In this light, we hope that this study's approach and observations will prove a useful contribution toward efforts to protect freshwater biodiversity.

Methods for Assessing the Biological Distinctiveness of Freshwater Ecoregions

The biological distinctiveness index (BDI) was designed to provide an objective measure of the degree to which the biodiversity of an ecoregion is distinctive over a range of biogeographic scales. We used four main criteria to evaluate biological distinctiveness: (1) species richness; (2) species endemism; (3) rare ecological and evolutionary phenomena; and (4) global rarity of habitat type.

The points awarded for these criteria were summed to yield a biological distinctiveness score for each ecoregion. The ecoregions were then assigned to one of four categories, which reflect the distinctiveness of the ecoregion's biodiversity at different biogeographic scales: globally outstanding, continentally (e.g., within a biogeographic realm or regionally) outstanding, bioregionally (e.g., Pacific Coast) outstanding, or nationally important. General explanations and the rationale for the four criteria are presented in Chapter 2: Approach. The design of the index is summarized in figure A.1, and below we describe the evaluation methods in detail. The information sources used to evaluate biological distinctiveness are listed in table A.1.

Species Presence/Absence and Endemism Data

We collected published and unpublished range and distribution data for North American species in five taxonomic groups, comprising over 2,200 species. The groups were fish, crayfish, unionid mussels, amphibians dependent on aquatic habitats, and aquatic and semiaquatic reptiles. Amphibians and reptiles have been grouped together as herpetofauna within this study (their distributions were not used to help delineate freshwater ecoregion boundaries). Only amphibians with life stages requiring freshwater were included, as were only aquatic or semiaquatic reptiles. Adequate biogeographic data on aquatic plants and other invertebrates were unsynthesized, difficult to locate, or of poor quality.

For fish and herpetofauna, the data were in the form of range maps. We compared the range map of each species with the ecoregion map and recorded the species as being present or absent in each ecoregion, as well as whether the species was endemic to the ecoregion (see below for endemism decision rules). Where range maps were composed of occurrence data, species were counted as being present in an ecoregion if an occurrence was located within the ecoregion boundary. Where range maps were composed of area rather than point data, any range overlap with an ecoregion was recorded as species presence within that ecoregion. If multiple range maps existed for a given species, maps were compared to achieve the greatest accuracy. Descriptions of species occurrences, which in some cases were the only available data source, were similarly employed.

Mexican species presented some challenges, as there are currently no comprehensive data sources available for the entire country. Fish distributions were determined using museum collection data combined with published species lists for specific basins. Fortunately, drainage basins in Mexico are well defined because of the prevalence of interior (endorheic) catchments. For herpetofauna, the best data source was the dissertation of Dr. Oscar Flores-Villela (1991). However, this document maps only amphibians and reptiles that are restricted to Mexico. The intra-Mexico distributions of species also found elsewhere were determined using an assortment

TABLE A.1 Data Sources for Biological Distinctiveness Indicators.

Indicator	Data Source/Type		
	Canada	U.S.	Mexico
Species richness and endemism			
Fishes	1–5	1, 6–21	1, 11, 14, 19, 20, 22–29, 30
Unionid mussels	31, 32	31	no data
Crayfish	33	33	30, 33
Amphibians	34–35	34–36	34–35, 37–50
Aquatic reptiles	34–35, 51–54	34–36, 51–54	34–35, 37–39, 42–43, 45–50, 52–54
Rare ecological or evolutionary phenomena	expert assessment	expert assessment	expert assessment
Higher-level taxonomic diversity	see richness/endemism	see richness/endemism	see richness/endemism
Rare habitat type	expert assessment, 55	expert assessment, 55	expert assessment, 55

1) Page and Burr 1991
2) Scott and Crossman 1973
3) Underhill 1986
4) Crossman and McAllister 1986
5) Lindsey and McPhail 1986
6) Burr and Page 1986
7) Conner and Suttkus 1986
8) Cross et al. 1986
9) Hocutt et al. 1986
10) McPhail and Lindsey 1986
11) Minckley et al. 1986
12) Robison 1986
13) Schmidt 1986
14) Smith and Miller 1986
15) Starnes and Etnier 1986
16) Swift et al. 1986
17) Underhill 1986
18) Mayden et al. 1992
19) Minckley and Brown 1994

20) Miller 1977
21) Burr, pers. comm.
22) Contreras-Balderas 1969
23) Espinosa-Pérez et al. 1993b
24) Hendrickson 1983
25) Lozano Vilano and Contreras Balderas 1987
26) Miller and Smith 1986
27) Minckley 1977
28) Obregón-Barboza et al. 1994
29) The University of Michigan 1995
30) Contreras Balderas, pers. comm.
31) The Nature Conservancy 1997
32) Clarke 1981
33) Taylor 1997
34) Conant and Collins 1991
35) Stebbins 1985
36) Degenhart et al. 1996
37) Dunn 1926
38) Conant 1977

39) Duellman 1961
40) Duellman 1963
41) Duellman 1964
42) Duellman 1965a
43) Duellman 1965b
44) Duellman 1970
45) Flores-Villela 1991
46) Flores-Villela 1993a
47) Hardy and McDiarmid 1969
48) Martin 1958
49) Smith and Taylor 1966
50) Scudday 1977
51) Ernst and Barbour 1989
52) Ernst et al. 1994
53) Gloyd and Conant 1990
54) Rossman et al. 1996
55) Olson et al. 1997

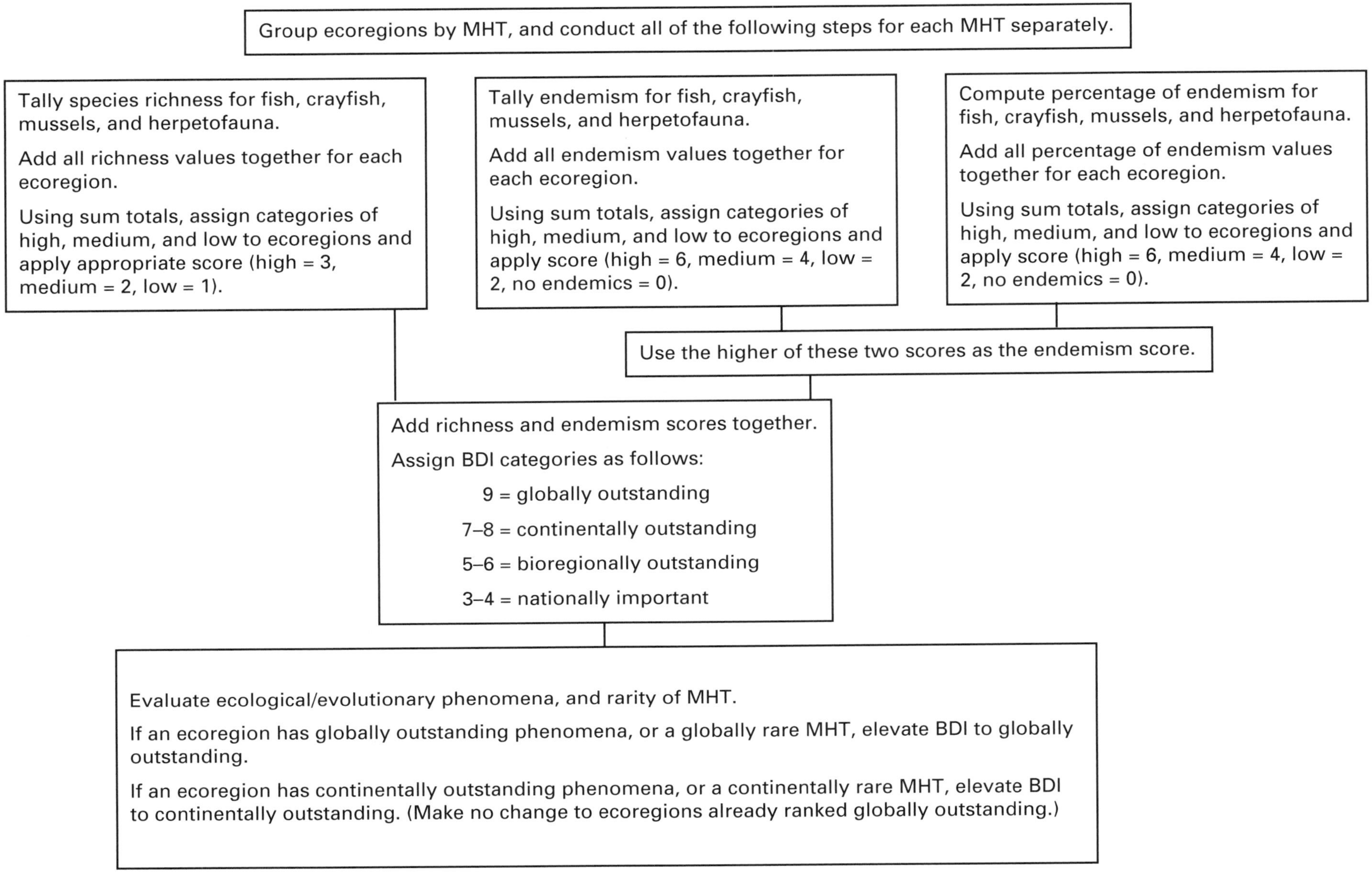

Group ecoregions by MHT, and conduct all of the following steps for each MHT separately.

Tally species richness for fish, crayfish, mussels, and herpetofauna.

Add all richness values together for each ecoregion.

Using sum totals, assign categories of high, medium, and low to ecoregions and apply appropriate score (high = 3, medium = 2, low = 1).

Tally endemism for fish, crayfish, mussels, and herpetofauna.

Add all endemism values together for each ecoregion.

Using sum totals, assign categories of high, medium, and low to ecoregions and apply score (high = 6, medium = 4, low = 2, no endemics = 0).

Compute percentage of endemism for fish, crayfish, mussels, and herpetofauna.

Add all percentage of endemism values together for each ecoregion.

Using sum totals, assign categories of high, medium, and low to ecoregions and apply score (high = 6, medium = 4, low = 2, no endemics = 0).

Use the higher of these two scores as the endemism score.

Add richness and endemism scores together.

Assign BDI categories as follows:

9 = globally outstanding

7–8 = continentally outstanding

5–6 = bioregionally outstanding

3–4 = nationally important

Evaluate ecological/evolutionary phenomena, and rarity of MHT.

If an ecoregion has globally outstanding phenomena, or a globally rare MHT, elevate BDI to globally outstanding.

If an ecoregion has continentally outstanding phenomena, or a continentally rare MHT, elevate BDI to continentally outstanding. (Make no change to ecoregions already ranked globally outstanding.)

Figure A.1   Steps for evaluating biological distinctiveness of ecoregions.

of additional sources, many of which were descriptive. Where there was uncertainty as to a species' presence in an ecoregion, it was counted as being present. This liberal approach was adopted following the assumption that overcounting of species would in part compensate for the large numbers of yet undescribed or unrecorded species.

The state of knowledge of freshwater taxa is such that new species are continually being "discovered," and many await formal description before they can be accepted as recognized species. Because more scientific investigation of freshwater species has occurred in Canada and the United States than in Mexico, it seems reasonable to assume that the number of undescribed species of fish, crayfish, mussels, amphibians, and reptiles in Mexico is substantially larger. Lists of species often include taxa that are not yet formally named but whose distributions are fairly well known. These species were not included in our biodiversity assessment because most of them occur in the United States and Canada, but the ratio of unnamed to undiscovered species is certainly much higher in Mexico. Additionally, only full species (as opposed to subspecies) were counted in the assessment, although many subspecies are considered at risk. This decision was made to standardize the assessment methodology. If, on review of the index scores, experts felt that an ecoregion's biological distinctiveness was undervalued because of the exclusion of unnamed species or subspecies, this was noted.

Species were recorded as being present in, endemic to, or absent from each ecoregion. For native crayfish and unionid mussels, no published range data were available. Crayfish data were supplied by Christopher Taylor of the Illinois Natural History Survey. Mussel data (excluding distributions of mussels in Mexico and Canada) were compiled jointly by James D. Williams and TNC. Mussel data for Canada were derived from the distribution maps published in Clarke (1981). The lack of mussel data for Mexico required that mussels be excluded from the analysis of richness (and endemism) of endorheic, xeric, and subtropical coastal MHTs.

For a species to be considered endemic to an ecoregion, in the strictest sense, would mean it occurs there and nowhere else on earth. We have relaxed this definition of endemism to better achieve conservation goals in two important

ways. First, we treated a near-endemic as endemic, so that the importance of conserving the species in the ecoregion containing most of its range is captured. Therefore, if most of a species' global range (≥ 75 percent) was found only within one ecoregion, it was counted as endemic to that ecoregion. Second, species with highly restricted ranges occasionally display distributions that straddle ecoregion boundaries, with no single ecoregion containing ≥ 75 percent of their ranges. In order not to lose restricted-range species that are at high risk of extinction from habitat loss, we scored species with such restricted global ranges ($<50,000$ km^2) as endemic to all ecoregions containing them. Based on these modifications, the assignment of endemism was made following the set of decision rules below.

Endemism Decision Rules

1. Total species range $> 50,000$ km^2:
   a. A single ecoregion contains 75–100 percent of the species' range: species is recorded as being endemic to that ecoregion and as being present in all others containing it.
   b. No single ecoregion contains >75 percent of the species' total range: species is recorded as being present in all ecoregions where it occurs.
2. Species range $< 50,000$ km^2:
   a. Species occurs in five or fewer ecoregions: species is recorded as being endemic to all ecoregions where it occurs.
   b. Species occurs in more than five ecoregions (many small, disjunct areas): species is recorded as being present in all ecoregions where it occurs.

We chose the 50,000 km^2 threshold for narrow-range endemics following Birdlife International's classification for endemic species (Bibby 1992). The 75 percent threshold for the proportion of a range within a single ecoregion is an arbitrary threshold but represents most of the total range. The point here was to highlight ecoregions that represent the only practical opportunity to conserve a certain species in the wild. An ecoregion containing over 75 percent of a species range fits that description much more than one containing less than 25 percent. A species that fits the criteria for endemism in the mapped area, but whose range extends outside North America, was *not* treated as endemic. An example would be a species found in the arctic areas of Eurasia as well as in a North American ecoregion. For diadromous species, only freshwater habitats were considered. Species that are considered marine but are commonly found within coastal rivers and streams were counted in this analysis but were not considered endemic to any freshwater ecoregion.

Richness

We compared species richness among ecoregions sharing the same major habitat type (MHT) by summing the totals for each of the four taxonomic groups analyzed and using the combined value as a proxy for the richness of the entire biota. Although fish species richness is nearly an order of magnitude higher than crayfish, mussel, or herpetofauna richness in certain ecoregions, the disparity between richness values was not judged so large as to bias the results in favor of fish protection over that of other groups. This method gives all species equal value; in other words, we consider the protection of one crayfish species as important as the protection of one fish. Alternative methods of evaluating richness, such as by ranking ecoregions according to the numbers of species, were not selected because they tended to obscure large differences between ecoregions.

Once the richness values for each faunal group were summed, the totals were used to define biodiversity richness categories of high, medium, and low. Separate category threshold points were obtained for each MHT. The total richness values for each MHT were plotted in increasing order, and threshold values were set where there was a sharp increase in slope between points (figure A.2). In the rare case where there were no such increases, we used a quartile method to set threshold values. After each ecoregion was assigned a richness level, we assigned point values to each level: high = 3 points, medium = 2 points, and low = 1 point.

Endemism

We collected endemism data in the same manner and from the same sources as richness. Conserving areas with the highest absolute levels of

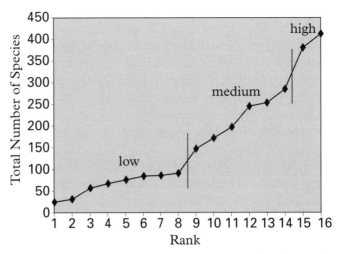

Figure A.2   Example of threshold determination for species richness in the temperate coastal rivers and lakes MHT.

endemism is desirable, but many species-poor ecoregions have a high number of endemics relative to their small number of total species. To capture both absolute and relative endemism, we employed two endemism measures. The first is simply the sum of all endemic species across the four selected faunal groups. The second is a calculation of percentage of endemism, the proportion of an ecoregion's richness that is endemic to that ecoregion, using the equation:

$$E\ (\%) = \frac{\text{number of endemic species by taxon}}{\text{number of species by taxon}} \times 100$$

For both endemism and percentage of endemism, we followed the same methodology as for species richness: ecoregions were separated by MHT, then the values for all faunal groups summed, threshold values chosen, and categories of high, medium, and low assigned. Next, for each ecoregion we compared the endemism and percentage of endemism categories and took whichever was the higher value. We felt that endemism was a highly significant factor in determining an ecoregion's distinctive biodiversity value, and accordingly gave it two times as much weight in the rankings as richness. The point values assigned for each level were as follows: high = 6, medium = 4, low = 2, no endemics or endemism = 0.

Rare Ecological or Evolutionary Phenomena

Ecological and evolutionary phenomena also contribute to the distinctiveness of an ecoregion's biodiversity. These criteria are included to capture important biodiversity features that are otherwise difficult to measure using quantitative methods. Distinctive evolutionary phenomena include globally outstanding centers of evolutionary radiation, higher-level taxonomic diversity, and unique species assemblages. Examples of rare ecological phenomena are large-scale migrations of fish and extraordinary seasonal concentrations of wildlife. We also noted high levels of beta-diversity which reflect both unusual ecological and evolutionary phenomena (e.g., high beta diversity is often associated with pronounced local endemism).

At the expert workshop, regional working groups were asked to list any appropriate phenomena occurring within their assigned regions of expertise. They also were asked to describe the level of rarity of each phenomenon, emphasizing those that are globally or continentally outstanding. We reviewed all the phenomena described by the experts and compared them with information gathered for similar phenomena around the world (Olson and Dinerstein 1998). We then placed ecoregions into one of three categories: globally outstanding phenomena, continentally outstanding phenomena, or not applicable.

Ecoregions that were judged to contain globally outstanding ecological or evolutionary phenomena were automatically categorized as globally outstanding in the overall BDI. Therefore, ecoregions without extraordinarily high richness or endemism but that nevertheless contained globally important and rare phenomena could be categorized as globally outstanding. Ecoregions that were categorized as continentally outstanding were automatically categorized as continentally outstanding in the overall BDI. Ecoregions awarded points for rare ecological and evolutionary phenomena are noted in table 3.3.

Rare Habitat Type

The global rarity of habitat types was evaluated to identify broad habitat types that offer few opportunities for conservation. This criterion encompasses ecological and evolutionary phenomena but addresses these characteristics at the scale of whole ecosystems and biotas, accounting for structural features of ecosystems and habitats as well.

The rarity of an ecoregion's general habitat type was evaluated based on expert assessment and a recently completed global analysis (Olson and Dinerstein 1998). Ecoregions were placed into one of three categories—globally rare, continentally rare, or not applicable—based on the following decision rules:

1. Fewer than eight ecoregions occur globally that contain the habitat type in question: globally rare.
2. Fewer than three ecoregions occur in North America that contain the habitat type in question: continentally rare.
3. Otherwise: not applicable.

As with the rare ecological and evolutionary phenomena criterion, an ecoregion that was judged to contain globally rare or continentally rare habitat type(s) was automatically categorized as globally or continentally outstanding in the overall BDI. No habitats were assessed as being globally rare; continentally rare habitats and their associated ecoregions are listed in table 3.3.

Final Biological Distinctiveness Categorization

The point totals for the four component criteria were added to yield a grand total for the BDI. Based on these totals, ecoregions were placed into one of four categories:

Globally outstanding:	9 points or more
Continentally outstanding:	7 or 8 points
Bioregionally outstanding:	5 or 6 points
Nationally important:	4 points or less

Methods for Assessing the Conservation Status of Freshwater Ecoregions

The conservation status index (CSI) was designed to measure different degrees of habitat alteration and spatial patterns of remaining natural habitats across landscapes as indicators for the ecological integrity of ecosystems. The index reflects how, with increasing habitat loss, degradation, and fragmentation, ecological processes cease to function naturally, or at all, and major components of biodiversity are steadily eroded. Here we assessed the conservation status of ecoregions in the tradition of the IUCN Red Data Books categories for threatened and endangered species (critical, endangered, and vulnerable), except that we estimated the state of whole biotas, ecological processes, and ecosystems. The conservation status categories we created for the analysis are as follows: critical, endangered, vulnerable, relatively stable, and relatively intact.

Although ample data are available documenting the degree to which freshwater systems are degraded, there are no easily accessed data sets that cover all of North America, are of appropriate spectral resolution, and address all the indicators that comprise our conservation status index. Instead of relying solely on data sets, we employed the knowledge and experience of experts to evaluate the criteria. This approach has been used in prior conservation assessments for both terrestrial and freshwater ecosystems (O'Keeffe et al. 1987; Dinerstein et al. 1995; Olson and Dinerstein 1998) and has proved useful as a quantitative tool that circumscribes strict information requirements. The pooled knowledge of the participants in our freshwater workshop formed an authoritative source from which to evaluate the criteria of the CSI, with available data sets used to provide supplementary information. In designing the CSI, we have tried to keep the evaluation process objective and transparent to facilitate reexamination of results and to analyze relationships among variables (figure B.1).

The criteria outlined below were used to assess the snapshot conservation status for each ecoregion. The snapshot assessment characterizes current (as of 1997) patterns associated with loss of diversity. A threat analysis subsequently modified the results, if necessary, to produce the final conservation status for the ecoregion. The CSI indicators described below were applied to all ecoregions, but scores were subsequently weighted according to MHT (see figure 2.4). MHTs were weighted because they vary in terms of their resiliency and response to disturbance, and the types of ecological processes that occur. For example, degraded water quality may have a much more serious biologic impact in lakes than in large rivers. Criteria and weightings were derived by the authors and modified according to the recommendations of experts.

Points were assigned for each indicator on a scale from 0 (least degraded) to 4 (most degraded). Weightings were then applied to the scores, and the total score for each ecoregion was converted to the percentage of the total possible score for the given MHT. All percentages can be compared across MHTs to evaluate the degree to which ecoregions have been disturbed and degraded, and the corresponding threat to biodiversity. Indicators are detailed below.

Score each ecoregion from 0 (least degraded) to 4 (most degraded) following criteria for each indicator.

Weight each indicator score according to appropriate factor for given MHT.

Total all indicator scores for each ecoregion.

Divide total score by total possible score for MHT, and multiply by 100 to get percentage.

Use percentage score to get snapshot assessment, according to the following scale:

percentage	rank	category
86 to 100	5	critical
71 to 85	4	endangered
51 to 70	3	vulnerable
31 to 50	2	relatively stable
0 to 30	1	relatively intact

Assess the likelihood of future threats to ecoregions as high, medium, or low.

Raise snapshot assessment category by one level if likelihood of future threats is considered high, to attain final conservation status category.

Figure B.1   Steps for evaluating conservation status of ecoregions.

Degree of Land Cover (Catchment) Alteration

This indicator assesses the degree to which land cover throughout an ecoregion's catchment(s) has been altered from pre-settlement conditions.

Percentage of Ecoregion Area Altered	Points
90 to 100	4
50 to 89	3
25 to 49	2
10 to 24	1
0 to 9	0

Water Quality Degradation

Water quality degradation includes, but is not limited to, changes in parameters such as pH, turbidity, DO, nutrients, pesticides, heavy metals, suspended solids, hydrocarbons, and temperature. The water quality score is determined as follows:

Percentage of Surface Water Degraded	Points
90 to 100	4
50 to 89	3
25 to 49	2
10 to 24	1
0 to 9	0

Alteration of Hydrographic Integrity

This indicator includes changes to flow regimes and water levels due to impoundments, channelization, groundwater or surface water extractions or diversions, or other modifications.

Percentage of Surface Water Altered	Points
90 to 100	4
50 to 89	3
25 to 49	2
10 to 24	1
0 to 9	0

Degree of Habitat Fragmentation

Fragmentation is defined here as the degree to which human-induced breaks in habitat act as barriers to the movement of aquatic organisms (e.g., dams, channelization, silted or dredged streambeds, areas of poor water quality).

Percentage of Original Habitat Fragmented		Points
very high:	core habitat highly affected; barriers preclude dispersal for most taxa	4
high:	core habitat significantly affected; barriers restrict short and/or long-distance dispersal for some species	3
medium:	core habitats moderately affected; barriers to some taxa for long-distance dispersal	2
low:	relatively contiguous, higher connectivity; long-distance dispersal still possible	1
none:	no habitat fragmentation	0

Additional Losses of Intact Original Habitat

This indicator is intended to account for habitat losses not measured by other indicators. Intact habitat represents relatively undisturbed areas characterized by the maintenance of most original ecological processes and by communities with most of their original suite of native species.

Percentage of Original Habitat Lost from Other Causes	Points
90 to 100	4
50 to 89	3
25 to 49	2
10 to 24	1
0 to 9	0

Effects of Introduced Species

This indicator should take into account both the magnitude of successful species introductions and the impact of these species on native biota.

Impact of Introductions		Points
very high:	major changes in species abundances or native community composition (includes extirpation or extinction)	4
high:	significant changes in species abundances or native community composition (may include extirpation)	3
medium:	moderate changes in species abundances; anticipated changes in community composition	2
low:	anticipated changes in species abundances	1
none:	no introduced species	0

Direct Species Exploitation

This indicator assesses the effect of overfishing, mussel extraction, or other exploitation on both the target species and other biota.

Impact of Exploitation		Points
very high:	major changes in species abundances or native community composition (includes extirpation or extinction)	4
high:	observable changes in species abundances and native community composition	3
medium:	moderate changes in species abundances; anticipated changes in community composition	2
low:	anticipated changes in species abundances	1
none:	no effect of species exploitation	0

Snapshot Total

To standardize the totals among all MHTs, a percentage is calculated by dividing the total by the maximum possible points. This total does not take into account impending future threats.

Conservation Status Ranking

A conservation status ranking is assigned to the percentages, as follows:

Percentage	Rank	Category
86 to 100	5	Critical
71 to 85	4	Endangered
51 to 70	3	Vulnerable
31 to 50	2	Relatively Stable
0 to 30	1	Relatively Intact

Likelihood of Future Threats

Some freshwater systems may face impending threats (within the next twenty years) that, if they occurred, would raise an ecoregion's conservation status ranking.

Threats could include adjacent intensive logging and associated road-building, agricultural expansion, clearing for development, intensive grazing, mining, dam-building and channelization, pollution, introduced species, excessive recreational impacts, and overfishing or unsustainable mussel extraction.

Threats are assessed by experts as high, medium, or low before all other points are tallied.

Projected (Final) Conservation Status, As Modified by Future Threat

The snapshot conservation status is revised on the basis of the threat analysis to develop final conservation status assessments.

Ecoregions with high threat rankings have their conservation status raised by one level. The final conservation status rankings, as with the snapshot status, range from 1 to 5 and correspond to the same categories.

Biological Distinctiveness Data and Scores

Species Richness

Ecoregion	Major Habitat Type (MHT)	No. of Native Fish Species	No. of Native Mussel Species	No. of Native Crayfish Species	No. of Native Herpetofauna Species	Total Number of Species*	Richness Points Assigned (by MHT)
South Hudson [53]	ARL	43	3	2	7	55	3
East Hudson [54]	ARL	32	2	2	7	43	2
Yukon [55]	ARL	39	1	0	0	40	2
Lower Mackenzie [56]	ARL	41	2	0	3	46	2
Upper Mackenzie [57]	ARL	41	3	0	8	52	3
North Arctic [58]	ARL	22	0	0	0	22	1
East Arctic [59]	ARL	18	0	0	0	18	1
Arctic Islands [60]	ARL	9	0	0	0	9	1
Superior [43]	LTL	60	7	3	21	91	1
Michigan-Huron [44]	LTL	123	35	10	29	197	3
Erie [45]	LTL	120	42	13	32	207	3
Ontario [46]	LTL	109	33	8	26	176	2
Central Prairie [28]	THL	143	60	26	45	274	2
Ozark Highlands [29]	THL	157	48	21	44	270	2
Ouachita Highlands [30]	THL	145	53	34	51	283	2
Southern Plains [31]	THL	83	20	10	44	157	1
Teays-Old Ohio [34]	THL	206	122	49	60	437	3
Tennessee-Cumberland [35]	THL	231	125	65	55	476	3
Canadian Rockies [49]	THL	23	1	0	5	29	1
Upper Saskatchewan [50]	THL	41	4	1	10	56	1
Lower Saskatchewan [51]	THL	38	4	1	5	48	1
English-Winnipeg Lakes [52]	THL	79	13	2	20	114	1
Colorado [12]	LTR	26	3	0	26	55	1
Upper Rio Grande/Bravo [15]	LTR	28	2	1	24	55	1
Lower Rio Grande/Bravo [20]	LTR	84	13	2	60	159	2
Mississippi [24]	LTR	164	53	13	54	284	3
Mississippi Embayment [25]	LTR	206	63	57	68	394	3
Upper Missouri [26]	LTR	50	5	3	19	77	1
Middle Missouri [27]	LTR	98	40	8	36	182	2
Bonneville [8]	ERLS	21	2	1	10	34	1
Lahontan [9]	ERLS	16	2	0	12	30	1
Oregon Lakes [10]	ERLS	8	0	1	6	15	1
Death Valley [11]	ERLS	9	0	0	11	20	1
Guzmán [16]	ERLS	16	1	1	52	70	2
Mapimí [19]	ERLS	26		1	57	84	3
Llanos el Salado [66]	ERLS	9†		3	46	58	2
Lerma [69]	ERLS	48		2†	62	112	3
South Pacific Coastal [7]	XRLS	26	0	0	58	84	3
Vegas-Virgin [13]	XRLS	11	1	0	8	20	1
Gila [14]	XRLS	19	1	0	27	47	1
Rio Conchos [17]	XRLS	47†		0	46	93	3
Pecos [18]	XRLS	54	6	0	24	84	3
Rio Salado [21]	XRLS	41†	(3)	1	38	80	3
Cuatro Ciénegas [22]	XRLS	19†		0	35	54	2
Rio San Juan [23]	XRLS	44†	(4)	1	57	102	3
Chapala [65]	XRLS	28		2	28	58	2
Rio Verde Headwaters [67]	XRLS	11		1	25	37	1
Sonoran [61]	XRLS	13†		0	53	66	2
North Pacific Coastal [1]	TCRL	58	4	4	19	85	1
Columbia Glaciated [2]	TCRL	34	5	4	13	56	1
Columbia Unglaciated [3]	TCRL	35	6	5	21	67	1
Upper Snake [4]	TCRL	14	4	5	9	32	1
Pacific Mid-Coastal [5]	TCRL	55	4	4	27	90	1
Pacific Central Valley [6]	TCRL	49	5	4	27	85	1
East Texas Gulf [32]	TCRL	115	46	29	78	268	2
Mobile Bay [36]	TCRL	187	65	60	77	389	3
Apalachicola [37]	TCRL	104	38	28	71	241	2
Florida Gulf [38]	TCRL	106	30	23	70	229	2
South Atlantic [40]	TCRL	177	59	56	82	374	3
Chesapeake Bay [41]	TCRL	95	22	14	53	184	2

(continues)

Species Richness

Ecoregion	Major Habitat Type (MHT)	No. of Native Fish Species	No. of Native Mussel Species	No. of Native Crayfish Species	No. of Native Herpetofauna Species	Total Number of Species*	Richness Points Assigned (by MHT)
North Atlantic [42]	TCRL	98	12	9	40	159	2
Lower St. Lawrence [47]	TCRL	93	20	6	28	147	2
North Atlantic-Ungava [48]	TCRL	20	1	0	4	25	1
West Texas Gulf [33]	SCRL	73	15	9	28	125	2
Florida [39]	SCRL	110	26	36	65	237	3
Sinaloan Coastal [62]	SCRL	18		2	81	102	1
Santiago [63]	SCRL	45		3	84	132	2
Manantlan-Ameca [64]	SCRL	25[†]		1	54	80	1
Tamaulipas-Veracruz [68]	SCRL	94[†]		20	108	222	3
Balsas [70]	SCRL	15[†]		5	97	117	2
Papaloapan [71]	SCRL	70[†]		6	104	177	2
Catemaco [72]	SCRL	10[†]		0	48	58	1
Coatzacoalcos [73]	SCRL	58[†]		2	71	131	2
Tehuantepec [74]	SCRL	29[†]		0	125	154	2
Grijalva-Usumacinta [75]	SCRL	70[†]		1	82	153	2
Yucatán [76]	SCRL	49[†]		1	44	94	1

*Mussel data not included for xeric, endorheic, and subtropical MHTs.

[†]Data modified by Dr. Salvador Contreras-Balderas.

Number of Endemics

Ecoregion	Major Habitat Type (MHT)	No. of Endemic Fish Species	No. of Endemic Mussel Species	No. of Endemic Crayfish Species	No. of Endemic Herpetofauna Species	Total Number of Endemic Species*	Endemism Points Assigned (by MHT)
South Hudson [53]	ARL	0	0	0	0	0	0
East Hudson [54]	ARL	0	0	0	0	0	0
Yukon [55]	ARL	4	1	0	0	5	4
Lower Mackenzie [56]	ARL	0	0	0	0	0	0
Upper Mackenzie [57]	ARL	0	0	0	0	0	0
North Arctic [58]	ARL	0	0	0	0	0	0
East Arctic [59]	ARL	0	0	0	0	0	0
Arctic Islands [60]	ARL	0	0	0	0	0	0
Superior [43]	LTL	0	0	0	0	0	0
Michigan-Huron [44]	LTL	1	0	0	0	1	2
Erie [45]	LTL	0	0	0	0	0	0
Ontario [46]	LTL	0	0	0	0	0	0
Central Prairie [28]	THL	8	1	13	1	23	4
Ozark Highlands [29]	THL	10	3	15	0	28	4
Ouachita Highlands [30]	THL	9	3	21	1	34	4
Southern Plains [31]	THL	3	0	2	0	5	2
Teays-Old Ohio [34]	THL	24	17	23	3	67	6
Tennessee-Cumberland [35]	THL	67	20	40	8	135	6
Canadian Rockies [49]	THL	0	0	0	0	0	0
Upper Saskatchewan [50]	THL	0	0	0	0	0	0
Lower Saskatchewan [51]	THL	0	0	0	0	0	0
English-Winnipeg Lakes [52]	THL	0	0	0	0	0	0
Colorado [12]	LTR	7	0	0	1	8	4
Upper Rio Grande/Bravo [15]	LTR	1	0	0	1	2	2
Lower Rio Grande/Bravo [20]	LTR	16	1	1	0	18	4
Mississippi [24]	LTR	1	1	1	0	3	2
Mississippi Embayment [25]	LTR	11	2	33	2	48	6
Upper Missouri [26]	LTR	0	0	0	0	0	0
Middle Missouri [27]	LTR	0	0	0	0	0	0
Bonneville [8]	ERLS	7	0	0	0	7	4
Lahontan [9]	ERLS	7	0	0	1	8	4
Oregon Lakes [10]	ERLS	3	0	0	0	3	2
Death Valley [11]	ERLS	6	0	0	4	10	4
Guzmán [16]	ERLS	9[†]	0	1	0	10	4
Mapimí [19]	ERLS	13		1	1	15	4
Llanos el Salado [66]	ERLS	8[†]		1	0	9	4
Lerma [69]	ERLS	30		2[†]	14	46	6
South Pacific Coastal [7]	XRLS	5	0	0	1	6	2
Vegas-Virgin [13]	XRLS	6	0	0	1	7	4
Gila [14]	XRLS	7	0	0	1	8	4
Rio Conchos [17]	XRLS	12[†]		0	0	12	6
Pecos [18]	XRLS	7	0	0	0	7	4
Rio Salado [21]	XRLS	6[†]		0	0	6	2
Cuatro Ciénegas [22]	XRLS	8		0	1	9	4
Rio San Juan [23]	XRLS	6[†]		1	1	8	4
Chapala [65]	XRLS	26[†]		2	0	28	6
Rio Verde Headwaters [67]	XRLS	9[†]		0	0	9	4
Sonoran [61]	XRLS	8		0	0	8	4
North Pacific Coastal [1]	TCRL	2	0	0	1	3	2
Columbia Glaciated [2]	TCRL	0	0	0	0	0	0
Columbia Unglaciated [3]	TCRL	4	0	1	0	5	2
Upper Snake [4]	TCRL	1	0	0	0	1	2
Pacific Mid-Coastal [5]	TCRL	16	0	0	2	18	4
Pacific Central Valley [6]	TCRL	14	1	1	5	21	4
East Texas Gulf [32]	TCRL	6	5	12	13	36	4
Mobile Bay [36]	TCRL	47	17	39	4	107	6
Apalachicola [37]	TCRL	8	7	17	3	35	4
Florida Gulf [38]	TCRL	6	7	13	4	30	4
South Atlantic [40]	TCRL	48	19	39	6	112	6
Chesapeake Bay [41]	TCRL	7	4	0	0	11	2

(continues)

Number of Endemics

Ecoregion	Major Habitat Type (MHT)	No. of Endemic Fish Species	No. of Endemic Mussel Species	No. of Endemic Crayfish Species	No. of Endemic Herpetofauna Species	Total Number of Endemic Species*	Endemism Points Assigned (by MHT)
North Atlantic [42]	TCRL	2	0	0	1	3	2
Lower St. Lawrence [47]	TCRL	2	0	0	0	2	2
North Atlantic-Ungava [48]	TCRL	0	0	0	0	0	0
West Texas Gulf [33]	SCRL	8	0	5	0	13	4
Florida [39]	SCRL	9	(8)	29	3	49	6
Sinaloan Coastal [62]	SCRL	2		1	0	3	2
Santiago [63]	SCRL	5		1	1	7	2
Manantlan-Ameca [64]	SCRL	14		0	1	15	4
Tamaulipas-Veracruz [68]	SCRL	29[†]		17	16	62	6
Balsas [70]	SCRL	7[†]		3	16	26	4
Papaloapan [71]	SCRL	12		2	27	41	6
Catemaco [72]	SCRL	8[†]		0	1	9	2
Coatzacoalcos [73]	SCRL	18[†]		0	2	20	4
Tehuantepec [74]	SCRL	6		0	42	48	6
Grijalva-Usumacinta [75]	SCRL	29[†]		0	12	41	6
Yucatán [76]	SCRL	12[†]		0	3	15	4

*Mussel data not included for xeric, endorheic, and subtropical MHTs.
[†]Data modified by Dr. Salvador Contreras-Balderas.

Percent Endemism

Ecoregion	Major Habitat Type (MHT)	Percent Endemism of Fish Species	Percent Endemism of Mussel Species	Percent Endemism of Crayfish Species	Percent Endemism of Herpetofauna Species	Total Percent Endemism of Species*	Endemism Points Assigned (by MHT)
South Hudson [53]	ARL	0	0	0	0	0	0
East Hudson [54]	ARL	0	0	0	0	0	0
Yukon [55]	ARL	10	100	0	0	110	6
Lower Mackenzie [56]	ARL	0	0	0	0	0	0
Upper Mackenzie [57]	ARL	0	0	0	0	0	0
North Arctic [58]	ARL	0	0	0	0	0	0
East Arctic [59]	ARL	0	0	0	0	0	0
Arctic Islands [60]	ARL	0	0	0	0	0	0
Superior [43]	LTL	0	0	0	0	0	0
Michigan-Huron [44]	LTL	1	0	0	0	1	2
Erie [45]	LTL	0	0	0	0	0	0
Ontario [46]	LTL	0	0	0	0	0	0
Central Prairie [28]	THL	6	2	50	2	59	4
Ozark Highlands [29]	THL	6	6	71	0	84	4
Ouachita Highlands [30]	THL	6	6	62	2	75	4
Southern Plains [31]	THL	2	0	20	0	22	2
Teays-Old Ohio [34]	THL	12	14	47	5	78	4
Tennessee-Cumberland [35]	THL	29	16	62	15	121	6
Canadian Rockies [49]	THL	0	0	0	0	0	0
Upper Saskatchewan [50]	THL	0	0	0	0	0	0
Lower Saskatchewan [51]	THL	0	0	0	0	0	0
English-Winnipeg Lakes [52]	THL	0	0	0	0	0	0
Colorado [12]	LTR	27	0	0	4	31	4
Upper Rio Grande/Bravo [15]	LTR	4	0	0	4	8	2
Lower Rio Grande/Bravo [20]	LTR	19	8	50	0	77	6
Mississippi [24]	LTR	1	2	8	0	10	2
Mississippi Embayment [25]	LTR	5	3	58	3	69	6
Upper Missouri [26]	LTR	0	0	0	0	0	0
Middle Missouri [27]	LTR	0	0	0	0	0	0
Bonneville [8]	ERLS	33	0	0	0	33	2
Lahontan [9]	ERLS	44	0	0	8	52	4
Oregon Lakes [10]	ERLS	38	0	0	0	38	2
Death Valley [11]	ERLS	67	0	0	36	103	6
Guzmán [16]	ERLS	50	0	100	0	150	6
Mapimí [19]	ERLS	50		100	2	152	6
Llanos el Salado [66]	ERLS	89		33	0	122	6
Lerma [69]	ERLS	63		100	23	185	6
South Pacific Coastal [7]	XRLS	19	0	0	2	21	2
Vegas-Virgin [13]	XRLS	55	0	0	13	67	4
Gila [14]	XRLS	37	0	0	4	41	4
Rio Conchos [17]	XRLS	25		0	0	25	2
Pecos [18]	XRLS	13	0	0	0	13	2
Rio Salado [21]	XRLS	15		0	0	15	2
Cuatro Ciénegas [22]	XRLS	42		0	3	45	4
Rio San Juan [23]	XRLS	14		100	2	116	6
Chapala [65]	XRLS	93		100	0	193	6
Rio Verde Headwaters [67]	XRLS	82		0	0	82	4
Sonoran [61]	XRLS	62		0	0	62	4
North Pacific Coastal [1]	TCRL	3	0	0	5	9	2
Columbia Glaciated [2]	TCRL	0	0	0	0	0	0
Columbia Unglaciated [3]	TCRL	11	0	20	0	31	4
Upper Snake [4]	TCRL	7	0	0	0	7	2
Pacific Mid-Coastal [5]	TCRL	29	0	0	7	36	4
Pacific Central Valley [6]	TCRL	29	20	25	19	93	6
East Texas Gulf [32]	TCRL	5	11	41	17	74	6
Mobile Bay [36]	TCRL	25	26	65	5	121	6
Apalachicola [37]	TCRL	8	18	61	4	91	6
Florida Gulf [38]	TCRL	6	23	57	6	91	6
South Atlantic [40]	TCRL	27	32	70	7	136	6
Chesapeake Bay [41]	TCRL	7	18	0	0	25	4

(continues)

Percent Endemism

Ecoregion	Major Habitat Type (MHT)	Percent Endemism of Fish Species	Percent Endemism of Mussel Species	Percent Endemism of Crayfish Species	Percent Endemism of Herpetofauna Species	Total Percent Endemism of Species*	Endemism Points Assigned (by MHT)
North Atlantic [42]	TCRL	2	0	0	3	5	2
Lower St. Lawrence [47]	TCRL	2	0	0	0	2	2
North Atlantic-Ungava [48]	TCRL	0	0	0	0	0	0
West Texas Gulf [33]	SCRL	11	0	56	0	67	4
Florida [39]	SCRL	8	31	81	5	124	6
Sinaloan Coastal [62]	SCRL	11		50	0	61	4
Santiago [63]	SCRL	11		33	1	46	4
Manantlan-Ameca [64]	SCRL	56		0	2	58	4
Tamaulipas-Veracruz [68]	SCRL	31		85	15	131	6
Balsas [70]	SCRL	47		60	16	123	6
Papaloapan [71]	SCRL	17		33	26	76	4
Catemaco [72]	SCRL	80		0	2	82	4
Coatzacoalcos [73]	SCRL	31		0	3	34	2
Tehuantepec [74]	SCRL	21		0	34	54	4
Grijalva-Usumacinta [75]	SCRL	41		0	15	56	4
Yucatán [76]	SCRL	24		0	7	31	2

*Mussel data not included for xeric, endorheic, and subtropical MHTs.

Biological Distinctiveness Ranking

Ecoregion	Major Habitat Type (MHT)	Richness Points	Endemism Points	Total of Richness & Endemism Points	Biological Distinctiveness Category	Phenomenon Category	Final Biological Distinctiveness Category
South Hudson [53]	ARL	3	0	3	NI		NI
East Hudson [54]	ARL	2	0	2	NI		NI
Yukon [55]	ARL	2	6	8	CO		CO
Lower Mackenzie [56]	ARL	2	0	2	NI		NI
Upper Mackenzie [57]	ARL	3	0	3	NI		NI
North Arctic [58]	ARL	1	0	1	NI	CO	CO
East Arctic [59]	ARL	1	0	1	NI		NI
Arctic Islands [60]	ARL	1	0	1	NI	CO	CO
Superior [43]	LTL	1	0	1	NI	CO	CO
Michigan-Huron [44]	LTL	3	2	5	BO	CO	CO
Erie [45]	LTL	3	0	3	NI		NI
Ontario [46]	LTL	2	0	2	NI		NI
Central Prairie [28]	THL	2	4	6	BO		BO
Ozark Highlands [29]	THL	2	4	6	BO		BO
Ouachita Highlands [30]	THL	2	4	6	BO		BO
Southern Plains [31]	THL	1	2	3	NI		NI
Teays-Old Ohio [34]	THL	3	6	9	GO		GO
Tennessee-Cumberland [35]	THL	3	6	9	GO		GO
Canadian Rockies [49]	THL	1	0	1	NI	CO	CO
Upper Saskatchewan [50]	THL	1	0	1	NI		NI
Lower Saskatchewan [51]	THL	1	0	1	NI	CO	CO
English-Winnipeg Lakes [52]	THL	1	0	1	NI	CO	CO
Colorado [12]	LTR	1	4	5	BO	CO	CO
Upper Rio Grande/Bravo [15]	LTR	1	2	3	NI	CO	CO
Lower Rio Grande/Bravo [20]	LTR	2	6	8	CO	CO	CO
Mississippi [24]	LTR	3	2	5	BO		BO
Mississippi Embayment [25]	LTR	3	6	9	GO	CO	GO
Upper Missouri [26]	LTR	1	0	1	NI		NI
Middle Missouri [27]	LTR	2	0	2	NI		NI
Bonneville [8]	ERLS	1	4	5	BO	CO	CO
Lahontan [9]	ERLS	1	4	5	BO	CO	CO
Oregon Lakes [10]	ERLS	1	2	3	NI		NI
Death Valley [11]	ERLS	1	6	7	CO	CO	CO
Guzmán [16]	ERLS	2	6	8	CO	CO	CO
Mapimí [19]	ERLS	2	6	8	CO	CO	CO
Llanos el Salado [66]	ERLS	2	6	8	CO	CO	CO
Lerma [69]	ERLS	3	6	9	GO	GO	GO
South Pacific Coastal [7]	XRLS	3	2	5	BO	CO	CO
Vegas-Virgin [13]	XRLS	1	4	5	BO	CO	CO
Gila [14]	XRLS	1	4	5	BO	CO	CO
Rio Conchos [17]	XRLS	3	6	9	GO	CO	GO
Pecos [18]	XRLS	3	4	7	CO	CO	CO
Rio Salado [21]	XRLS	3	2	5	BO	CO	CO
Cuatro Ciénegas [22]	XRLS	2	4	6	BO	GO	GO
Rio San Juan [23]	XRLS	3	6	9	GO	CO	GO
Chapala [65]	XRLS	2	6	8	CO	GO	GO
Rio Verde Headwaters [67]	XRLS	1	4	5	BO	GO	GO
Sonoran [61]	XRLS	2	4	6	BO	CO	CO
North Pacific Coastal [1]	TCRL	1	2	3	NI	CO	CO
Columbia Glaciated [2]	TCRL	1	0	1	NI		NI
Columbia Unglaciated [3]	TCRL	1	4	5	BO	CO	CO
Upper Snake [4]	TCRL	1	2	3	NI		NI
Pacific Mid-Coastal [5]	TCRL	1	4	5	BO	CO	CO
Pacific Central Valley [6]	TCRL	1	6	7	CO	GO	GO
East Texas Gulf [32]	TCRL	2	6	8	CO	CO	CO
Mobile Bay [36]	TCRL	3	6	9	GO		GO
Apalachicola [37]	TCRL	2	6	8	CO		CO
Florida Gulf [38]	TCRL	2	6	8	CO		CO
South Atlantic [40]	TCRL	3	6	9	GO	CO	GO
Chesapeake Bay [41]	TCRL	2	4	6	BO	CO	CO

(continues)

Biological Distinctiveness Ranking

Ecoregion	Major Habitat Type (MHT)	Richness Points	Endemism Points	Total of Richness & Endemism Points	Biological Distinctiveness Category	Pheno-menon Category	Final Biological Distinctiveness Category
North Atlantic [42]	TCRL	2	2	4	NI		NI
Lower St. Lawrence [47]	TCRL	2	2	4	NI	CO	CO
North Atlantic-Ungava [48]	TCRL	1	0	1	NI		NI
West Texas Gulf [33]	SCRL	1	4	5	BO	CO	CO
Florida [39]	SCRL	3	6	9	GO	GO	GO
Sinaloan Coastal [62]	SCRL	1	4	5	BO		CO
Santiago [63]	SCRL	1	4	5	BO	CO	CO
Manantlan-Ameca [64]	SCRL	1	4	5	BO	GO	GO
Tamaulipas-Veracruz [68]	SCRL	3	6	9	GO	GO	GO
Balsas [70]	SCRL	1	6	7	CO	CO	CO
Papaloapan [71]	SCRL	2	6	8	CO	CO	CO
Catemaco [72]	SCRL	1	4	5	BO	GO	GO
Coatzacoalcos [73]	SCRL	2	4	6	BO	CO	CO
Tehuantepec [74]	SCRL	2	6	8	CO		CO
Grijalva-Usumacinta [75]	SCRL	2	6	8	CO	CO	CO
Yucatán [76]	SCRL	1	4	5	BO	CO	CO

APPENDIX D

Conservation Status Assessment and Scores

Ecoregion	Major Habitat Type (MHT)	Degree of Land-cover (Catchment) Alteration (0–4)	Water Quality Degradation (0–4)	Alteration of Hydrographic Integrity (0–4)	Degree of Habitat Fragmentation (0–4)	Additional Losses of Original Intact Habitat (0–4)	Effects of Introduced Species (0–4)	Direct Species Exploitation (0–4)
South Hudson [53]	ARL	2	1	1.5	2	1	0	1
East Hudson [54]	ARL	3	1	4	3.5	1	0	1
Yukon [55]	ARL	2	2	1	1	2	0	1.5
Lower Mackenzie [56]	ARL	1	1	0	0	1	0	2
Upper Mackenzie [57]	ARL	2.5	2	0	2.5	2	1	2
North Arctic [58]	ARL	1	1	0	0	1	0	2
East Arctic [59]	ARL	1	1	0	0	1	0	1
Arctic Islands [60]	ARL	1	1	0	0	1	0	1
Superior [43]	LTL	3	1	2	2	1	4	2
Michigan-Huron [44]	LTL	3	4	2	3	2	4	3
Erie [45]	LTL	4	4	2	3	1	4	3.5
Ontario [46]	LTL	4	4	2	3	2.5	4	3
Central Prairie [28]	THL	2.5	2	2	1	0	1	0
Ozark Highlands [29]	THL	1.5	1	2	1	0	1	0
Ouachita Highlands [30]	THL	2	1	3	3	0	1	0
Southern Plains [31]	THL	4	4	4	4	0	1	0
Teays-Old Ohio [34]	THL	2.5	2.5	3	2	1	2	1.5
Tennessee-Cumberland [35]	THL	2.5	2	3.5	2.5	2	1.5	1.5
Canadian Rockies [49]	THL	3	1	2	2	1	1.5	2
Upper Saskatchewan [50]	THL	3	2.5	3.5	3.5	2	1.5	2
Lower Saskatchewan [51]	THL	3	2.5	2.5	2.5	1	0	2
English-Winnipeg Lakes [52]	THL	3	3	2.5	3	1	1.5	3
Colorado [12]	LTR	3	4	4	4	4	4	1
Upper Rio Grande/Bravo [15]	LTR	4	2	4	4	4	3	1
Lower Rio Grande/Bravo [20]	LTR	4	4	4	3	4	4	1
Mississippi [24]	LTR	4	3	3.5	3.5	0	2	2
Mississippi Embayment [25]	LTR	3	3	4	4	3	1	2
Upper Missouri [26]	LTR	4	3	4	4	0	0	0
Middle Missouri [27]	LTR	4	4	3	3	0	1	0
Bonneville [8]	ERLS	2	3	4	3	3	3	1
Lahontan [9]	ERLS	3	3	4	3	4	4	1
Oregon Lakes [10]	ERLS	2	3	3	2	4	4	0
Death Valley [11]	ERLS	3	2	3	2	3	3	0
Guzmán [16]	ERLS	3	3	4	4	4	3	1
Mapimí [19]	ERLS	3	3	3	4	4	3	1
Llanos el Salado [66]	ERLS	4	4	4	4	0	4	4
Lerma [69]	ERLS	4	4	3	3	2	3	1
South Pacific Coastal [7]	XRLS	4	4	4	4	3	3	1
Vegas-Virgin [13]	XRLS	3	4	2	2	3	4	0
Gila [14]	XRLS	4	4	4	3	4	4	1
Rio Conchos [17]	XRLS	3	2	4	4	4	3	1
Pecos [18]	XRLS	2	3	3	3	4	3	1
Rio Salado [21]	XRLS	3	2	4	2	2	2	0
Cuatro Ciénegas [22]	XRLS	2	1	1	1	2	4	0
Rio San Juan [23]	XRLS	4	4	4	2	4	2	0
Sonoran [61]	XRLS	3	4	4	2	0	2	1
Chapala [65]	XRLS	4	4	1	1	0	4	3
Rio Verde Headwaters [67]	XRLS	1	1	1	1	0	3	3
North Pacific Coastal [1]	TCRL	2.5	1.5	2.5	2.5	1.5	1	3
Columbia Glaciated [2]	TCRL	3	3	4	3	3	2	3
Columbia Unglaciated [3]	TCRL	3	3	4	3	3	2	3
Upper Snake [4]	TCRL	3	3	4	3	3	3	2
Pacific Mid-Coastal [5]	TCRL	4	2	2	3	3	3	3
Pacific Central Valley [6]	TCRL	4	3	4	4	4	4	3
East Texas Gulf [32]	TCRL	3	1	4	4	0	1	1
Mobile Bay [36]	TCRL	4	3	4	4	3	1	1
Apalachicola [37]	TCRL	4	4	4	4	3	2	1
Florida Gulf [38]	TCRL	4	3	2	1	2	1	1
South Atlantic [40]	TCRL	4	3	4	4	2	2	1
Chesapeake Bay [41]	TCRL	3	3	3	3	2	1	1
North Atlantic [42]	TCRL	3	2.5	3	3	2	1	1

(continues)

Ecoregion	Major Habitat Type (MHT)	Degree of Land-cover (Catchment) Alteration (0–4)	Water Quality Degradation (0–4)	Alteration of Hydrographic Integrity (0–4)	Degree of Habitat Fragmentation (0–4)	Additional Losses of Original Intact Habitat (0–4)	Effects of Introduced Species (0–4)	Direct Species Exploitation (0–4)
Lower St. Lawrence [47]	TCRL	3	4	1	3	2.5	3	2.5
North Atlantic-Ungava [48]	TCRL	2	2	3	2	0	1	2
West Texas Gulf [33]	SCRL	3.5	1	4	4	0	1	1
Florida [39]	SCRL	4	3	4	4	4	2.5	1
Sinaloan Coastal [62]	SCRL	3	4	3	2	0	2	1
Santiago [63] *	SCRL							
Manantlan-Ameca [64]	SCRL	2	2	2	1	3	1	1
Tamaulipas-Veracruz [68]	SCRL	3	2	2	2	2	2	2
Balsas [70]	SCRL	3	3	3	3	2	4	1
Papaloapan [71]	SCRL	2	3	3	2	2	2	2
Catemaco [72]	SCRL	1	2	1	1	1	2	3
Coatzacoalcos [73]	SCRL	3	3	3	2	3	2	2
Tehuantepec [74]	SCRL	2	2	2	2	3	2	2
Grijalva-Usumacinta [75]	SCRL	3	3	2	2	2	3	2
Yucatán [76]	SCRL	3	2	2	1	3	1	1

* No conservation status data available; rank taken from Olson et al. (1997).

Ecoregion	Major Habitat Type (MHT)	Degree of Land-cover (Catchment) Alteration (Weighted)	Water Quality Degradation (Weighted)	Alteration of Hydrographic Integrity (Weighted)	Degree of Habitat Fragmentation (Weighted)	Additional Losses of Original Intact Habitat (Weighted)	Effects of Introduced Species (Weighted)	Direct Species Exploitation (Weighted)
South Hudson [53]	ARL	2	1	1.5	2	1	0	1
East Hudson [54]	ARL	3	1	4	3.5	1	0	1
Yukon [55]	ARL	2	2	1	1	2	0	1.5
Lower Mackenzie [56]	ARL	1	1	0	0	1	0	2
Upper Mackenzie [57]	ARL	2.5	2	0	2.5	2	1	2
North Arctic [58]	ARL	1	1	0	0	1	0	2
East Arctic [59]	ARL	1	1	0	0	1	0	1
Arctic Islands [60]	ARL	1	1	0	0	1	0	1
Superior [43]	LTL	3	2	2	2	1	8	2
Michigan-Huron [44]	LTL	3	8	2	3	2	8	3
Erie [45]	LTL	4	8	2	3	1	8	3.5
Ontario [46]	LTL	4	8	2	3	2.5	8	3
Central Prairie [28]	THL	5	2	4	2	0	2	0
Ozark Highlands [29]	THL	3	1	4	2	0	2	0
Ouachita Highlands [30]	THL	4	1	6	6	0	2	0
Southern Plains [31]	THL	8	4	8	8	0	2	0
Teays-Old Ohio [34]	THL	5	2.5	6	4	1	4	1.5
Tennessee-Cumberland [35]	THL	5	2	7	5	2	3	1.5
Canadian Rockies [49]	THL	6	1	4	4	1	3	2
Upper Saskatchewan [50]	THL	6	2.5	7	7	2	3	2
Lower Saskatchewan [51]	THL	6	2.5	5	5	1	0	2
English-Winnipeg Lakes [52]	THL	6	3	5	6	1	3	3
Colorado [12]	LTR	6	4	8	8	4	4	1
Upper Rio Grande/Bravo [15]	LTR	8	2	8	8	4	3	1
Lower Rio Grande/Bravo [20]	LTR	8	4	8	6	4	4	1
Mississippi [24]	LTR	8	3	7	7	0	2	2
Mississippi Embayment [25]	LTR	6	3	8	8	3	1	2
Upper Missouri [26]	LTR	8	3	8	8	0	0	0
Middle Missouri [27]	LTR	8	4	6	6	0	1	0
Bonneville [8]	ERLS	4	6	12	6	6	6	2
Lahontan [9]	ERLS	6	6	12	6	8	8	2
Oregon Lakes [10]	ERLS	4	6	9	4	8	8	0
Death Valley [11]	ERLS	6	4	9	4	6	6	0
Guzmán [16]	ERLS	6	6	12	8	8	6	2
Mapimí [19]	ERLS	6	6	9	8	8	6	2
Llanos el Salado [66]	ERLS	8	8	12	8	0	8	8
Lerma [69]	ERLS	8	8	9	6	4	6	2
South Pacific Coastal [7]	XRLS	8	12	12	8	6	6	2
Vegas-Virgin [13]	XRLS	6	12	6	4	6	8	0
Gila [14]	XRLS	8	12	12	6	8	8	2
Rio Conchos [17]	XRLS	6	6	12	8	8	6	2
Pecos [18]	XRLS	4	9	9	6	8	6	2
Rio Salado [21]	XRLS	6	6	12	4	4	4	0
Cuatro Ciénegas [22]	XRLS	4	3	3	2	4	8	0
Rio San Juan [23]	XRLS	8	12	12	4	8	4	0
Sonoran [61]	XRLS	6	12	12	4	0	4	2
Chapala [65]	XRLS	8	12	3	2	0	8	6
Rio Verde Headwaters [67]	XRLS	2	3	3	2	0	6	6
North Pacific Coastal [1]	TCRL	5	1.5	5	5	1.5	1	6
Columbia Glaciated [2]	TCRL	6	3	8	6	3	2	6
Columbia Unglaciated [3]	TCRL	6	3	8	6	3	2	6
Upper Snake [4]	TCRL	6	3	8	6	3	3	4
Pacific Mid-Coastal [5]	TCRL	8	2	4	6	3	3	6
Pacific Central Valley [6]	TCRL	8	3	8	8	4	4	6
East Texas Gulf [32]	TCRL	6	1	8	8	0	1	2
Mobile Bay [36]	TCRL	8	3	8	8	3	1	2
Apalachicola [37]	TCRL	8	4	8	8	3	2	2
Florida Gulf [38]	TCRL	8	3	4	2	2	1	2
South Atlantic [40]	TCRL	8	3	8	8	2	2	2
Chesapeake Bay [41]	TCRL	6	3	6	6	2	1	2
North Atlantic [42]	TCRL	6	2.5	6	6	2	1	2

(continues)

Ecoregion	Major Habitat Type (MHT)	Degree of Land-cover (Catchment) Alteration (Weighted)	Water Quality Degradation (Weighted)	Alteration of Hydrographic Integrity (Weighted)	Degree of Habitat Fragmentation (Weighted)	Additional Losses of Original Intact Habitat (Weighted)	Effects of Introduced Species (Weighted)	Direct Species Exploitation (Weighted)
Lower St. Lawrence [47]	TCRL	6	4	2	6	2.5	3	5
North Atlantic-Ungava [48]	TCRL	4	2	6	4	0	1	4
West Texas Gulf [33]	SCRL	7	1	8	8	0	1	2
Florida [39]	SCRL	8	3	8	8	4	2.5	2
Sinaloan Coastal [62]	SCRL	6	4	6	4	0	2	2
Santiago [63] *	SCRL							
Manantlan-Ameca [64]	SCRL	4	2	4	2	3	1	2
Tamaulipas-Veracruz [68]	SCRL	6	2	4	4	2	2	4
Balsas [70]	SCRL	6	3	6	6	2	4	2
Papaloapan [71]	SCRL	4	3	6	4	2	2	4
Catemaco [72]	SCRL	2	2	2	2	1	2	6
Coatzacoalcos [73]	SCRL	6	3	6	4	3	2	4
Tehuantepec [74]	SCRL	4	2	4	4	3	2	4
Grijalva-Usumacinta [75]	SCRL	6	3	4	4	2	3	4
Yucatán [76]	SCRL	6	2	4	2	3	1	2

* No conservation status data available; rank taken from Olson et al. (1997).

Ecoregion	Major Habitat Type (MHT)	Sum of Weighted Conservation Status Indicators	"Snapshot" Total (% Possible)	Conservation Status Ranking (1–5)	Likelihood of Future Threats (H–M–L)	Final Conservation Status, Modified by Future Threat (1–5)	Final Conservation Status Category
South Hudson [53]	ARL	8.5	26.6	1	M	1	Rel. Intact
East Hudson [54]	ARL	13.5	42.2	2	M	2	Rel. Stable
Yukon [55]	ARL	9.5	29.7	1	M	1	Rel. Intact
Lower Mackenzie [56]	ARL	5	15.6	1	L	1	Rel. Intact
Upper Mackenzie [57]	ARL	12	37.5	2	H	3	Vulnerable
North Arctic [58]	ARL	5	15.6	1	L	1	Rel. Intact
East Arctic [59]	ARL	4	12.5	1	L	1	Rel. Intact
Arctic Islands [60]	ARL	4	12.5	1	L	1	Rel. Intact
Superior [43]	LTL	20	55.6	3	M	3	Vulnerable
Michigan-Huron [44]	LTL	29	80.6	4	H	5	Critical
Erie [45]	LTL	29.5	81.9	4	H	5	Critical
Ontario [46]	LTL	30.5	84.7	4	H	5	Critical
Central Prairie [28]	THL	15	34.1	2	M	2	Rel. Stable
Ozark Highlands [29]	THL	12	27.3	1	L	1	Rel. Intact
Ouachita Highlands [30]	THL	19	43.2	2	M	2	Rel. Stable
Southern Plains [31]	THL	30	68.2	3	L	3	Vulnerable
Teays-Old Ohio [34]	THL	24	54.5	3	L	3	Vulnerable
Tennessee-Cumberland [35]	THL	25.5	58.0	3	H	4	Endangered
Canadian Rockies [49]	THL	21	47.7	2	H	3	Vulnerable
Upper Saskatchewan [50]	THL	29.5	67.0	3	H	4	Endangered
Lower Saskatchewan [51]	THL	21.5	48.9	2	L	2	Rel. Stable
English-Winnipeg Lakes [52]	THL	27	61.4	3	M	3	Vulnerable
Colorado [12]	LTR	35	87.5	5	H	5	Critical
Upper Rio Grande/Bravo [15]	LTR	34	85.0	4	H	5	Critical
Lower Rio Grande/Bravo [20]	LTR	35	87.5	5	H	5	Critical
Mississippi [24]	LTR	29	72.5	4	M	4	Endangered
Mississippi Embayment [25]	LTR	31	77.5	4	H	5	Critical
Upper Missouri [26]	LTR	27	67.5	3	L	3	Vulnerable
Middle Missouri [27]	LTR	25	62.5	3	L	3	Vulnerable
Bonneville [8]	ERLS	42	70.0	3	H	4	Endangered
Lahontan [9]	ERLS	48	80.0	4	H	5	Critical
Oregon Lakes [10]	ERLS	39	65.0	3	M	3	Vulnerable
Death Valley [11]	ERLS	35	58.3	3	M	3	Vulnerable
Guzmán [16]	ERLS	48	80.0	4	H	5	Critical
Mapimí [19]	ERLS	45	75.0	4	H	5	Critical
Llanos el Salado [66]	ERLS	52	86.7	5	H	5	Critical
Lerma [69]	ERLS	43	71.7	4	H	5	Critical
South Pacific Coastal [7]	XRLS	54	84.4	4	H	5	Critical
Vegas-Virgin [13]	XRLS	42	65.6	3	H	4	Endangered
Gila [14]	XRLS	56	87.5	5	H	5	Critical
Rio Conchos [17]	XRLS	48	75.0	4	H	5	Critical
Pecos [18]	XRLS	44	68.8	3	H	4	Endangered
Rio Salado [21]	XRLS	36	56.3	3	H	4	Endangered
Cuatro Ciénegas [22]	XRLS	24	37.5	2	H	3	Vulnerable
Rio San Juan [23]	XRLS	48	75.0	4	M	4	Endangered
Sonoran [61]	XRLS	40	62.5	3	H	4	Endangered
Chapala [65]	XRLS	39	60.9	3	H	4	Endangered
Rio Verde Headwaters [67]	XRLS	22	34.4	2	M	2	Rel. Stable
North Pacific Coastal [1]	TCRL	25	56.8	3	H	4	Endangered
Columbia Glaciated [2]	TCRL	34	77.3	4	M	4	Endangered
Columbia Unglaciated [3]	TCRL	34	77.3	4	M	4	Endangered
Upper Snake [4]	TCRL	33	75.0	4	M	4	Endangered
Pacific Mid-Coastal [5]	TCRL	32	72.7	4	H	5	Critical
Pacific Central Valley [6]	TCRL	41	93.2	5	H	5	Critical
East Texas Gulf [32]	TCRL	26	59.1	3	M	3	Vulnerable
Mobile Bay [36]	TCRL	33	75.0	4	H	5	Critical
Apalachicola [37]	TCRL	35	79.5	4	H	5	Critical
Florida Gulf [38]	TCRL	22	50.0	2	M	2	Rel. Stable
South Atlantic [40]	TCRL	33	75.0	4	H	5	Critical
Chesapeake Bay [41]	TCRL	26	59.1	3	H	4	Endangered
North Atlantic [42]	TCRL	25.5	58.0	3	H	4	Endangered

(continues)

Ecoregion	Major Habitat Type (MHT)	Sum of Weighted Conservation Status Indicators	"Snapshot" Total (% Possible)	Conservation Status Ranking (1–5)	Likelihood of Future Threats (H–M–L)	Final Conservation Status, Modified by Future Threat (1–5)	Final Conservation Status Category
Lower St. Lawrence [47]	TCRL	28.5	64.8	3	H	4	Endangered
North Atlantic-Ungava [48]	TCRL	21	47.7	2	H	3	Vulnerable
West Texas Gulf [33]	SCRL	27	61.4	3	M	3	Vulnerable
Florida [39]	SCRL	35.5	80.7	4	H	5	Critical
Sinaloan Coastal [62]	SCRL	24	54.5	3	H	4	Endangered
Santiago [63] *	SCRL			3	H	4	Endangered
Manantlan-Ameca [64]	SCRL	18	40.9	2	M	2	Rel. Stable
Tamaulipas-Veracruz [68]	SCRL	24	54.5	3	H	4	Endangered
Balsas [70]	SCRL	29	65.9	3	H	4	Endangered
Papaloapan [71]	SCRL	25	56.8	3	H	4	Endangered
Catemaco [72]	SCRL	17	38.6	2	L	2	Rel. Stable
Coatzacoalcos [73]	SCRL	28	63.6	3	H	4	Endangered
Tehuantepec [74]	SCRL	23	52.3	3	M	3	Vulnerable
Grijalva-Usumacinta [75]	SCRL	26	59.1	3	M	3	Vulnerable
Yucatán [76]	SCRL	20	45.5	2	M	2	Rel. Stable

* No conservation status data available; rank taken from Olson et al. (1997).

Statistical Analyses of Biological Distinctiveness and Conservation Status Data

Patterns of Biodiversity, by MHT

To determine whether MHTs exhibited different patterns of species richness and endemism, we ran one-way (fixed-effects) analyses of variance (ANOVAs) for fish, crayfish, and herpetofauna species richness, endemism, and percentage of endemism, with MHT as the classification factor. For fish richness and percentage of endemism there were significant differences ($p < 0.00005$ for both) among MHTs; for fish endemism, the difference was nonsignificant at $\alpha = 0.05$ ($p = 0.2441$). There was a significant difference among MHTs for crayfish richness ($p = 0.0115$), mussel richness ($p = 0.0100$), and herpetofauna richness ($p < 0.00005$), as well as for herpetofauna endemism ($p = 0.0055$) and herpetofauna percentage of endemism ($p = 0.0321$), but not for crayfish endemism ($p = 0.0950$), crayfish percentage of endemism ($p = 0.1753$), mussel endemism ($p = 0.2796$), or mussel percentage of endemism ($p = 0.6183$).

These results confirm that, for the BDI analysis, a direct comparison of all ecoregions regardless of MHTs would likely undervalue most or all ecoregions of certain MHTs (e.g. xeric and arctic habitat types) because they support low numbers of species. They also suggest that the percentage of endemism statistic (number of endemic species divided by total number of species) often gives information different from the number of endemic species taken alone. This is confirmed by correlations between fish richness, fish endemism, and percentage of fish endemism. There was a moderately strong positive relationship between fish richness and endemism ($r = 0.4988$, $p < 0.0005$), but there was a significant negative correlation

between fish richness and percentage of endemism ($r = {}^-0.3463$, $p = 0.002$). The ratio of endemics to the total number of fish species, then, tends to be high where species richness is low. This finding reinforces the use of percentage of endemism as an indicator for the BDI, as high endemism within species-poor ecoregions would be lost otherwise.

The same patterns were not observed for crayfish, mussels, or herpetofauna; however, for crayfish, the correlation between richness and percentage of endemism was significant and positive (rather than negative, as for fish), and substantially weaker than the positive relationship between richness and endemism ($r = 0.9656$, $p < 0.0005$). Similarly, for herpetofauna there were positive and significant relationships between richness and both endemism ($r = 0.6613$, $p < 0.0005$) and percentage of endemism ($r = 0.4565$, $p < 0.0005$). The same general pattern held for mussels, with a strong positive relationship between richness and endemism ($r = 0.7861$, $p < 0.0005$) but a nonsignificant relationship between richness and percentage of endemism ($p = 0.129$).

To further explore the differences among MHTs, *post-facto* mean pairwise comparisons were examined. Using the conservative Tukey-HSD test with a significance level of 0.05, significant differences between any two MHTs were found only for fish species richness, fish percentage of endemism, mussel richness, herpetofauna species richness, and herpetofauna endemism. For fish species richness, endorheic rivers, lakes, and springs (with low richness) were significantly less species rich than were temperate coastal rivers and lakes and large temperate rivers. In addition, temperate headwaters and lakes were significantly more speciose than were xeric rivers, lakes, and springs; arctic rivers and lakes; and subtropical coastal rivers and lakes. For fish percentage of endemism, endorheic and xeric MHTs (with high means) were significantly different from all other MHTs (except subtropical coastal systems and xeric ecoregions), and the subtropical coastal MHT had significantly higher values than did arctic and temperate headwaters and lakes MHTs.

For mussel richness, temperate headwaters and lakes had significantly more species than arctic rivers and lakes (endorheic and xeric MHTs were not included in the analysis because there were no data). For herpetofauna richness, subtropical coastal rivers and lakes had significantly higher numbers of species than all other MHTs, and the arctic ecoregions had, on average, lower richness than large rivers and temperate coastal MHTs. For herpetofauna endemism, subtropical coastal rivers and lakes had higher numbers of endemics than the arctic, xeric, large rivers, temperate coastal, and headwaters MHTs.

These pairwise comparison tests serve to highlight the most striking differences among MHTs. They show that, on average, xeric and endorheic systems are far less speciose for fish than are most other MHTs, but that these habitats exhibit extremely high rates of endemism. Crayfish biodiversity appears to be more evenly distributed among MHTs. Herpetofauna richness and endemism, not surprisingly, is highest in subtropical ecoregions, due largely to diverse assemblages of frogs. Patterns are less easily discerned for mussels, given the lack of data for Mexico, though it appears obvious that temperate headwaters and lakes support the greatest mussel diversity.

A direct examination of the mean MHT values for each BDI indicator suggests again that the same measures of biodiversity are not equally appropriate to all MHTs. Rankings of MHTs according to species richness, endemism, and percentage of endemism for fish, crayfish, and herpetofauna show that ranks change markedly depending on the indicator (table 3.1). For instance, large temperate lakes stand out for high numbers of fish species but rank low in most other categories. In contrast, endorheic rivers, lakes, and streams have low numbers of species but rank high for percentage of endemism in fish and herpetofauna. The ranks also show that, in terms of strict numbers of species, temperate headwaters and lakes have the

highest diversity in most categories, and arctic rivers and lakes have the lowest numbers.

Relationships of Diversity Among Faunal Groups

Because of a lack of distributional data for poorly known groups, this study relied on fish, crayfish, mussels, and selected amphibians and reptiles as the basis for its baseline biodiversity index data. We made the assumption that these groups serve as a reasonable proxy for the distribution of other taxa, particularly those with entirely aquatic life cycles. While we cannot directly test the validity of this assumption (because we have no distributional data for unmapped taxa), we were able to test the correspondence between the distributions of our four mapped groups.

Pearson correlation correspondence analyses showed strong associations between fish and crayfish richness ($r = 0.8835$, $p < 0.0005$), fish and mussel richness ($r = 0.9330$, $p < 0.0005$), mussel and crayfish richness ($r = 0.8873$, $p < 0.0005$), and mussel and herpetofauna richness ($r = 0.6291$, $p < 0.0005$); and between fish and crayfish endemism ($r = 0.6925$, $p < 0.0005$), fish and mussel endemism ($r = 0.8778$, $p < 0.0005$), crayfish and mussel endemism ($r = 0.8758$, $p < 0.0005$), and mussel and herpetofauna endemism ($r = 0.6319$, $p < 0.0005$). Herpetofauna richness showed a weaker correlation with both fish and crayfish richness ($r = 0.2893$, $p < 0.0005$; $r = 0.4087$, $p < 0.0005$), and herpetofauna endemism was even more weakly associated with fish and crayfish endemism ($r = 0.2893$, $p = 0.011$; $r = 0.1485$, $p = 0.200$). This suggests that general habitat conditions favoring high fish and crayfish diversity may also favor amphibians and reptiles, but that different mechanisms may have caused localized speciation in aquatic herpetofauna.

Relationship of Ecoregion Size to Biodiversity Index Scores

Ecoregion boundaries were drawn without regard for ecoregion size, resulting in a wide range in ecoregion areas (table E.1). The mean ecoregion area was 278,584 km^2, with a minimum of 192 km^2 and a maximum of 1,567,884 km^2. Because, all things being equal, we might expect that a larger area would contain more species, we tested the assumption that ecoregion size was not significantly associated with number of species.

All Pearson correlation relationships between ecoregion area and fish, crayfish, and herpetofauna richness, endemism, and percentage of endemism were negative (mussels were not used because there were no data for Mexico). In other words, as ecoregion area increased, all of these measures decreased. However, correlations with fish richness, crayfish richness, crayfish endemism, and herpetofauna endemism were nonsignificant. A moderately strong negative correlation between ecoregion area and herpetofauna species richness ($r = {}^-0.4416$, $p < 0.00005$) is explained by the facts that arctic ecoregions were significantly larger than those in all other MHTs (with an average size of 908,005 km^2), and that these ecoregions had the lowest numbers of herpetofauna species. Significant correlations between ecoregion area and fish endemism ($r = {}^-0.2899$, $p = 0.011$), percentage of fish endemism ($r = {}^-0.4399$, $p < 0.0005$), percentage of crayfish endemism ($r = {}^-0.3001$, $p = 0.008$), and percentage of herpetofauna endemism ($r = {}^-0.2323$, $p = 0.043$) are explained by the delineation of unusually small ecoregions where species exhibited highly localized distributions. These tests allowed us to conclude that our

TABLE E.1   Area and Latitude of Ecoregions.

Ecoregion	Area (sq. km)	Latitude (°)	Ecoregion	Area (sq. km)	Latitude (°)
Yukon [55]	1,567,884	65	Bonneville [8]	155,863	40
Arctic Islands [60]	1,407,571	83	Tennessee-Cumberland [35]	152,292	36
Lower Mackenzie [56]	1,209,314	63	Grijalva-Usumacinta [75]	136,290	18
North Pacific Coastal [1]	928,013	61	Tamaulipas-Veracruz [68]	132,524	22
Lower St. Lawrence [47]	855,097	49	Superior [43]	128,464	50
Upper Mackenzie [50]	702,806	58	Sinaloan Coastal [62]	126,886	25
Upper Missouri [26]	678,741	46	Mobile Bay [36]	115,514	33
English-Winnipeg Lakes [52]	639,319	50	Ouachita Highlands [30]	115,370	34
North Atlantic-Ungava [48]	626,555	56	Pecos [18]	114,782	33
East Arctic [59]	624,339	64	Balsas [70]	114,547	19
South Hudson [53]	601,216	52	Pacific Mid-Coastal [5]	108,888	42
North Arctic [58]	598,863	67	Santiago [63]	106,988	22
Middle Missouri [27]	594,095	42	Guzmán [16]	97,442	30
East Hudson [54]	552,051	63	Tehuantepec [74]	93,703	17
Lower Saskatchewan [51]	523,809	55	Upper Snake [4]	92,848	43
Colorado [12]	507,245	37	Ozark Highlands [29]	85,600	36
Mississippi [24]	478,549	43	Death Valley [11]	80,707	36
Southern Plains [31]	417,780	36	Yucatán [76]	79,602	20
East Texas Gulf [32]	408,405	32	Erie [45]	79,527	43
Teays-Old Ohio [34]	373,887	39	West Texas Gulf [33]	75,596	28
North Atlantic [42]	335,412	48	Rio Conchos [17]	74,885	28
Columbia Unglaciated [3]	322,518	45	Lerma [69]	69,780	20
Upper Saskatchewan [50]	304,749	52	Ontario [46]	67,573	44
South Atlantic [40]	295,608	35	Papaloapam [71]	55,870	18
Mississippi Embayment [25]	258,675	33	Llanos el Salado [66]	55,330	23
Columbia Glaciated [2]	257,332	49	Apalachicola [37]	54,403	32
Michigan-Huron [44]	250,901	46	Manantlan-Ameca [64]	49,563	20
Lahontan [9]	213,566	40	Rio Salado [21]	48,675	28
Sonora [61]	203,936	30	Oregon Lakes [10]	47,925	43
Mapimí [19]	187,773	25	Rio San Juan [23]	46,677	26
Central Valley [6]	184,129	38	Canadian Rockies [49]	36,928	51
Chesapeake Bay [41]	179,243	40	Florida Gulf [38]	35,513	31
South Pacific Coastal [7]	170,320	29	Vegas-Virgin [13]	34,565	37
Lower Rio Grande/Bravo [20]	162,576	30	Coatzacoalcos [73]	29,294	17
Florida [39]	161,393	29	Chapala [65]	7,486	20
Gila [14]	159,875	33	Rio Verde Headwaters [67]	4,859	22
Central Prairie [28]	158,635	38	Cuatro Ciénegas [22]	492	27
Upper Rio Grande [15]	156,769	34	Catemaco [72]	192	18

assumption was valid and that ecoregion size had not biased our analysis of biodiversity. The same correlations were run within each MHT. The only significant relationships were for crayfish in subtropical ecoregions (richness: $r = 0.5611$, $p = 0.046$; endemism: $r = 0.5763$, $p = 0.039$; percentage of endemism: $r = 0.6272$, $p = 0.022$).

Relationship of Latitude to Species Richness

Because it is well established that for many taxa species richness increases toward the equator (see chapter 2), we investigated the relationship between latitude and species richness for our ecoregions. Using a GIS, we determined the latitude of each ecoregion's centroid. For ecoregions composed of multiple polygons (such as the Arctic Islands), the average of all polygon centroids was taken. The center latitudes of each ecoregion are listed in table E.1.

To investigate latitude as a potentially strong factor in determining species richness within ecoregions, we ran Pearson correlations between ecoregion latitude and fish, crayfish, and herpetofauna richness (mussel richness was not used, because there were no data for Mexico). The relationships between latitude and fish and crayfish richness were nonsignificant and negative. However, there was a strong negative correlation between herpetofauna species richness and latitude ($r = ^-0.7400$, $p < 0.0005$); this is not surprising, given the extraordinary amphibian diversity within central and southern Mexico. Although herpetofauna comprise only a fourth of the biodiversity index data, this finding justifies separation of subtropical from temperate coastal systems.

Because the associations between herpetofauna richness and ecoregion area and latitude were strong, a multiple linear regression was generated to further investigate these relationships. These variables together yielded a model with an R^2 value of 0.590 ($p < 0.0005$), whereas the simple linear regression using latitude alone produced a model with an R^2 of 0.548 ($p < 0.0005$). Herpetofauna, then, follow established biogeographic trends more strongly than do North American fish or crayfish, suggesting again that the processes shaping herpetofauna distributions may be more different than those for the other faunal groups considered. That differences in

fish and crayfish diversity are not similarly predicted by latitude and area lends support to the ecoregion approach used in this and other studies.

Within MHTs, as opposed to across all ecoregions, certain relationships between richness and latitude were significantly negative (arctic ecoregions: fish and mussel richness; headwater ecoregions: fish, mussel, crayfish, herpetofauna richness; endorheic ecoregions: herpetofauna richness; xeric ecoregions: crayfish richness; coastal ecoregions: mussel, crayfish herpetofauna richness; subtropical ecoregions: mussel richness). These findings are not unexpected, given the strong relationship between temperature and species distributions for freshwater systems. Because richness comprises only a part of our analysis and one of our objectives is the identification of ecoregions supporting high species richness, we do not feel that these relationships invalidate our approach. A detailed discussion of latitudinal effects and implications for our approach is found in Ricketts et al. (1999b).

Relationship of Species Imperilment Among Faunal Groups and MHTs

Because species distributions were available for fish and herpetofauna (as opposed to summary data for crayfish and mussels), imperiled species in these two groups could be mapped by ecoregions and their patterns examined. There was a significant positive relationship between the percentage of imperiled (endangered and threatened) fish and herpetofauna species within ecoregions ($r = 0.4515$, $p < 0.0005$). This suggests that, to a certain extent, the factors threatening fish may also be threatening aquatic reptiles and amphibians. However, an analysis of the relationship between patterns of imperilment and conservation status indicators did not offer strong evidence that the same threats, as assessed by workshop participants, were similarly affecting these taxa.

The degree of species imperilment was quite uneven among MHTs. For fish, a Tukey HSD test found significant differences in the percentage of endangered and threatened species between xeric and endorheic rivers, lakes, springs, and all other MHTs except large rivers.

For herpetofauna, the degree of imperilment was greater in endorheic ecoregions than in all other MHTs except xeric and subtropical coastal ecoregions, and imperilment in xeric and subtropical ecoregions was greater than in temperate coastal rivers and lakes, temperate headwaters and lakes, large temperate lakes, and arctic ecoregions. For the percentage of endemic species that were imperiled, there were no significant differences among MHTs for herpetofauna. For fish, xeric region and large river MHTs had significantly greater imperilment than did subtropical coastal ecoregions.

Relationship of U.S. EPA IWI Data and Conservation Status Assessment Scores, for U.S. Ecoregions

National Watershed Characterization data (U.S. EPA 1997), which describe water quality and vulnerability for U.S. watersheds, were aggregated for ecoregions for comparison with conservation status scores. For many of these watersheds, defined as USGS hydrologic units, there were no data available, and in some cases such watersheds comprised the majority of their ecoregions. Where there were no National Watershed Characterization data for 50 percent or more of the area of an ecoregion, the entire ecoregion was coded as "providing insufficient data." Where less than 50 percent of an ecoregion was without data, an area-weighted average score was computed for the ecoregion, with watersheds without data removed from the computation.

Correlations between the EPA scores and conservation status scores were run to evaluate the value of the expert assessment. There was a significant relationship between the EPA scores for water quality and the expert assessment scores for the water quality indicator ($r = 0.3146$, $p = 0.040$). The EPA water quality scores were also significantly correlated with expert assessment scores for additional losses of intact original habitat ($r = 0.3588$, $p = 0.018$) and for the effects of exotic species ($r = 0.3910$, $p = 0.010$). In other words, better water quality was associated with less habitat loss and fewer exotic species. These results do not imply cause and effect but merely suggest that areas with poor water quality also face these additional threats. On the other hand, threats may be related, such as where poor water quality encourages the establishment and proliferation of tolerant exotics.

More important, the EPA water quality data were significantly correlated with this study's snapshot conservation status total ($r = 0.4261$, $p = 0.004$). To some extent, it appears that ecoregions that are degraded from many threats tend to suffer poor water quality. This is reinforced by the strong correlation between the EPA water quality and EPA vulnerability scores ($r = 0.6341$, $p = 0.000$). EPA vulnerability scores were not significantly correlated with this study's snapshot total.

APPENDIX F

Integration Matrices for the Eight Major Habitat Types

INTEGRATION MATRIX: Arctic Rivers and Lakes

Final Conservation Status (Before Threats)

Biological Distinctiveness	Critical	Endangered	Vulnerable	Relatively Stable	Relatively Intact
Globally Outstanding					
Continentally Outstanding					Yukon [55], North Arctic [58], Arctic Islands [60]
Bioregionally Outstanding					
Nationally Important				East Hudson [54], Upper Mackenzie [57]*	South Hudson [53], Lower Mackenzie [56], East Arctic [59]

Final Conservation Status (With Threats)

Biological Distinctiveness	Critical	Endangered	Vulnerable	Relatively Stable	Relatively Intact
Globally Outstanding					
Continentally Outstanding					Yukon [55], North Arctic [58], Arctic Islands [60]
Bioregionally Outstanding					
Nationally Important			Upper Mackenzie [57]*	East Hudson [54]	South Hudson [53], Lower Mackenzie [56], East Arctic [59]

*Ecoregions that shift cells because of threats.

INTEGRATION MATRIX: Large Temperate Lakes

Final Conservation Status (Before Threats)

Biological Distinctiveness	Critical	Endangered	Vulnerable	Relatively Stable	Relatively Intact
Globally Outstanding					
Continentally Outstanding		Michigan-Huron [44]*	Superior [43]		
Bioregionally Outstanding					
Nationally Important		Erie [45]*, Ontario [46]*			

Final Conservation Status (With Threats)

Biological Distinctiveness	Critical	Endangered	Vulnerable	Relatively Stable	Relatively Intact
Globally Outstanding					
Continentally Outstanding	Michigan-Huron [44]*		Superior [43]		
Bioregionally Outstanding					
Nationally Important	Erie [45]*, Ontario [46]*				

*Ecoregions that shift cells because of threats.

INTEGRATION MATRIX: Temperate Headwaters and Lakes

Final Conservation Status (Before Threats)

Biological Distinctiveness	Critical	Endangered	Vulnerable	Relatively Stable	Relatively Intact
Globally Outstanding			Teays-Old Ohio [34], Tennessee-Cumberland [35]*		
Continentally Outstanding			English-Winnipeg Lakes [52]	Canadian Rockies [49]*, Lower Saskatchewan [51]	Ozark Highlands [29]
Bioregionally Outstanding				Central Prairie [28], Ouachita Highlands [30]	
Nationally Important			Southern Plains [31], Upper Saskatchewan [50]*		

Final Conservation Status (With Threats)

Biological Distinctiveness	Critical	Endangered	Vulnerable	Relatively Stable	Relatively Intact
Globally Outstanding		Tennessee-Cumberland [35]*	Teays-Old Ohio [34]		
Continentally Outstanding			Canadian Rockies [49]*, English-Winnipeg Lakes [52]	Lower Saskatchewan [51]	Ozark Highlands [29]
Bioregionally Outstanding				Central Prairie [28], Ouachita Highlands [30]	
Nationally Important		Upper Saskatchewan [50]*	Southern Plains [31]		

*Ecoregions that shift cells because of threats.

INTEGRATION MATRIX: Large Temperate Rivers

Final Conservation Status (Before Threats)

Biological Distinctiveness	Critical	Endangered	Vulnerable	Relatively Stable	Relatively Intact
Globally Outstanding	Mississippi Embayment [25]*				
Continentally Outstanding	Colorado [12], Lower Rio Grande/Bravo [20]	Upper Rio Grande/Bravo [15]*			
Bioregionally Outstanding		Mississippi [24]			
Nationally Important			Upper Missouri [26], Middle Missouri [27]		

Final Conservation Status (With Threats)

Biological Distinctiveness	Critical	Endangered	Vulnerable	Relatively Stable	Relatively Intact
Globally Outstanding	Mississippi Embayment [25]*				
Continentally Outstanding	Colorado [12], Upper Rio Grande/Bravo [15]*, Lower Rio Grande/Bravo [20]				
Bioregionally Outstanding		Mississippi [24]			
Nationally Important			Upper Missouri [26], Middle Missouri [27]		

*Ecoregions that shift cells because of threats.

INTEGRATION MATRIX: Endorheic Rivers, Lakes, and Springs

Final Conservation Status (Before Threats)

Biological Distinctiveness	Critical	Endangered	Vulnerable	Relatively Stable	Relatively Intact
Globally Outstanding		Lerma [69]*			
Continentally Outstanding	Llanos el Salado [66]	Lahontan [9]*, Guzmán [16]**, Mapimí [19]*	Bonneville [8]*, Death Valley [11]		
Bioregionally Outstanding					
Nationally Important			Oregon Lakes [10]		

Final Conservation Status (With Threats)

Biological Distinctiveness	Critical	Endangered	Vulnerable	Relatively Stable	Relatively Intact
Globally Outstanding	Lerma [69]*				
Continentally Outstanding	Lahontan [9]*, Guzmán [16]**, Mapimí [19]*, Llanos El Salado [66]	Bonneville [8]*	Death Valley [11]		
Bioregionally Outstanding					
Nationally Important			Oregon Lakes [10]		

*Ecoregions that shift cells because of threats.

160

INTEGRATION MATRIX: Xeric-Region Rivers, Lakes and Springs

Final Conservation Status (Before Threats)

Biological Distinctiveness	Critical	Endangered	Vulnerable	Relatively Stable	Relatively Intact
Globally Outstanding		Rio Conchos [17]*, Rio San Juan [23]	Chapala [65]*	Cuatro Ciénegas [22]*, Rio Verde Headwaters [67]	
Continentally Outstanding	Gila [14]	South Pacific Coastal [7]*	Vegas-Virgin [13]*, Pecos [18]*, Rio Salado [21]*, Sonoran [61]*		
Bioregionally Outstanding					
Nationally Important					

Final Conservation Status (With Threats)

Biological Distinctiveness	Critical	Endangered	Vulnerable	Relatively Stable	Relatively Intact
Globally Outstanding	Rio Conchos [17]*	Rio San Juan [23], Chapala [65]*	Cuatro Ciénegas [22]*	Rio Verde Headwaters [67]	
Continentally Outstanding	South Pacific Coastal [7]*, Gila [14]	Vegas-Virgin [13]*, Pecos [18]*, Rio Salado [21]*, Sonoran [61]*			
Bioregionally Outstanding					
Nationally Important					

*Ecoregions that shift cells because of threats.

INTEGRATION MATRIX: Temperate Coastal Rivers and Lakes

Final Conservation Status (Before Threats)

Biological Distinctiveness	Critical	Endangered	Vulnerable	Relatively Stable	Relatively Intact
Globally Outstanding	Pacific Central Valley [6]	Mobile Bay [36]*, South Atlantic [40]*			
Continentally Outstanding		Columbia Unglaciated [3], Pacific Mid-Coastal [5]*, Apalachicola [37]*	North Pacific Coastal [1], East Texas Gulf [32], Chesapeake Bay [41], Lower St. Lawrence [47]*,	Florida Gulf [38]	
Bioregionally Outstanding					
Nationally Important		Columbia Glaciated [2], Upper Snake [4]	North Atlantic [42]*	North Atlantic-Ungava [48]*	

Final Conservation Status (With Threats)

Biological Distinctiveness	Critical	Endangered	Vulnerable	Relatively Stable	Relatively Intact
Globally Outstanding	Pacific Central Valley [6], Mobile Bay [36]*, South Atlantic [40]*				
Continentally Outstanding	Pacific Mid-Coastal [5]*, Apalachicola [37]*	Columbia Unglaciated [3], Chesapeake Bay [41], Lower St. Lawrence [47]*	North Pacific Coastal [1], East Texas Gulf [32]	Florida Gulf [38]	
Bioregionally Outstanding					
Nationally Important		Columbia Glaciated [2], Upper Snake [4], North Atlantic [42]*	North Atlantic-Ungava [48]*		

*Ecoregions that shift cells because of threats.

INTEGRATION MATRIX: Subtropical Coastal Rivers and Lakes

Final Conservation Status (Before Threats)

Biological Distinctiveness	Critical	Endangered	Vulnerable	Relatively Stable	Relatively Intact
Globally Outstanding		Florida [39]*	Tamaulipas-Veracruz [68]*	Manantlan-Ameca [64], Catemaco [72]	
Continentally Outstanding			West Texas Gulf [33], Sinaloan Coastal [62]*, Santiago [63]*, Balsas [70]*, Papaloapan [71]*, Coatzacoalcos [73]*, Tehuantepec [74], Grijalva-Usumacinta [75]	Yucatán [76]	
Bioregionally Outstanding					
Nationally Important					

Final Conservation Status (With Threats)

Biological Distinctiveness	Critical	Endangered	Vulnerable	Relatively Stable	Relatively Intact
Globally Outstanding	Florida [39]*	Tamaulipas-Veracruz [68]*		Manantlan-Ameca [64], Catemaco [72]	
Continentally Outstanding		Sinaloan Coastal [62]*, Santiago [63]*, Balsas [70]*, Papaloapan [71]*, Coatzacoalcos [73]*	West Texas Gulf [33], Tehuantepec [74], Grijalva-Usumacinta [75]	Yucatán [76]	
Bioregionally Outstanding					
Nationally Important					

*Ecoregions that shift cells because of threats.

Ecoregion Descriptions

Ecoregion Number: **1**
Ecoregion Name: **North Pacific Coastal**
Major Habitat Type: **Temperate Coastal Rivers, Lakes, and Springs**
Ecoregion Size: **928,013 km²**
Biological Distinctiveness: **Continentally Outstanding**
Conservation Status: **Snapshot—Vulnerable Final—Endangered**

Introduction

This ecoregion extends from southeastern Alaska through the southwestern portion of the Yukon Territory and the western and central portions of British Columbia to northwestern Washington. The Queen Charlotte Islands, Vancouver Island, and the islands in the Tongass National Forest in Alaska fall within this ecoregion as well. Among the major rivers are the Copper River in Alaska, the Fraser River in British Columbia, and the Skagit River in Washington. This is a cool, high-rainfall area, formerly covered by rain forests on the coast and drier forests and grasslands in the interior.

Biological Distinctiveness

Fifty-eight freshwater fish species are native to this ecoregion. The Canadian portion of this region was once home to rich runs of five species of anadromous salmon, plus cutthroat and rainbow trout. Endemic sticklebacks (Enos, Texada, and giant sticklebacks, *Gasterosteus* spp.) are found in three locations in British Columbia, and the Olympic mudminnow (*Novumbra*

hubbsi) is in Washington. Two endemic fish species are known from the drainage basin of the Puget Sound. The Nooksack dace (*Rhinichthys cataractae* ssp.) inhabits small streams in southern British Columbia and western Washington, and the salish sucker (*Catostomus* sp.) is known from fewer locations in the same area. A blind cave-dwelling species of amphipod, *Stygobromus* sp., has been described from Vancouver Island.

Because parts of Washington and British Columbia escaped glaciation during the Pleistocene, further study may reveal other endemics besides those already described. For instance, hot springs and caves, particularly on Vancouver Island, likely support many undescribed and potentially endangered invertebrates, including mites, mollusks, amphipods, and isopods (Scudder 1996).

Conservation Status

Threats to this ecoregion's biota include expanding agriculture and associated water quality impacts; development of gravel pits; poisoning campaigns in lakes to benefit introduced game fish; damming of rivers for hydroelectric projects; pipeline construction; clear-cut logging, mining, and overfishing; genetic dilution from hatchery fish; loss of floodplain and riparian habitats from forest clearing; urbanization and industrialization along the lower reaches of rivers; and catchment modification for development (McAllister et al. 1985; Mosquin et al. 1995; Environment Canada 1996; Gregory and Bisson 1997; Northcote and Atagi 1997; Reisenbichler 1997). Freshwater mollusk habitat has been largely destroyed or degraded, having been dredged, diverted into sewers, and extensively polluted, particularly by agricultural activities (Scudder 1996). A few rare freshwater species are apparently well protected, such as the giant stickleback (*Gasterosteus* sp.), known only from Mayer Lake in the Queen Charlotte Islands, and for which the B.C. provincial government has established a natural history preserve (McAllister et al. 1985). In general, though, most freshwater species and their habitats are highly endangered, with destructive logging practices constituting the largest and most extensive threat throughout this ecoregion.

Suite of Priority Activities to Enhance Biodiversity Conservation

- Replace clear-cut logging with sustainable forestry practices. Protection of riparian areas is essential.
- Find alternatives to building dams.
- Establish protected areas, especially for endemics and migratory fish. Restore important habitats that have been degraded by logging and other activities.
- Ensure that international treaties provide for the persistence of native fish stocks throughout their range.
- Within streams supporting migratory fish runs, protect seasonally high flows that create essential spawning and rearing habitats (American Rivers 1999).
- Pursue alternatives to logging and associated road construction across the Copper River Delta (The Wilderness Society 1999).
- Contain urban sprawl from Seattle and develop and implement a water conservation program for the city (American Rivers 1999).
- Prevent the establishment and reopening of copper and gold mines. Clean up old and abandoned mines that continue to pollute streams with acid effluent.

Conservation Partners

For contact information, please see appendix H.

- American Rivers
- British Columbia Ministry of Environment, Lands, and Parks
- Environment Canada

- Friends of the Earth
- The Nature Conservancy (Alaska and Washington Field Offices)
- The Nature Conservancy of Canada (British Columbia Office)
- The Sierra Club, Cascade Chapter
- Southeast Alaska Conservation Council
- The Wilderness Society
- World Wildlife Fund Canada

Ecoregion Number:	**2**
Ecoregion Name:	**Columbia Glaciated**
Major Habitat Type:	**Temperate Coastal Rivers, Lakes, and Springs**
Ecoregion Size:	**257,332 km²**
Biological Distinctiveness:	**Nationally Important**
Conservation Status:	**Snapshot—Endangered**
	Final—Endangered

Introduction

This ecoregion marks the glaciated portion, or upper third, of the Columbia River Basin (McPhail and Lindsey 1986). It includes most of eastern Washington, the northern portion of Idaho, the northwestern corner of Montana, and southeastern British Columbia. Among tributaries to the Columbia are the Yakima, Okanagan, Spokane, and Kootenai Rivers. The ecoregion is mountainous and heavily forested, with cold, high-gradient streams and several lakes, primarily in British Columbia (McPhail and Lindsey 1986).

Biological Distinctiveness

As a result of Pleistocene glaciation, this area is relatively depauperate in freshwater species. The fish fauna, consisting of thirty-four native species, is derived largely from the lower Columbia [3] ecoregion and ecoregions to the east; consequently, there are no fish truly endemic to this ecoregion. A number of eastern fish found in this ecoregion are likely absent from the lower Columbia [3] ecoregion because of low water temperature; these include the pygmy whitefish (*Prosopium coulteri*), lake chub (*Couesius plumbeus*), longnose sucker (*Catostomus catosto-*

mus), burbot (*Lota lota*), and slimy sculpin (*Cottus cognatus*), all of which are widespread in other ecoregions. Historically, the upper Columbia [2] ecoregion also supported substantial runs of *Oncorhynchus nerka*, *O. tshawytscha*, *O. clarki*, and *O. mykiss* (McPhail and Lindsey 1986).

Conservation Status

Discussions about the conservation status of the upper Columbia [2] ecoregion tend to focus on the well-documented decline of anadromous fish populations. Dams and overfishing are widely held to be the major causes of these declines, but agricultural development and associated irrigation, mining, grazing, logging, and urbanization have doubtless contributed as well (McPhail and Lindsey 1986). Streams that were once clear are now turbid. Loss of riparian cover has been cited as an important contributor to the disappearance of salmon, as lack of cover can lead to increased temperatures, decreased organic matter inputs, loss of materials for in-stream habitat, increased bank erosion, modified channel morphology, and altered hydrology (Beschta 1997). Distinct salmon populations may also be disappearing as a result of interbreeding with hatchery-raised fish (Reisenbichler 1997).

Suite of Priority Activities to Enhance Biodiversity Conservation

- Modify existing dams and dam operations to release flows that mimic the natural spring surge and allow fish to spill over the dams.
- Regulate surrounding land use to restore riparian zones and minimize catchment-wide impacts of agriculture, grazing, and logging.
- Develop and promote alternative energy sources and less destructive agriculture and forestry practices.
- Protect Hanford Reach, the only free-flowing stretch of the mainstem Columbia River in the United States from damming and irrigation development. Federal ownership of the surrounding land should be retained, and the settlement agreement between utilities and fishing interests that regulates flows for salmon should be upheld (Geist 1995). The National Park Service has recommended creating a 102,000-acre national wildlife refuge

around the reach and giving the river a "Wild and Scenic Rivers" designation; this would help to protect the reach from harm caused by agricultural expansion and other development (American Rivers 1997, 1998).

Conservation Partners

For contact information, please see appendix H.

- American Rivers
- Columbia River Inter-Tribal Fish Commission
- Confederated Tribes of the Umatilla Indian Reservation
- National Audubon Society
- The Nature Conservancy (Idaho, Montana, and Washington Field Offices)
- The Nature Conservancy of Canada (British Columbia Office)
- Northwest Power Planning Council
- Washington Environmental Council
- World Wildlife Fund Canada

Ecoregion Number:	**3**
Ecoregion Name:	**Columbia Unglaciated**
Major Habitat Type:	**Temperate Coastal Rivers, Lakes, and Springs**
Ecoregion Size:	**322,518 km²**
Biological Distinctiveness:	**Continentally Outstanding**
Conservation Status:	**Snapshot—Endangered Final—Endangered**

Introduction

The Columbia River is the second-largest river in the United States. The Columbia Unglaciated [3] ecoregion, together with the Upper Snake [4] ecoregion, marks the portion of the larger Columbia River Basin that was never glaciated. This ecoregion covers most of eastern and northern Oregon, reaches just into Washington along the southern border of the state, spans central and southeastern Idaho, and covers small portions of northern Nevada and western Montana. The major river in this ecoregion is the Columbia, from just above the point where it becomes the Washington-Oregon border to its confluence in the Pacific. Other important rivers include the lower half of the Snake, its tributary the Salmon River in Idaho, and the John Day and Deschutes Rivers in Oregon.

The climate in the ecoregion is varied, with the Dalles Dam separating what some have termed the lower and middle Columbia Rivers. Above the dam the climate is arid, with high summer temperatures and periodic droughts; below the dam the climate is mild and water is abundant (McPhail and Lindsey 1996).

Biological Distinctiveness

This ecoregion is home to thirty-five native species of fish, twelve of which are found both in the lower and middle Columbia. Four fish species are endemic: the relict sand roller (*Percopsis transmontana*), the shorthead sculpin (*Cottus confusus*), the marginated sculpin (*C. marginatus*), and the Oregon chub (*Oregonichthys crameri*). The chub is also found in the Umpqua River drainage of the adjacent Mid-Coastal ecoregion [5]. Equally important, the ecoregion supports seven species of salmon (*Oncorhynchus* spp.), whose historic migrations made the Columbia one of the biggest salmon-producing rivers in North America (McPhail and Lindsey 1986).

Within this ecoregion mollusk endemism is high and often localized; discrete areas of endemism are found in the lower Columbia River, the lower Deschutes River drainage, the Columbia Gorge, the Blue Mountains, the lower Salmon River, the middle Snake River, the Clearwater River drainage, and Hells Canyon. Frest and Johannes (1995) list sixty-three freshwater snail and clam species and subspecies of special concern, twenty-five of which are apparently restricted to single locations.

Conservation Status

The degradation of freshwater habitats and their species in the entire Columbia Basin has been the subject of much attention in recent years because of the visible decline of the basin's native salmonid migrations. More than 200 salmon runs have disappeared from the basin, and 76 more are at risk of extinction (American Rivers 1996). The causes are numerous: unregulated grazing and logging, leading to loss of riparian cover and subsequent destruction of in-stream habitat and impaired water quality; stream chan-

nelization; genetic losses as a result of interbreeding with hatchery fish; irrigation diversions for agriculture and subsequent dewatering of streams; nonpoint source pollution from agriculture; and hydroelectric dams that alter river flows and block migration routes (McPhail and Lindsey 1986; American Rivers 1996; Kostow 1997; Reisenbichler 1997). Of these, hydroelectric dams have likely had the greatest impact. There are nineteen federal power and irrigation dams on the Columbia and Snake Rivers (ecoregions [2], [3], and [4]), and these have virtually eliminated the spring surge of water that historically flushed juvenile salmon down to the ocean (American Rivers 1996). All Snake River salmon runs are now listed under the Endangered Species Act (American Rivers 1996).

Suite of Priority Activities to Enhance Biodiversity Conservation

- Manage dams better to mimic natural river flows and allow for fish spillover.
- Restore and protect riparian zones by regulating livestock grazing and logging.
- Reduce irrigation diversions from streams prone to dewatering, such as the John Day River in Oregon.
- Protect watersheds by incorporating aquatic conservation needs into regional land management plans.
- Remove Condit Dam on the White Salmon River. The White Salmon River's once-abundant salmon and steelhead are at risk of extinction, and spring chinook have already disappeared. The dam is presently under review for FERC relicensing (American Rivers 1996).
- Partially remove four dams on the lower Snake River that currently block passage for migrating salmon (American Rivers 1999).

Conservation Partners

For contact information, please see appendix H.

- American Rivers
- Central Cascades Alliance
- Friends of the White Salmon
- Idaho Rivers United
- The Nature Conservancy (Idaho, Montana, Nevada, Oregon, and Washington Field Offices)

- Oregon Natural Desert Association
- Oregon Natural Resources Council
- Pacific Rivers Council
- Save Our Wild Salmon

Ecoregion Number: 4

Ecoregion Name:	**Upper Snake**
Major Habitat Type:	**Temperate Coastal Rivers, Lakes, and Springs**
Ecoregion Size:	**92,848 km²**
Biological Distinctiveness:	**Nationally Important**
Conservation Status:	**Snapshot—Endangered Final—Endangered**

Introduction

This ecoregion is defined by the Snake River above the 30,000-to-60,000-year-old Shoshone Falls, which serve as a total barrier to the upstream movement of fish (McPhail and Lindsey 1986). The ecoregion boundary has been placed about 50 km downstream of the falls to include the Wood River, a tributary to the Snake. Only 35 percent of the fish fauna of the Snake River above Shoshone Falls, and 40 percent of the Wood River fish fauna, are shared with the lower Snake River (McPhail and Lindsey 1986). The ecoregion is predominantly restricted to the southeastern part of Idaho but also extends into eastern Wyoming, northeastern Nevada, and the extreme northwestern corner of Utah. Other major freshwater habitats include Jackson Lake and other lakes of Grand Teton National Park.

Biological Distinctiveness

Fourteen fish species are found in this ecoregion, many of which are found nowhere else in the Columbia River Basin but are shared with the adjacent Bonneville ecoregion [8]. The Snake River sucker (*Chasmistes muriei*) is known from a single specimen collected from the Snake River below Jackson Lake, Wyoming, and is apparently extinct (McPhail and Lindsey 1986). The Wood River contains the endemic Wood River sculpin (*Cottus leiopomus*), and in the Snake River between the Wood River and Shoshone Falls lives the endemic Shoshone sculpin (*C. greenei*) (McPhail and Lindsey

1986). Additionally, the Upper Snake [4] ecoregion is an area of high freshwater mollusk endemism; Frest and Johannes (1995) identify twenty-one snail and clam species and subspecies of special concern, fifteen of which are apparently restricted to single clusters.

Conservation Status

Threats to the Upper Snake [4] ecoregion are similar to those facing other Columbia Basin ecoregions. Dams and diversions have altered the Snake River's flow and water levels, and some stretches are dewatered at times. Watershed-wide grazing and logging, and expanding agriculture, have taken their toll on in-stream habitat. Pollution from agriculture, fish farms, and municipalities has resulted in water quality that often fails to meet minimum state standards. Exotic fish species outnumber native species in the upper Snake River as a result of game fish introductions, though their impacts on native species are yet to be determined. All salmon runs in the Snake River are listed under the federal Endangered Species Act, as are five mollusk species (American Rivers 1997).

Suite of Priority Activities to Enhance Biodiversity Conservation

- Restore flows to the Snake River by reducing irrigation diversions and reoperating dams.
- Remove hydroelectric dams through the FERC relicensing process if power companies do not implement measures to mitigate water quality impacts.
- Develop new water quality standards and enforce them.
- Prevent development of the Auger Falls hydroelectric project at one of the last remaining falls on the Snake River (American Rivers 1997).

Conservation Partners

For contact information, please see appendix H.

- American Rivers
- Idaho Rivers United
- The Nature Conservancy (Idaho, Nevada, and Wyoming Field Offices)
- Pacific Rivers Council

Ecoregion Number: **5**
Ecoregion Name: **Pacific Mid-Coastal**
Major Habitat Type: **Temperate Coastal Rivers, Lakes, and Springs**
Ecoregion Size: **108,888 km²**
Biological Distinctiveness: **Continentally Outstanding**
Conservation Status: **Snapshot—Endangered Final—Critical**

Introduction

This ecoregion extends along the Pacific Coast of Oregon and California to the northern shore of San Francisco Bay, including the western portion of the San Francisco peninsula south to Santa Cruz. In southern Oregon and Northern California, the ecoregion reaches further inland, encompassing the western drainages of the Klamath and Siskiyou Mountains. Important rivers include the Umpqua, Mad, and Klamath. The ecoregion contains several lakes, including Upper Klamath Lake in southern Oregon.

Biological Distinctiveness

This ecoregion has plentiful water resources because of its location on the slopes of four mountain ranges—the Cascades, Coastal Ranges, Klamath Mountains, and Siskiyou Mountains. Its mild climate and rainfall have made this ecoregion a haven for terrestrial and aquatic species. Stream capture, the process by which the headwaters of a neighboring watershed are "captured," has also contributed to this ecoregion's remarkable biodiversity; many species found in the Pacific Mid-Coastal [5] ecoregion were originally found in more inland watersheds.

The Pacific Mid-Coastal [5] ecoregion is the southernmost habitat for many of the Pacific Coast anadromous species, including coho salmon (*Oncorhynchus kisutch*) and cutthroat salmon (*O. clarki*). While this ecoregion contains no endemic mussel or crayfish species, sixteen of the region's fifty-five fish species are endemic, including the Umpqua squawfish (*Ptychocheilus umpquae*), the Klamath smallscale sucker (*Catostomus rimiculus*), and the Klamath largescale sucker (*C. snyderi*).

The ecoregion includes a number of restricted-range freshwater mollusks and crustaceans, including the Great Basin rams-horn (*Helisoma newberryi newberryi*) (Frest and

Johannes 1995) and species of endemic syncarid shrimp (*Syncaris* spp.).

Conservation Status

Industrial logging on the western slopes of the ranges has removed and altered natural vegetation over vast landscapes. Removal of ground cover means that the plentiful rain washes valuable topsoil into the high-gradient streams, clogging them with sediment. Loss of riparian vegetation or inadequate riparian buffer widths contribute to increased sedimentation, higher temperatures, lower dissolved oxygen, and increased severity and frequency of floods. Road-building associated with logging also leads to increased sedimentation.

Agriculture, grazing, mining, and urbanization also threaten the aquatic biodiversity of this region. Nonpoint source and industrial pollution disrupt the balance of these watersheds and harm the diversity of naturally present species.

Of special interest for this ecoregion are salmonid hatcheries. These are high-density, controlled hatcheries for the production of salmon for food. Impacts associated with hatcheries include nutrient loading and eutrophication of streams from hatchery waste products, and genetic pollution of native salmon populations when hatchery fish escape.

Suite of Priority Activities to Enhance Biodiversity Conservation

- Direct logging activities away from biologically rich and sensitive streams, especially headwaters areas. Where logging activities do occur, ensure that stream buffers are wide enough to reduce siltation and runoff.
- Introduce "true" watershed management for conserving terrestrial and aquatic habitats (reducing logging, protecting riparian areas).
- Enforce stricter control over aquaculture operations to prevent escapes and pollution.
- Enforce strict controls over other exotic species releases.
- Maintain roadless areas.

Conservation Partners

For contact information, please see appendix H.

- American Rivers
- The California Native Plant Society

- Headwaters Environmental Center
- Humboldt State University
- Klamath Forest Alliance
- The Nature Conservancy (California and Oregon Field Offices)
- Northcoast Environmental Center
- Siskiyou Regional Education Project
- The Wildlands Project

Ecoregion Number:	**6**
Ecoregion Name:	**Pacific Central Valley**
Major Habitat Type:	**Temperate Coastal Rivers, Lakes, and Springs**
Ecoregion Size:	**184,129 km²**
Biological Distinctiveness:	**Globally Outstanding**
Conservation Status:	**Snapshot—Critical**
	Final—Critical

Introduction

This ecoregion lies almost entirely within the state of California, encircling the Central Valley and including the western drainages of the Sierra Nevada mountains. A tiny portion also lies just inside the northeastern corner of Nevada. Within the ecoregion, the main freshwater systems are the Sacramento-San Joaquin Rivers, the Pit River, Clear Lake, the Pajaro-Salinas system, and the upper Kern River.

Biological Distinctiveness

Four to five million years ago, when the predecessors of the Sierra Nevada and coastal ranges had been eroded, fishes invaded the Sacramento-San Joaquin region from the ancient Columbia River system. Subsequent mountain-building isolated the fishes, and the Sacramento-San Joaquin became the center for evolution within the larger Central Valley region (Moyle 1976). Today, the Pacific Central Valley [6] is one of the richest ecoregions in North America west of the Rockies in terms of fish species.

In total, the ecoregion supports forty-nine native freshwater fish, including ten anadromous species. The anadromous species are made up of two lampreys, two sturgeons, a smelt, and five salmonids. Fish groups with high representation are lampreys, sturgeons, smelt, salmonids,

cyprinids, and suckers; additionally, there are the threespine stickleback (*Gasterosteus aculeatus*), Sacramento perch (*Archoplites interruptus*), tule perch (*Hysterocarpus traski*), and tidewater goby (*Eucyclogobius newberryi*). The Sacramento perch is the only centrarchid to occur west of the Rockies, and the tule perch is the only freshwater member of the surfperch family (Embiotocidae).

Fourteen fish species are endemic to the ecoregion, and many of these have localized distributions. Species with historically small ranges include the extinct Clear Lake splittail (*Pogonichthys ciscoides*), and the Modoc sucker (*Catostomus microps*) and rough sculpin (*Cottus asperrimus*), the two latter species from the Pit River drainage. The primitive Sacramento perch was once widespread throughout the Sacramento-San Joaquin system but today inhabits a much reduced range because of its inability to compete with introduced centrarchids (Moyle 1976).

Conservation Status

The water of the Central Valley has been channeled, pumped, diverted, and extracted since the gold rush, and the native freshwater biota has suffered substantially. The severity of the problems has been described by Moyle (1976):

> Most of California's major inland waterways today bear little resemblance to the streams and lakes encountered by the first white explorers and settlers . . . the giant lakes of the San Joaquin Valley are today vast grain farms. The Sacramento-San Joaquin Delta, once an enormous tule marsh dissected by meandering river channels, has been transformed into islands of farmland protected by high levees from the water that flows by in straight, dredged channels. Almost every stream of any size has been dammed at least once to control its flow. Thus, it is not surprising that habitat modification is the major cause of the changes in California's fish fauna.

These changes have resulted in the extinction of native forms (the Clear Lake splittail and thicktail chub, *Gila crassicauda*) and the precarious situation of others—eight additional fish species are state or federally listed. Other species at risk include the California freshwater shrimp

(*Syncaris pacifica*), giant garter snake (*Thamnophis couchii gigas*), limestone salamander (*Hydromantes brunus*), and Shasta crayfish (*Pacifastacus fortis*) (Steinhart 1990).

Stream-channel alterations, dam and reservoir construction, dewatering of streams and lakes, pollution, and watershed changes have touched nearly every freshwater habitat in the ecoregion (Moyle 1976). Modification of freshwater environments has opened the door to introduced species that are better suited to the new conditions. At least thirty-six species of non-native fish now inhabit the Pacific Central Valley [6] ecoregion, and they exist in far greater numbers than native species. The majority of these introduced species were intentionally introduced from eastern North America as game fish. The Shasta crayfish may lose its battle with two exotic species of crayfish introduced into the Pit River drainage (Steinhart 1990)

Suite of Priority Activities to Enhance Biodiversity Conservation

*   Promote whole watershed conservation and restoration to preserve both terrestrial and aquatic habitats.
*   Consider removal of dams and restoration of natural flow regimes to the region. Pursue dam re-operation through FERC relicensing.
*   Enforce strict pollution controls on urban and agricultural polluters.
*   Respond to rapid growth and development with land use and water conservation plans (American Rivers 1999).

Conservation Partners

For contact information, please see appendix H.

*   California Department of Fish and Game
*   California Native Grass Association
*   California Native Plant Society
*   National Audubon Society
*   The Nature Conservancy (California Field Office)
*   Sacramento River Preservation Trust
*   Sierra Club
*   U.S. Bureau of Land Management

Ecoregion Number:	**7**
Ecoregion Name:	**South Pacific Coastal**
Major Habitat Type:	**Xeric-Region Rivers, Lakes, and Springs**
Ecoregion Size:	**170,320 km²**
Biological Distinctiveness:	**Continentally Outstanding**
Conservation Status:	**Snapshot—Endangered**
	Final—Critical

Introduction

This coastal ecoregion begins just south of Monterey, California, and encompasses southwestern California and the entire Baja Peninsula in Mexico. One of the few major rivers is the Cuyama River.

Biological Distinctiveness

The South Pacific Coastal [7] ecoregion, sometimes grouped with the Sonoran Desert [61] ecoregion, is characterized by extreme aridity. There are almost no permanent watercourses in Baja California, and most of the endemism of the region is found in the Los Angeles Basin, in California. There are five known endemic fish species in the ecoregion, and one endemic snake, *Thamnopis hammondii*. There are no known endemic crayfish. Endemic fish that occur in the Los Angeles Basin include the Santa Ana sucker (*Catostomus santaanae*), Santa Ana speckled dace (*Rhinichthys osculus* ssp.), tidewater goby (*Eucyclogobius newberryi*), and California killifish (*Fundulus parvipinnis*). The arroyo chub (*Gila orcutti*) is native to Malibu Creek and to the Santa Margarita River drainage (Page and Burr 1991).

This ecoregion is the southernmost range for some anadromous fish, such as steelhead (*Oncorhynchus mykiss*) and the Pacific lamprey (*Lampetra tridentata*).

Conservation Status

The ecology of the South Pacific Coastal [7] ecoregion is under pressure from several sources. Perhaps the most severe is the diversion of the meager water resources of the area from natural channels to provide water for metropolitan centers such as Los Angeles, San Diego, Tijuana, and Ensenada. Other ecological pressures include overgrazing and resultant loss of native riparian vegetation, and habitat loss from pollution of springs.

Suite of Priority Activities to Enhance Biodiversity Conservation

- Provide immediate protection for the remaining freshwater habitats of the Los Angeles Basin, including protecting water quantity and natural flow regimes.
- Provide immediate protection for coastal lagoons, habitat for killifish and gobies.
- Control grazing in the watershed.
- Manage the region on a watershed basis.

Conservation Partners

For contact information, please see appendix H.

- Centro de Investigaciones Biológicas del Noreste, S.C. (CIBNOR)
- Desert Fishes Council
- Friends of the Los Angeles River
- The Nature Conservancy (Headquarters and California Field Office)
- Universidad Autónoma de Baja California (UABC)
- Universidad Autónoma de Baja California Sur (UABCS)
- Universidad Nacional Autónoma de México (UNAM)
- World Wildlife Fund México

Ecoregion Number:	**8**
Ecoregion Name:	**Bonneville**
Major Habitat Type:	**Endorheic Rivers, Lakes, and Springs**
Ecoregion Size:	**155,863 km²**
Biological Distinctiveness:	**Continentally Outstanding**
Conservation Status:	**Snapshot—Vulnerable**
	Final—Endangered

Introduction

This ecoregion corresponds to the Bonneville Basin, the single largest interior drainage in the Great Basin (Minckley et al. 1986). The ecoregion occupies much of western Utah with extensions into eastern Nevada, southeastern Idaho, and a small portion of southwestern Wyoming.

The Butte and Pequop Mountains constitute the western boundary of this ecoregion, while in Utah the eastern boundary lies west of the Wasatch Plateau and north of Dixie National Forest. The Sevier and Bear are the two largest rivers in the ecoregion. Major lakes include the Great Salt Lake, Bear Lake, Utah Lake, and Sevier Lake, though the Great Salt Lake cannot be considered a true freshwater habitat because of its prohibitive salinity.

Biological Distinctiveness

Much of the Bonneville [8] ecoregion's distinctive biodiversity is harbored in its lakes. Bear Lake, located high in the mountains near the borders of Utah, Idaho, and Wyoming, has four endemic fish species, all remnants of the Pleistocene-era Lake Bonneville fauna—the Bear Lake sculpin (*Cottus extensus*), Bear Lake whitefish (*Prosopium abyssicola*), Bonneville whitefish (*P. spilonotus*), and Bonneville cisco (*P. gemmiferum*). Utah Lake, located southeast of the Great Salt Lake and distinguished as the largest freshwater lake west of the Mississippi, historically supported two endemic fish, also Lake Bonneville remnants—the June sucker (*Chasmistes liorus*), found also in the lake's tributaries, and the now-extinct Utah Lake sculpin (*Cottus echinatus*). The Bonneville Basin shares the leatherside chub (*Gila copei*) and the Utah sucker (*Catostomus ardens*) with the Upper Snake ecoregion [4], but the species can be considered endemic to both ecoregions. Finally, the Bonneville [8] ecoregion is home to the endemic least chub (*Iotichthys phlegethontis*), historically found in the Great Salt Lake marshes, in streams along the Wasatch Front, and in the Great Salt Desert Springs; and to two subspecies of speckled dace (*Rhinichthys osculus adobe* and *R. o. carringtoni;* Minckley et al. 1986; Sigler and Sigler 1994). Only two species of mussel and one species of crayfish, *Pacifastacus gambelii*, are native to this ecoregion, and neither is endemic (Johnson 1986).

Conservation Status

Like the other endorheic ecoregions of the Great Basin, the Bonneville [8] ecoregion's freshwater species are highly susceptible to habitat modification and introduced species. The least chub's distribution has been reduced to two locations in the basin, a result of predation by introduced fishes, bullfrogs, and birds as well as habitat destruction; the species is now protected in Utah. The June sucker presents an interesting case, since the form (*Chasmistes liorus liorus*) from which the species was described went extinct in the 1930s when its Utah Lake habitat was devastated by drought and agricultural practices. Today, a cross between the Utah sucker (*Catostomus ardens*) and *C. liorus liorus* is present in Utah Lake and is listed as endangered. The fish of Bear Lake do not appear to be imperiled at present.

The greatest threats facing this arid ecoregion's freshwater habitats and biota may be water diversions and groundwater pumping. According to American Rivers (1999), Salt Lake City's over-consumption of water is one of the single largest threats, particularly to the Bear River.

Suite of Priority Activities to Enhance Biodiversity Conservation

- Purchase water rights where possible.
- Reduce pumping of groundwater.
- Reduce agricultural consumption of water.
- Guard against introduction of new exotic species.

Conservation Partners

For contact information, please see appendix H.

- American Rivers
- Desert Fishes Council
- The Nature Conservancy (Idaho, Nevada, and Utah Field Offices)
- Sierra Club
- Southern Utah Wilderness Alliance
- Utah Rivers Council

Ecoregion Number:	**9**
Ecoregion Name:	**Lahontan**
Major Habitat Type:	**Endorheic Rivers, Lakes, and Springs**
Ecoregion Size:	**213,566 km²**
Biological Distinctiveness:	**Continentally Outstanding**
Conservation Status:	**Snapshot—Endangered**
	Final—Critical

Introduction

This ecoregion occupies almost the entire state of Nevada, portions of eastern California east of the Sierra Nevada Mountains, and two small disjunct areas of southeastern Oregon. Major watersheds include those of the Humboldt, Walker, Carson, and Truckee Rivers, all of which flow east and originate as high-elevation, high-gradient streams. Large lakes in this ecoregion include Lake Tahoe on the Nevada-California border, Mono Lake to the east of Yosemite in California, Eagle Lake to the east of Lassen National Forest in California, Pyramid Lake on the Pyramid Lake Indian Reservation in western Nevada, and Walker Lake just south of the Walker River Indian Reservation in western Nevada. Both Pyramid and Walker Lakes are remnants of Lake Lahontan, a late-Pleistocene-era lake that at its maximum size approximated that of Lake Erie. Lake Tahoe and Eagle Lake are both surrounded by coniferous forests, in contrast with the other desert lakes (Minckley et al. 1986). Lake Tahoe is also distinguished by being one of the largest high-mountain lakes in the world, with a surface area of 304 km², a maximum depth of 501 m, and an altitude of 1,899 m above sea level (Moyle 1976). Numerous small playa lakes, as well as both warm and cold springs, are also within the ecoregion. The ecoregion is characterized by extreme aridity that has reduced the native freshwater fauna to a handful of species from a much greater number in pluvial times (Hubbs et al. 1974).

One endorheic basin of note is Railroad Valley, located in central Nevada and encompassing approximately 9,233 km². Within Railroad Valley, a series of cool and thermal springs discharges relatively high volumes of water, creating in some cases large spring pools and outflow creeks (Williams et al. 1985).

Mono Lake, on the border with the Death Valley ecoregion [11], represents a rare habitat type. It is too saline to support fish but contains abundant populations of the alkali fly (*Ephydra hians*) and an endemic brine shrimp (*Artemia monica*), which in turn support a diverse migratory bird fauna (Moyle 1976).

Biological Distinctiveness

The harsh conditions of the Lahontan [9] ecoregion, with its few perennial freshwater habitats that become torrents during infrequent storms, have excluded all but the hardiest and most adaptable fish species. Those that remain have, in some cases, become extraordinarily differentiated within isolated habitats; for instance, six subspecies each of tui chub (*Gila bicolor*) and five of speckled dace (*Rhinichthys osculus*) are recognized in the ecoregion (Hubbs et al. 1974; Williams et al. 1985; Sigler and Sigler 1994). Other endemic fish are the Lahontan redside (*Richardsonius egregius*), cui-ui (*Chasmistes cujus*), desert dace (*Erimichthys acros*), and Tahoe sucker (*Catostomus tahoensis*) (Moyle 1976; Page and Burr 1991; Sigler and Sigler 1994). The cui-ui was historically found only in Pyramid and Winnemucca Lakes, but today the latter is dry (Page and Burr 1991). The desert dace, a relict species found only in thermal spring habitats of Soldier Meadow, Nevada, is distinguished by its ability to inhabit much hotter water than any other minnow species. Soldier Meadow is also home to at least four undescribed species of hydrobiid springsnails and an endemic plant, the basalt cinquefoil (*Potentilla basaltica*) (Vinyard 1996).

Railroad Valley supports a highly endemic fauna, with five subspecies of tui chub (*Gila bicolor* ssp.), the Railroad Valley springfish (*Crenichthys nevadae*), and at least five hydrobiid snails, all found nowhere else. Each of the tui chub subspecies is restricted to single localities (Kate Spring, Butterfield Spring, Blue Eagle Spring, Bull Creek, Green Springs, and Duckwater Creek), with no apparent overlap. A sixth subspecies of tui chub is shared with two other nearby valleys (Williams et al. 1985).

Conservation Status

The status of the Lahontan [9] ecoregion's freshwater species is similar to that of other endorheic desert ecoregions in North America: a combination of flow modification, overgrazing of riparian habitats, and introduced species has led to the extinction and extirpation of a number of native forms and seriously threatens those that remain. The highly localized endemic species, in particular, have no escape from exotics and no alternate home when their habitats dry up. The rare ecosystem of Mono Lake is imperiled by lowered water levels, a consequence of water diversions for consumption by Los Angeles residents. With no new freshwater source, evaporative loss has lowered water levels by about a foot per year since the diversion of the lake's tributary streams fifty years ago. As a result, salt concentrations have increased, making the lake unsuitable for its historic flocks of migratory birds (Hart 1996).

The cui-ui is now restricted to Pyramid Lake and is federally listed as endangered. The population made spawning runs up the Truckee River until 1905, when Derby Dam was built 38.5 miles upstream of the lake and began diverting most of the river water to the Carson River Basin. In low-water years, the water level of Pyramid Lake was so diminished that fish were unable to traverse the resulting delta at the mouth of the river, and by 1938 the population had virtually disappeared (Sigler and Sigler 1994; Page and Burr 1991). Lake Winnemucca, which received its water from Pyramid Lake, had dried up by 1938 as well, when the surface elevation of Pyramid Lake dropped below Winnemucca's inlet level. Similarly, water levels in Walker Lake have declined as a result of water diversions (Minckley et al. 1986). In each of these cases, lower water levels lead to increased salinity.

The desert dace is listed as threatened, its decline a result of habitat degradation (grazing in riparian areas by cattle, feral horses, and burros; recreational use of the Soldier Meadow area; diversion of water to irrigation ditches and associated impoundments) and introductions of exotic fishes (Sigler and Sigler 1994; Vinyard 1996). Likewise, various endemic subspecies of tui chub and speckled dace are threatened by exotics and habitat modification (the Independence Valley tui chub—*G. b. mohavensis*—is extinct, probably from predation by largemouth bass; Sigler and Sigler 1994).

According to one assessment (Williams et al. 1985), all the endemic Railroad Valley freshwater fish species are rare, vulnerable, or their status indeterminate. The six subspecies of tui chub are variously threatened by exotic goldfish (*Carassius auratus*), carp (*Cyprinus carpio*), channel catfish (*Ictalurus punctatus*), and guppies (*Poecilia reticulata*); channelization; overgrazing of riparian zones; diversions and groundwater withdrawal; and construction of aquaculture projects for channel catfish. Future threats include petroleum development and agricultural expansion (Williams et al. 1985).

Suite of Priority Activities to Enhance Biodiversity Conservation

- Protect springs and associated pools and out-flow creeks in Railroad Valley from further modifications and from surrounding land development, and guard springs with no exotics from future introductions.
- To protect the desert dace, the springs and flowing waters in Soldier Meadow must be protected from overgrazing, and exotic species must not be introduced. Water should also be restored to natural stream channels to improve habitat.
- Purchase water rights where possible.
- Restrict groundwater pumping.
- Regulate grazing in riparian zones.
- Work with Native American stakeholders, such as the Pyramid Lake Paiute Tribe, to develop and implement water quality models and standards.

Conservation Partners

For contact information, please see appendix H.

- Desert Fishes Council
- Mono Lake Committee
- The Nature Conservancy (California and Nevada Field Offices)
- Pyramid Lake Paiute Tribe
- Sierra Club
- The Wilderness Society

Ecoregion Number:	**10**
Ecoregion Name:	**Oregon Lakes**
Major Habitat Type:	**Endorheic Rivers, Lakes, and Springs**
Ecoregion Size:	**47,925 km²**
Biological Distinctiveness:	**Nationally Important**
Conservation Status:	**Snapshot—Vulnerable**
	Final—Vulnerable

Introduction

This ecoregion covers inland basins of central-southern Oregon, a small part of northeastern California, and parts of northwestern Nevada. These desert basins are largely defined by alkaline lakes. Large lakes include Goose Lake,

which straddles the California-Oregon border, and Lake Abert, Summer Lake, and Harney Lake in Oregon. There are numerous smaller lakes, several of which are far more important to the ecoregion's biological distinctiveness.

Biological Distinctiveness

This ecoregion, like other desert areas, has low numbers of freshwater species but relatively high endemism. Of the ecoregion's eight native fish species, three are endemic. These are the Alvord chub (*Gila alvordensis*), limited to the warm waters of the Alvord Basin in Oregon; the Borax Lake chub (*G. boraxobius*), a dwarf species found in Borax Lake, Lower Borax Lake, and associated ponds and marshes, together comprising approximately 260 hectares; and the Warner sucker (*Catostomus warnerensis*), found in ephemeral lakes, sloughs, lower-gradient streams, and three permanent lakes in the Warner Basin in Oregon and Nevada (Sigler and Sigler 1994; Williams 1995a, 1995b). There is also a native redband trout subspecies (*Oncorhynchus mykiss* ssp.).

This Great Basin ecoregion also includes a number of restricted-range freshwater mollusks, including the turban pebblesnail (*Fluminicola turbiniformis*), which is dependent on oligotrophic springs. Known endemic snails associated with the alkaline lakes of this ecoregion include the lamb rams-horn (*Planorbella oregonensis*) and Harney Lake springsnail (*Pyrgulopsis hendersoni*). There are likely others as well (Frest and Johannes 1995).

Conservation Status

The threats that imperil the biota of other Great Basin ecoregions—water diversion, poor land-use practices, and exotic species—are also present in the Oregon Lakes [10] ecoregion. The Warner sucker was once abundant and widely distributed but is now listed as federally threatened. Spawning migrations have been cut off by many small agricultural diversion dams in the Warner Lake Basin, and introduced brown bullhead (*Ameiurus nebulosis*) and crappie (*Pomoxis* spp.) are now abundant in the lakes instead (Williams 1995a). Drought since the late 1980s has placed an additional stress on remaining fish. As part of recovery actions recommended for the species, fencing to restore riparian vegetation has been erected, ephemeral lake habitat has been acquired, and a fishway over a diversion dam on Twentymile Creek was built (Williams 1995a). A number of adult fish have also been transplanted to nearby Summer Lake Wildlife Management Area.

The Borax Lake chub is federally listed as endangered, a designation that resulted from proposed geothermal energy development near Borax Lake. The species is also threatened by damage to the fragile lake shoreline by overgrazing and off-road vehicle use, and by apparently natural fluctuations in the lake temperature that exceed the species' critical thermal maximum (Williams 1995b). Today, protection of Borax Lake by a TNC land lease offers the species partial protection. The endemic Alvord chub is considered imperiled by TNC because of habitat degradation and loss (Sigler and Sigler 1994; Natural Heritage Center Databases 1997).

Suite of Priority Activities to Enhance Biodiversity Conservation

- Permanently acquire and protect 260 hectares of private lands designated as critical habitat for the Borax Lake chub, and close this area to vehicle, livestock, mining, and energy development activities (Williams 1995b).
- Restore Lower Borax Lake and adjacent marshes (Williams 1995b).
- Construct additional fishways to allow movement of fish over dams (Williams 1995a).
- Identify and remove deteriorating dams (Williams 1995a).
- Restore natural stream channels (Williams 1995a).

Conservation Partners

For contact information, please see appendix H.

- The Nature Conservancy (Oregon Field Office)
- Oregon Lakes Association
- Oregon Trout

Ecoregion Number:	**11**
Ecoregion Name:	**Death Valley**
Major Habitat Type:	**Endorheic Rivers, Lakes, and Springs**
Ecoregion Size:	**80,707 km²**
Biological Distinctiveness:	**Continentally Outstanding**
Conservation Status:	**Snapshot—Vulnerable**
	Final—Vulnerable

Introduction

The Death Valley [11] ecoregion, comprising the southwest corner of the Great Basin, is an endorheic basin defined by the drainages of the Owens, Amargosa, and Mojave Rivers (Sada et al. 1995). This ecoregion dominates central-southern California and reaches into southwestern Nevada. Included in the ecoregion are the eastern slopes of the southern portion of the Sierra Nevada Mountains and the northeastern slopes of the Transverse Range. This area has few large lakes or perennially flowing rivers and streams, but abundant springs rising along faults provide much of the habitat available to freshwater species. Water flowing from these springs is 8,000 to 12,000 years old and originates in southern and eastern Nevada (Minckley et al. 1986). This ecoregion contains some of the most extreme conditions inhabited by freshwater life and its biota has been extensively studied.

Ash Meadows, covering an area of about 756 km², is of particular interest, as its more than thirty springs and seeps create an oasis in the middle of the desert (Williams et al. 1985). Devils Hole is the highest in elevation of these springs, at 732 m. With increasing elevation, springs have been isolated from each other for a longer time, and springs only a kilometer apart may have been isolated for thousands of years (Williams et al. 1985). In this arid area groundwater recharge is so slow that the aquifers supplying springs such as Devils Hole contain fossil water (Pister 1990).

Biological Distinctiveness

Like other arid ecoregions of North America, the Death Valley [11] ecoregion exhibits low species diversity but extraordinarily high endemism. Within the ecoregion, there are twenty-two endemic mollusks (all snails), six endemic aquatic insect species, and six endemic fish species. The number of endemic fish forms is increased to nineteen if subspecies are included, which some authors argue is appropriate given the degree of differentiation observed in forms associated with distinct, persistent water bodies (Minckley et al. 1986; Sada et al. 1995). With the exception of the fish in the Owens River and Mojave River Basins, all of the endemic freshwater species in this ecoregion are associated with springs or spring margins. Given the minute amount of fresh water available in this ecoregion, its biodiversity is truly impressive.

The endemic fish come from four families, with subspecies of two minnows (*Rhinichthys osculus* and *Gila bicolor*), one sucker species (*Catostomus fumeiventris*), two springfish (the extinct *Empetrichthys merriami* and the extirpated *E. latos*), and four pupfish (*Cyprinodon radiosus*, *C. diabolis*, *C. nevadensis*, and *C. salinus*). Of these, five subspecies of *C. nevadensis* and two of *C. salinus* are recognized, as are three subspecies of speckled dace (*R. osculus*) (Sada et al. 1995). The Devils Hole pupfish (*C. diabolis*) is distinguished by having the smallest range of any vertebrate species—23 yd^2 in a spring-fed, limestone cavern in Ash Meadows (Williams et al. 1985; Sada et al. 1995). The species is tiny, rarely exceeding 20 mm standard length, and populations fluctuate seasonally from 150 to 400 individuals (Williams et al. 1985). The two forms of the Salt Creek pupfish (*C. salinus*) live between 180 and 240 ft below sea level, where temperatures can reach 130°F (Sigler and Sigler 1994).

Conservation Status

The USFWS lists nine of the ecoregion's fish species and subspecies as endangered, and the American Fisheries Society has proposed addition of a tenth (Sigler and Sigler 1994). Many of these species gain their status from being found in only a handful of locations, in habitats that are extremely vulnerable to disturbance.

In Ash Meadows, diversion, channelization, impoundments, flow reduction, elimination of riparian habitat, and dredging of spring systems have destroyed much of the native freshwater habitat. Native species also suffer from competition and/or predation from the exotic sailfin molly (*Poecilia latipinna*), mosquitofish (*Gambusia affinis*), bullfrogs (*Rana catesbeiana*), and crayfish (*Procambarus clarkii*). The Oriental snail (*Melanoides tuberculatus*) also threatens native snail populations (Williams et al. 1985).

The Devils Hole pupfish, known from a single population in Ash Meadows, is threatened by groundwater pumping and resultant lowered water levels. Following a 1976 U.S. Supreme Court decision, water levels have been maintained to protect the pupfish, and Devils Hole is now part of the Ash Meadows National Wildlife Refuge. Two transplanted populations of the species have been established to protect against loss of the species to a catastrophic event, but one of the populations has undergone morphometric change (Williams et al. 1985; Pister 1990). In a similar situation, the Owens pupfish (*C. radiosus*) was saved from its desiccating pond when approximately 400 individuals were transplanted to a temporary refugium, later to be introduced into the Owens Valley Native Fish Sanctuary (Pister 1990).

The Ash Meadows poolfish (*E. merriami*) was known from five separate springs and apparently went extinct due to interactions with introduced species. Three subspecies of the Pahrump poolfish (*E. latos*) were once found in springs of the Pahrump Valley, in Nye County, Nevada; groundwater pumping eliminated the habitat of two, and the third (*E. latos latos*) was transplanted to a spring outside its native range before the destruction of its habitat as a result of mosquito-control measures (Sigler and Sigler 1994). Efforts to restore the native habitat have been unsuccessful, but as of 1990 the transplanted population was stable (Pister 1990). As the sole remaining representative of its genus, the Pahrump poolfish is an important conservation target.

Suite of Priority Activities to Enhance Biodiversity Conservation

- Protect remaining habitat of endemic fish, mollusks, and insects from desiccation, pollution, and other forms of disturbance.
- Restrict groundwater pumping in sensitive areas.
- Guard against the introduction of non-native species and control existing populations of exotics.
- Conduct *ex situ* breeding programs for species found in single locations.
- Conduct ecological and taxonomic studies of the invertebrate fauna (Williams et al. 1985).
- Restore damaged spring and marsh areas (Williams et al. 1985).

- Restrict residential or commercial development.
- Purchase water rights where possible.

Conservation Partners

For contact information, please see appendix H.

- Desert Fishes Council
- The Nature Conservancy (California and Nevada Field Offices)
- The Wilderness Society

Ecoregion Number: **12**
Ecoregion Name: **Colorado**
Major Habitat Type: **Large Temperate Rivers**
Ecoregion Size: **507,245 km²**
Biological Distinctiveness: **Continentally Outstanding**
Conservation Status: **Snapshot—Critical**
Final—Critical

The Colorado is probably the most utilized, controlled, and fought over river in the world. It flows through lands of incomparable beauty and includes nearly seven percent of the nation's contiguous land mass, including parts of seven states. From the time of the early settlers to the present, the water of the Colorado River has been the key to development of the arid region.
—Crawford and Peterson (1974)

Introduction

Stretching from southwestern Wyoming to the northeastern tip of Baja California in Mexico, this ecoregion covers a portion of southeastern Nevada, parts of western and northern Arizona, the northwestern corner of New Mexico, most of eastern Utah, and western Colorado. The ecoregion primarily lies within the physical province of the Colorado Basin and includes portions of the Wyoming Basin and the Sonoran Desert, making it one of the most arid ecoregions in North America (Minckley et al. 1986).

The drainage area of the Colorado River, which flows for 2,282 km, is typically divided into an upper and a lower basin. The major tributaries to the upper Colorado River are the Green, Gunnison, Dolores, and San Juan

Rivers, and those in the lower basin are the Little Colorado, Virgin, Bill Williams, and Gila Rivers (Williams et al. 1985). The Gila River contains a biota distinct enough to warrant a separate ecoregion [14], as does the Virgin River [13] ecoregion, which includes the now-disjunct White River.

Historically, this ecoregion's freshwater habitats were dominated by warm, silt-laden rivers with highly variable flows. These large rivers were fed by cold, clear montane creeks as well as springs. Mountain lakes have also supplied cold-water habitats. At the other extreme, habitats of the Colorado Delta were characterized by extremely high and variable tides, temperatures, salinities, and siltation loads (Minckley et al. 1986).

Biological Distinctiveness

Because of its long isolation from neighboring river systems, the Colorado [12] ecoregion is not rich in freshwater species; it has a total of twenty-six fish species (Behnke and Benson 1983). However, many of its endemic fish species are so distinct in form that they are instantly recognizable. The assemblage of large-river fish species historically found in the Colorado and its main tributaries (including, in some cases, the Gila) is truly extraordinary. The humpback chub (*Gila cypha*), bonytail (*G. elegans*), Colorado squawfish (*Ptychocheilus lucius*), roundtail chub (*G. robusta*), and razorback sucker (*Xyrauchen texanus*) all display morphological adaptations for life in turbid, fast-flowing habitats. The Colorado squawfish is of particular interest as a top carnivore and the largest cyprinid in North America. Other large-river endemic or near-endemic fishes include the flannelmouth sucker (*Catostomus latipinnis*), Little Colorado River sucker (*C.* sp.), Sonora sucker (*C. insignis*), and desert sucker (*C. clarki*), as well as subspecies of cutthroat trout (*Oncorhynchus clarki stomias*, *O. c. pleuriticus*) and bluehead sucker (*C. discobolus yarrowi*) (Page and Burr 1991, Sigler and Sigler 1994).

Rather than relatively large-bodied minnows and suckers, this ecoregion's streams and creeks support a suite of endemic fishes and other taxa adapted to small freshwater habitats. The Apache trout (*Oncorhynchus apache*) was historically found in the clear, cool mountain headwaters and lakes in the upper Salt River and Little

Colorado River systems (Page and Burr 1991). The desert pupfish (*Cyprinodon macularius*) is endemic to springs, marshes, slow-flowing streams, and backwater areas along large rivers in the Lower Colorado River Basin (Williams et al. 1985). The Little Colorado spinedace (*Lepidomeda vittata*), found only in creeks and small rivers of eastern Arizona, shares its genus with just two other species in the neighboring Vegas-Virgin [13] ecoregion (Page and Burr 1991). The Las Vegas dace (*Rhinichthys deaconi*), now extinct, was historically found in springs along Las Vegas Creek, a small tributary to the Colorado River in Nevada. The extinct Las Vegas frog (*Rana fisheri*) was found in springs, marshes, and creeks in the same area (Williams et al. 1985; Sigler and Sigler 1994). Springs in this ecoregion also support endemic spring snails, including the Overton assiminea (*Assiminea* sp.) and Grand Wash springsnail (*Fontelicella* sp.), found in separate springs in the vicinity of Lake Mead (Williams et al. 1985). Like the ecoregion's large river habitats, these springs and associated small freshwater habitats also support a number of endemic subspecies, such as the Kendall Warm Springs dace (*Rhinichthys osculus thermalis*), found only in a small tributary to the Green River (Williams et al. 1985).

Some forty marine fish species have been confirmed in the Colorado Delta or farther upstream, and an additional thirty species are reported in the historical literature (Minckley et al. 1986).

Conservation Status

Gross habitat change in combination with introductions of exotic species (Pister 1990) have devastated the Colorado River biota. The river was historically a warm, silt-laden stream with major variations in flow, but today a series of dams and diversions has placed most of the river's water in twenty huge, cold-water reservoirs into which more than fifty exotic fish species have been introduced (Pister 1981). Whereas flow in the Colorado naturally varied with the seasons and precipitation, today it is stabilized and fluctuates unseasonally as a result of water diversions and dam operations (Williams et al. 1985). In most years, water diversions from the impoundments result in the total cessation of flow out of the Colorado, and flows reach the Gulf of California only in

extremely wet years when reservoirs become overfull. The water that is released from the deep reservoirs generally comes from the hypolimnion (the cold, lower layer), resulting in cold temperatures downstream that favor cold-water exotics (Williams et al. 1985). The reservoirs also trap sediment and dissolved nutrients that would naturally flow downstream to the Colorado Delta, where riparian forests and scrublands have been replaced by bare hypersaline plains (Minckley et al. 1986). An additional threat is pollution from historic and current gold mining and coal and gas development in the basin.

Virtually every Colorado [12] ecoregion endemic fish species is considered imperiled. The federally endangered razorback sucker, once found throughout the upper and lower basins but now restricted to river segments not affected by dams, is seriously threatened by hybridization and predation of young by introduced fishes (Williams et al. 1985; Pister 1990). The bonytail and humpback chub, both federally endangered, have been severely affected by dam construction, exotic species, water diversion, pollution, and hybridization with other *Gila* species (Williams et al. 1985). Small stocks of both are being maintained at the Dexter National Fish Hatchery in New Mexico (Pister 1990). The federally endangered Colorado squawfish, once found in the Colorado River and its tributaries throughout the upper and lower basins into Mexico, today has an extremely truncated range as a result of dam construction and associated habitat loss, blockage of spawning and migration routes by dams, lowered water temperatures below impoundments, and interactions with exotic species (Behnke and Benson 1983; Williams et al. 1985). The federally endangered desert pupfish is reduced to three introduced populations in Arizona and a single native population in Santa Clara Slough, Mexico; original populations were reduced by habitat desiccation and interactions with exotics (as of 1985, an exotic tilapia appeared to be replacing portions of the Mexico population). The Las Vegas dace and Las Vegas frog both went extinct in the 1950s as a result of increased water use and development in the Las Vegas metropolitan area (Williams et al. 1985).

Many of the impacts that have produced this grim state of affairs appear to be irreversible. Exotic species, once established, are virtually impossible to remove, and the effects of hybridization are basically permanent. Large-scale impoundments, such as the Hoover and Glen Canyon Dams, are unlikely to be removed. For these reasons, the conservation status of the Colorado region is assessed as critical. However, carefully implemented, localized actions, such as dam reoperation to create flows that resemble the natural hydrograph, may serve to mitigate the impacts of severe disturbance on the biota.

Suite of Priority Activities to Enhance Biodiversity Conservation

*   Maintain remaining free-flowing river areas and lobby for reoperation of existing dams so that flow resembles the natural hydrograph.
*   Actively control existing exotic species, especially to minimize hybridization with native trout and minnows, and guard against the introduction of any new species.
*   Prevent construction of the Animas-La Plata Project, which would impound and dewater the free-flowing Animas River (a tributary to the San Juan), dry up riparian wetlands, and likely increase selenium loading to freshwater habitats as a result of crop irrigation (American Rivers 1997).
*   Clean up historic gold-mining sites that continue to pollute Colorado Basin waters, such as the sites near Eagle on the Eagle River (a tributary to the upper Colorado River), along the Uncompahgre River from the San Juan Mountains south of Ouray, and on the San Miguel River (a tributary of the Dolores) (Colorado Plateau Forum 1998).
*   End federal subsidies of irrigation and municipal water taken from the lower Colorado River (American Rivers 1997).

Conservation Partners

For contact information, please see appendix H.

*   American Rivers
*   Colorado Rivers Alliance
*   Defenders of Wildlife
*   Desert Fishes Council
*   Environmental Defense Fund
*   Friends of the Animas River
*   Grand Canyon Trust

*   The Nature Conservancy (Headquarters, and Arizona, California, Colorado, New Mexico, Utah, and Wyoming Field Offices)
*   Sierra Club, Rocky Mountain Chapter
*   Southwest Center for Biological Diversity
*   World Wildlife Fund México

Ecoregion Number:	**13**
Ecoregion Name:	**Vegas-Virgin**
Major Habitat Type:	**Xeric-Region Rivers, Lakes, and Springs**
Ecoregion Size:	**34,565 km²**
Biological Distinctiveness:	**Continentally Outstanding**
Conservation Status:	**Snapshot—Vulnerable**
	Final—Endangered

Introduction

This ecoregion is centered in southeastern Nevada and also occupies the extreme northwestern corner of Arizona and the southwestern corner of Utah. Though nestled between ecoregions of the Great Basin, this ecoregion is part of the Colorado complex, as its waters historically drained into the Colorado River. The main watersheds are those of the White River, Moapa River, Meadow Valley Wash, and Virgin River. The White River was a tributary to the Colorado River during pluvial times but is now dry for much of its length. The Virgin River naturally flows into the Colorado but today flows into Lake Mead, as does the Moapa River (Williams et al. 1985). Many of the streams are ephemeral, and the more permanent habitats are associated with springs.

Three spring-fed habitats stand out in this ecoregion for their biotic distinctiveness and diverse habitats. The upper White River basin is largely desiccated, but a short streamcourse is fed by a series of springs and outflow creeks, each with its own characteristic physiochemistry and temperature. Pahranagat Valley, located farther south along the intermittent White River, is distinguished by Hiko, Crystal, and Ash Springs and their outflow creeks. The 40-km-long Moapa River flows from more than twenty warm springs, traverses through blackbrush and creosote communities, and supports important riparian zone biota (Williams et al. 1985).

Biological Distinctiveness

Like the ecoregions of the Great Basin that surround it, the Vegas-Virgin [13] ecoregion is characterized by low freshwater species richness but high endemism, especially at the subspecies level. Endemic fish historically present in the ecoregion's freshwater habitats are the White River spinedace (*Lepidomeda albivallis*) of the upper White River; the Virgin River spinedace (*L. mollispinis mollispinis*) of the Virgin River; the Big Spring spinedace (*L. mollispinis pratensis*) of Big Spring in Meadow Valley Wash; the Moapa speckled dace (*Moapa coriacea*) in the Moapa River headwaters; the woundfin (*Plagopterus argentissimus*) of the Virgin and Gila River Basins; the extinct Las Vegas dace (*Rhinichthys deaconi*) in springs and outflows along Las Vegas Creek; the Moorman springfish (*Crenichthys baileyi thermophilus*) in the warm springs of the White River and Moapa River Basins; and the Preston springfish (*C. b. albivallis*), White River springfish (*C. b. baileyi*), Hiko springfish (*C. b. grandis*), and Moapa springfish (*C. b. moapae*); This list accounts for six of the eleven native species that are endemic. Endemic subspecies of roundtail chub (*Gila robusta seminuda*), speckled dace (*Rhinichthys osculus velifer*), and desert sucker (*Catostomus clarki intermedius*) are recognized as well (Williams et al. 1985; Page and Burr 1991; Sigler and Sigler 1994).

The warm springs of this ecoregion also support an endemic invertebrate fauna that includes the Pahranagat pebblesnail (*Fluminicola merriami*), White River tryonia (*Tryonia clathrata*), Moapa riffle beetle (*Stenelmis calida moapa*), Moapa naucorid (*Usingerina moapensis*), Moapa pebblesnail (*Fluminicola avernalis*), Hot Creek pebblesnail (*Fluminicola* sp.), and an unnamed springsnail (*Fontelicella* sp.). These endemic invertebrates are restricted to spring systems in this ecoregion (Williams et al. 1985).

Conservation Status

Because nearly all of the Vegas-Virgins [13] ecoregion's freshwater species are endemic, and many are found in single localities, their rarity puts them at high risk from the many threats imposed on this desert ecoregion. To date, the most significant disturbance may have been the creation of Lake Mead by the Hoover Dam on the Colorado River. Because the lake extends thirty-five miles up the Virgin River, it has

destroyed precious river habitat and blocked connections between the Moapa River, the Virgin River, and the Colorado. Utah's Washington County, through which more than half of the Virgin River flows, is one of the United States' fastest-growing counties and is putting extreme pressure on the river's water resources and riparian habitat (Grand Canyon Trust 1997). Other more insidious disturbances have taken their toll on this ecoregion's species and habitats as well.

In the Pahranagat Valley, the Pahranagat roundtail chub (*Gila robusta jordani*), a federally endangered subspecies, is limited to one of its original three spring habitats as a result of water diversions, channelization, and interactions with exotic species (particularly carp, *Cyprinus carpio*; shortfin molly, *Poecilia mexicana*; mosquitofish, *Gambusia affinis*; convict cichlid, *Cichlasoma nigrofasciatum*, and Oriental snail, *Melanoides tuberculatus*). The Pahranagat spinedace (*Lepidomeda altivelis*) is extinct, apparently from the establishment of carp and mosquitofish. Likewise, these and other exotics have led to the near extinction of the White River desert sucker, the White River springfish, and the Hiko White River springfish. The Pahranagat pebblesnail and the White River tryonia are at risk from the Oriental snail.

In the Moapa River, the monotypic Moapa dace is federally endangered, having been reduced to low numbers in the Warm Springs area as a result of development of springs for landscaping, recreation, and tourism; water diversion for domestic and agricultural use; and establishment of exotic species. Other spring endemics in the Moapa River Basin face similar threats. Natives of the Moapa River proper have been reduced by channelization, siltation, removal of riparian vegetation, pollution, and exotics, as well as by the inundation of the river's lower reaches by Lake Mead (Williams et al. 1985).

In the upper White River, the White River spinedace is federally endangered as a result of major habitat loss and alteration following the installation of an irrigation pipe below the springhead; herbicide application within the spring outflow; and the establishment of exotic species. The White River springfish is likewise imperiled by interactions with exotics (Williams et al. 1985).

Suite of Priority Activities to Enhance Biodiversity Conservation

• Work to prevent construction of the Sand Hollow Reservoir on the Virgin River, which would dewater the last viable reaches of the river for endangered fish (American Rivers 1997).
• Following the example of a refuge established to protect the Moapa dace, establish similar protections for other imperiled endemics with highly restricted ranges (Williams et al. 1985).
• Conduct surveys of invertebrate fauna so that endemic species can be protected (Williams et al. 1985)
• Guard against the introduction of any new species.
• Protect Lund Town Spring, where the only spring pool within the upper White River remains relatively unaltered (Williams et al. 1985).
• Restore altered spring habitats where possible.
• Consider establishing captive populations of species known from single localities.

Conservation Partners

For contact information, please see appendix H.

• American Rivers
• Desert Fishes Council
• Grand Canyon Trust
• The Nature Conservancy (Nevada and Utah Field Offices)
• Southern Utah Wilderness Alliance

Ecoregion Number:	**14**
Ecoregion Name:	**Gila**
Major Habitat Type:	**Xeric-Region Rivers, Lakes, and Springs**
Ecoregion Size:	**159,875 km²**
Biological Distinctiveness:	**Continentally Outstanding**
Conservation Status:	**Snapshot—Critical Final—Critical**

Introduction

This ecoregion covers most of southern Arizona and part of southwestern New Mexico and extends into northern Sonora in Mexico. The major watershed in this ecoregion is that of the Gila River, a tributary to the lower Colorado River. Other important rivers include the San Pedro, Santa Cruz, and Salt Rivers, all tributaries to the Gila.

Biological Distinctiveness

As one would expect, the Gila [14] ecoregion shares many of its freshwater species with the Colorado [12] ecoregion, and the Gila and Colorado Rivers, in particular, share many distinctive large-river species. However, whereas the other major tributaries to the Colorado, with the exception of the Virgin River, do not have distinct biotas at the species level, the Gila does. As many as seven fish species that are not found in the Colorado [12] ecoregion's waters can be considered endemic to the Gila ecoregion; given a total of nineteen native species found in the Gila, this is an impressive number of endemics.

The woundfin (*Plagopterus argentissimus*) was historically found in seasonally hot and turbid habitats of the Gila River basin as well as in the Vegas-Virgin [13] ecoregion, but today it is limited to the Virgin River system (Minckley et al. 1986; Sigler and Sigler 1994). Related to the woundfin is the spikedace (*Meda fulgida*), found only in fragmented warm-water reaches of the Gila River. The loach minnow (*Tiaroga cobitis*) is restricted to warm-water reaches of the Gila and San Francisco (a tributary to the Gila) River systems, though it was formerly found more widely in the Gila Basin (Page and Burr 1991; Propst and Bestgen 1991). Both the woundfin and the loach minnow are significant in their representation of monotypic genera (though most ichthyologists now place the loach minnow in the genus *Rhinichthys;* Robins et al. 1991). The Gila trout (*Salmo gilae*) was historically restricted to the headwaters of the Gila and San Francisco Rivers in the Gila National Forest but has been translocated elsewhere in the ecoregion (Page and Burr 1991; Propst et al. 1992). Both the longfin dace (*Agosia chrysogaster*) and the Gila topminnow (*Poeciliopsis occidentalis*) can be considered endemic to the Gila [14] ecoregion, though their ranges extend southward into the Sonoran [61] ecoregion (Minckley et al. 1986; Page and Burr 1991). The Gila chub (*Gila intermedia*), whose taxonomy is uncertain, persists in one location in the ecoregion. Colorado squawfish (*Ptychocheilus lucius*) and razorback sucker (*Xyrauchen texanus*), endemic to the Colorado complex, may once have occurred in the Gila River, and roundtail chub (*Gila robusta*) still occurs in three separate reaches of the river.

Conservation Status

The Gila River is unusual in having no flow-controlling dams, and flow regimes in the basin are natural. The native freshwater fish fauna of the Gila [14] ecoregion is relatively intact, yet land- and water-use practices and introduced species pose substantial threats to all native species' persistence. The upper Gila River in New Mexico still provides important habitat for the federally threatened spikedace and loach minnow, which have been extirpated from much of their former habitat elsewhere in the basin (Minckley 1973; Propst et al. 1986; Propst et al. 1988). These species are primarily threatened by habitat modification in the form of channelization, dam construction, removal of riparian vegetation, and stream desiccation by water diversion, and by exotic species. Non-native trouts (rainbow, *Oncorhynchus mykiss*, and brown, *Salmo trutta*) pose perhaps the greatest threat to the federally endangered Gila trout, through competition, predation, and hybridization (Propst et al. 1992). This species, which now inhabits twelve sites thanks to intense recovery efforts, is also threatened by livestock grazing, illegal angling, and stochastic natural events. Black bullhead (*Ameiurus melas*), smallmouth bass (*Micropterus dolomieu*), and other exotics are responsible for the elimination of spikedace and loach minnow from much of their ranges. The roundtail chub, listed in New Mexico as endangered, has suffered from the establishment of exotic fishes as well as from habitat loss. The Gila topminnow has been extirpated from New Mexico, and the Gila chub may or may not persist in one location. Reductions in the ranges of all native fishes in the Gila River in the past fifty to seventy-five years has been documented.

Elsewhere in the ecoregion, freshwater species and habitats are threatened by the effects of tourism, pesticide runoff from agricultural land, and grazing of livestock in riparian zones. In the past, railroad accidents have spilled sulfuric acid into the Santa Cruz River, and the possibility of future spills exists.

Suite of Priority Activities to Enhance Biodiversity Conservation

- Reclaim and manage entire subdrainages with multiple tributaries in which populations of imperiled species persist, particularly in montane streams where regularly occurring natural events such as fire, floods, and drought threaten individual populations (Propst et al. 1992).
- Acquire land and/or establish conservation easements to protect essential aquatic and riparian habitats. Such areas should include the Tularosa River near Eagle Peak Road, the San Francisco River near Glenwood, and the upper East Fork Gila River in the vicinity of Cliff, New Mexico.
- Block plans to channelize the Gila River in the Cliff-Gila Valley reach.
- Work with land management agencies to sufficiently regulate potentially damaging activities on lands under their jurisdiction.
- Stop wildfire suppression activities, which result in more severe wildfire than might be expected under more natural fire cycles (wildfire and its associated impacts eliminated Gila trout from its type locality in 1989).
- Prevent the construction of flow-controlling dams on the Gila.

Conservation Partners

For contact information, please see appendix H.

- Desert Fishes Council
- The Nature Conservancy (Headquarters, and Arizona and New Mexico Field Offices)
- New Mexico Department of Game and Fish
- Sonoran Institute
- The Southwest Center for Biodiversity
- Universidad de Sonora (UNISON)
- World Wildlife Fund México

CONTRIBUTING AUTHOR: Dr. David L. Propst

Ecoregion Number:	**15**
Ecoregion Name:	**Upper Rio Grande/Río Bravo del Norte**
Major Habitat Type:	**Large Temperate Rivers**
Ecoregion Size:	**156,769 km²**
Biological Distinctiveness:	**Continentally Outstanding**
Conservation Status:	**Snapshot—Endangered**
	Final—Critical

Introduction

This long, narrow ecoregion begins in the middle portion of southern Colorado, extending through west-central New Mexico into western Texas. The watersheds of the Rio Grande, from its headwaters to the mouth of the Río Conchos in Mexico, form this ecoregion's boundary. The eastern extent of this ecoregion is partially defined by the Sangre de Cristo Mountains, while the western extent is defined by the series of ranges that creates the Continental Divide. The Rio Grande flows through New Mexico's major cities, and more than 80 percent of New Mexico's human population lives along the river (Forest Guardians 1998). In Mexico this river is known as the Río Bravo del Norte.

The Rio Grande's headwaters are in the San Juan Mountains in southern Colorado, and from there the river flows south through the San Luis Valley and the Rio Grande Depression. The character of the river changes markedly when it exits the depression below El Paso, Texas, and this change signals the break between the Upper and Lower Rio Grande. Streams in the upper portions of this basin tend to have relatively predictable flow regimes and contain fairly diverse habitat types. Farther downstream, tributaries are more widely spaced and their flow erratic, with the portion of the Rio Grande between El Paso, Texas, and the Río Conchos at times virtually dry (Smith and Miller 1986).

Major tributaries to the Upper Rio Grande are the Rio Chama, Rio Puerco, and Rio Paraje, and the Alamosa River. Dams have created the major reservoirs of Elephante Butte, Cochiti, Abiquiu, Angostrura, Isleta, San Acacia, and El Vado.

Biological Distinctiveness

With only twenty-eight native fish species, the Upper Rio Grande [15] ecoregion is faunistically depauperate compared with the Lower Rio Grande [20], Pecos [18] and Río Conchos [17]

ecoregions. This ecoregion is home to only one endemic fish species, the White Sands pupfish (*Cyprinodon tularosa*); it is the only fish species found in three restricted areas in the endorheic Tularosa Basin (Propst et al. 1985; Miller and Smith 1986). The Rio Grande silvery minnow (*Hybognathus amarus*) was once found in the Upper and Lower Rio Grande [20] ecoregion as well as the Pecos [18] ecoregion, but today it is restricted to a reach of the Rio Grande between Cochiti Dam and Elephant Butte Reservoir, in central New Mexico. Historically, the Upper Rio Grande [15] also shared with the Lower Rio Grande [20] the phantom shiner (*Notropis orca*) and the Rio Grande bluntnose shiner (*N. simus simus*), both of which are now extinct. No endemic unionid mussels or crayfish occur in the ecoregion, though there are a number of other endemic invertebrates. Within the springs of Socorro County, New Mexico, are the Socorro springsnail (*Fontelicella neomexicana*), the Chupadera springsnail (*Fontelicella* sp.), the Alamosa tryonia (*Tryonia* sp.), the Boquillas isopod (*Thermosphaeroma subequalum*), and the Socorro isopod (*T. thermophilum*). An endemic amphibian, the Sacramento mountain salamander (*Aneides hardii*), also is found in these springs.

Conservation Status

The Upper Rio Grande, as the main water supply for most of New Mexico's human population, has been critically altered by a range of impacts, many of which may be irreversible. Major diversion dams along the Rio Grande have been built in the last century to supply water for agriculture, and these have dewatered downstream reaches, flooded upstream habitat, and altered water quality and flow regimes. To expedite water delivery to agricultural areas, the river and its tributaries have also been heavily channelized. In dry years, virtually all of the Rio Grande's water may be diverted for irrigation. Pollution from agriculture, livestock grazing, urban runoff, industrial uses, logging, and resource extraction adds another important threat (Forest Guardians 1998). Into these altered habitats exotic species have been introduced and have flourished, further affecting native species populations. Such threats have led to the extirpation of the phantom shiner and Rio Grande bluntnose shiner in this ecoregion, and they threaten the native species that remain in reduced numbers. Big-river fish are suffering in particular; for instance, the blue sucker and shovelnose sturgeon, both native to this ecoregion, have likely been extirpated from the Upper Rio Grande (New Mexico Department of Game and Fish 1997). Spring habitats that support unique invertebrates are also easily disturbed by pollution and dewatering.

The White Sands pupfish and the Rio Grande silvery minnow, both endemic, have garnered attention as highly imperiled species. At present, the White Sands pupfish does not share its habitat with any other fish species. Alien mosquitofish (*Gambusia affinis*), goldfish (*Carassius auratus*), and largemouth bass (*Micropterus salmoides*) occur in ponds on White Sands Missile Range and Holloman Air Force Base in the Tularosa Basin, and their introduction into the habitats of the White Sands pupfish would likely cause its elimination. Feral horses have damaged banks at watering holes where White Sands pupfish live, causing loss of water. Other potential threats to the survival of this pupfish include pollution from pesticides, herbicides, toxic missile residues, chemical spills, and waste products of feral horses and oryx; physical destruction of habit due to vehicles, livestock, or missile damage; water removal via diversion or groundwater pumping; and lowering of the water table due to expanding populations of non-native vegetation (e.g., salt cedar) (New Mexico Department of Game and Fish 1997). The Rio Grande silvery minnow has been at the center of controversy, as it competes directly with agriculture for scarce water. State, federal, and international water rights laws conflict with each other and with the provisions of the Endangered Species Act, such that the minnow's listing as a federally endangered species far from ensures its survival.

Pollution in rivers with the most degraded water quality may be addressed by an agreement reached between New Mexico environmental groups and the state in 1997. According to the agreement, the state will clean up specified rivers according to a strict time schedule. Within this ecoregion, rivers identified for cleanup are the Middle Rio Grande and the Rio Puerco (Forest Guardians 1998).

Suite of Priority Activities to Enhance Biodiversity Conservation

- Work to protect "unique waters" identified in New Mexico House Bill 1298, introduced in 1997. Within the Upper Rio Grande [15] ecoregion, designated waters include:
  - Tesuque Creek from the headwaters to the Santa Fe National Forest boundary;
  - the Santa Fe River from the headwaters to McClure Reservoir;
  - Rio Frijoles from the headwaters to its confluence with the Rio Grande;
  - the East Fork of the Jemez from the Santa Fe National Forest boundary downstream to the confluence with the Jemez River;
  - the Rio Grande from the Colorado/New Mexico state line downstream to the Rio Arriba/Taos county line;
  - Rio Embudo from the Picuris Pueblo boundary downstream approximately 4.5 miles to where the river enters private land;
  - Las Huertas Creek from the headwaters to the Cibola National Forest boundary.
- Support clean-up of polluted surface waters, particularly the Rio Grande, by regulating municipal discharges and other point and nonpoint sources of pollution.
- Restore riparian vegetation along rivers and streams, with restriction of livestock grazing where possible.
- Operate dams along the Rio Grande to create flows that resemble natural regimes and benefit native fish species.
- Prohibit use of non-native minnows as bait for fishing, and guard against introduction of any new exotics.

Conservation Partners

For contact information, please see appendix H.

- American Rivers
- Desert Fishes Council
- Forest Guardians
- The Nature Conservancy (Colorado, New Mexico, and Texas Field Offices)
- Rio Grande Alliance
- Southwest Center for Biodiversity

Ecoregion Number:	**16**
Ecoregion Name:	**Guzmán**
Major Habitat Type:	**Endorheic Rivers, Lakes, and Springs**
Ecoregion Size:	**97,442 km²**
Biological Distinctiveness:	**Continentally Outstanding**
Conservation Status:	**Snapshot—Endangered**
	Final—Critical

Introduction

Part of the Rio Grande complex, this ecoregion covers most of the Mexican state of Chihuahua and extends into southwestern New Mexico. A small portion of the ecoregion is found in northeastern Sonora and extreme southeastern Arizona. The southwestern extent of this ecoregion is defined by the mountains of the Sierra Madre Occidental. Additionally, a small portion of the ecoregion lies on the western side of the Continental Divide within Arizona. Watersheds include the Rio San Pedro, Río Casa Grandes, Río Santa Maria, and the Mimbres River. Part of this ecoregion is the region known as Médanos Samalayuca, a harsh area of parched sand dunes that was once the bed of an ancient lake (Jagger 1998).

Biological Distinctiveness

As is typical of many endorheic habitats, the Guzmán [16] ecoregion is rich in endemic species, with 56 percent endemism in fish. These endemic fish species consist of at least one minnow (*Notropis* sp.), one chub (*Gila* sp.), one sucker (*Catostomos* sp.), and one pupfish (*Cyprinodon* sp.) (Arriaga et al. 1998). In addition, the beautiful shiner (*Cyprinella formosa*) is endemic to this region but has been extirpated from much of its former range (see below). One important site for the conservation of freshwater biodiversity in this ecoregion is the Mimbres River of New Mexico.

The fish fauna of the Mimbres River consists of just three confirmed species: the Chihuahua chub (*Gila nigrescens*), beautiful shiner (*Cyprinella formosa*), and Rio Grande sucker (*Catostomus plebeius*). The beautiful shiner was extirpated from the Mimbres River by the mid-1950s (the last verified specimen was collected in the early 1950s). The Chihuahua chub was believed extirpated until 1975 when a small population was found in a spring system associated with the Mimbres River. Since then, the range of the Chihuahua chub has been documented to consist of about 15 km of the Mimbres River between Allie Canyon and New Mexico Game and Fish property (purchased to provide habitat for the species). The Chihuahua chub is under federal and state protection as a threatened species. The Rio Grande sucker is common in warm-water reaches of the stream. An introduced population of Gila trout is present in McKnight Creek (a Mimbres tributary). Non-native rainbow trout (*Oncorhynchus mykiss*), speckled dace (*Rhinichthys osculus*), and longfin dace (*Agosia chrysogaster*) are present in portions of the Mimbres River occupied by the Chihuahua chub; all, particularly rainbow trout and longfin dace, may compete with the Chihuahua chub for habitat.

Conservation Status

Among the most serious problems in the Guzmán [16] ecoregion are those related to water demand. This is a very dry region, and even in the best circumstances water supplies are limited. As a result, aquifer drawdown is a major concern for the conservation of subterranean freshwater biodiversity. Also problematic is deforestation and the introduction of exotic species such as the blue tilapia (*Oreochromis aureas;* Arriaga et al. 1998).

Although there are no dams controlling flows on the Mimbres River, flows are greatly diminished in much of the river and eliminated in some reaches during the irrigation season (March through October) by numerous, small diversions. Habitats are further degraded by gravel pit operations, channelization, removal of woody riparian vegetation, row crop agricultural practices (tillage to edge of river), and sewage effluent (from septic fields in Mimbres Valley). Ash flows, associated with wildfire, greatly reduced fish abundance in the Mimbres and almost eliminated the Chihuahua chub from the river (except Moreno Spring) until 1995. If introduced rainbow trout occupy pool habitat that otherwise appears suitable, Chihuahua chub are almost always absent.

Suite of Priority Activities to Enhance Biodiversity Conservation

The New Mexico Department of Game and Fish acquired property encompassing about

1.2 km of stream habitat for the Chihuahua chub in 1985 and has since managed the property as a wildlife area for the species (all activities that might degrade habitat or disturb the Chihuahua chub are forbidden). TNC has recently purchased three properties (collectively comprising TNC Mimbres Reserve), in part to protect habitats critical to the Chihuahua chub.

- Because almost all of the Mimbres River that contains occupied or potential habitat for the Chihuahua chub is on private land, additional land purchases or conservation easements are needed to conserve the species in the United States of America.
- Stricter enforcement of Section 404 of the Clean Water Act is needed. Uplands of the watershed have been seriously overgrazed and contribute to the flashiness of floods. Although the U.S. Forest Service (USFS) has attempted to improve upland range conditions, more aggressive efforts are needed.

Conservation Partners

For contact information, please see appendix H.

- Desert Fishes Council
- The Nature Conservancy (Colorado, New Mexico, and Texas Field Offices)
- New Mexico Department of Game and Fish
- World Wildlife Fund México

CONTRIBUTING AUTHOR: Dr. David L. Propst

Ecoregion Number:	**17**
Ecoregion Name:	**Río Conchos**
Major Habitat Type:	**Xeric-Region Rivers, Lakes, and Springs**
Ecoregion Size:	**74,885 km²**
Biological Distinctiveness:	**Globally Outstanding**
Conservation Status:	**Snapshot—Endangered Final—Critical**

Introduction

Part of the Rio Grande complex, this ecoregion encompasses most of central and southern Chihuahua, extending into northern Durango. The major rivers in this xeric area are the Río San Pedro and the Río Conchos. These rivers drain into the Presa Madero and the Presa Boquillo, respectively.

Biological Distinctiveness

The Río Conchos [17] ecoregion exhibits remarkable endemism in its fish species, with twelve of its forty-seven native fish endemic. These endemics include the Chihuahua shiner (*Notropis chihuahua*), *N. santarosaliae*, *Cyprinella* sp., the ornate shiner (*Codoma ornata*), *Dionda* sp., *Lotriurus* sp., the blotched gambusia (*Gambusia senilis*), the crescent gambusia (*G. hurtadoi*), the yellowfin gambusia (*G. alvarezi*), the Conchos pupfish (*Cyprinodon eximius*), the bighead pupfish (*C. pachycephalus*), and the largescale pupfish (*C. macrolepis*). There is also a high degree of endemism among the ecoregion's aquatic herpetofauna; twelve of its forty-six native species are endemic.

The Río Conchos [17] is the only free-flowing, large-river environment left in the Rio Grande catchment basin. This system still harbors an impressive assemblage of species because its ecology has not been affected by channel modifications. This ecoregion is important not only for its support of surface water biota, but also for the fauna in its specialized spring and cave habitats. These spring habitats contribute most to the region's high endemism.

Conservation Status

This area is suffering under a wide variety of human impacts. Its water quality has been degraded by sewage (primarily from border communities); agricultural wastes such as nutrients, pesticides, and fertilizers; and industrial contamination, including chemical and thermal pollution. Also harmful to the species of the basin are poor land and water management practices, such as clear-cutting (in the upper Río Conchos), and inefficient use of groundwater resources. The spring biota are particularly sensitive to groundwater extraction, since their habitats are tightly linked to the water table (Rio Grande River Keeper 1994, quoted in Berger 1995).

Suite of Priority Activities to Enhance Biodiversity Conservation

- Enforce responsible land use and land management practices, such as terracing and drip irrigation.
- Control overharvesting of timber in the upper watershed of the Río Conchos.
- Complete a biological survey of the aquatic fauna of the watershed.
- Place strict limits on groundwater use.
- Study underground ecosystems and determine sensitivity to change.

Conservation Partners

For contact information, please see appendix H.

- Bioconservación, A.C.
- La Comisión Nacional para el Conocimiento y uso de la Biodiversida (CONABIO)
- The Nature Conservancy
- Universidad Autónoma de Chihuahua (UACH)
- Universidad Autónoma de Nuevo León (UANL)
- World Wildlife Fund México

Ecoregion Number:	**18**
Ecoregion Name:	**Pecos**
Major Habitat Type:	**Xeric-Region Rivers, Lakes, and Springs**
Ecoregion Size:	**114,782 km²**
Biological Distinctiveness:	**Continentally Outstanding**
Conservation Status:	**Snapshot—Vulnerable Final—Endangered**

Introduction

This ecoregion, part of the Rio Grande complex, extends from east-central and southeastern New Mexico into western Texas. The major river in this ecoregion is the Pecos, and its drainage area defines the extent of the ecoregion. The Pecos runs for 920 km from the glaciated Sangre de Cristo Mountains in north-central New Mexico to southwestern Texas, where it meets the Rio Grande (Williams et al. 1985; Minckley et al. 1986). The river contains a variety of habitats, including high-gradient, rocky-bottomed areas;

sluggish, soft-bottomed meanders; and rocky riffles. Major tributaries to the Pecos, virtually all of which arise in the mountains, include the North Seven River, Rio Penasco, Rio Felix, Rio Hondo and its tributary Rio Bonito, Ciénega del Macho, and Alamosa Creek. Five major reservoirs have been constructed on the Pecos: Los Esteros Reservoir, Lake Sumner, Lake McMillan, and Avalon Reservoir in New Mexico, and Red Bluff Lake in Texas. Amistad Reservoir, built on the Rio Grande, inundates 10 to 20 km of the Pecos River at its confluence with the Rio Grande (Williams et al. 1985).

Biological Distinctiveness

The Pecos [18] ecoregion is relatively species rich compared with the Upper Rio Grande [15] ecoregion and the other Rio Grande tributaries, particularly in terms of fish. Fifty-four native species occur here, seven of which (13 percent) are endemics or near-endemics. These are the proserpine shiner (*Cyprinella proserpina*), found only from the lower Pecos River and a few small localities in the Lower Rio Grande [20] ecoregion; the Leon Springs pupfish (*Cyprinodon bovinus*), known only from Leon Creek and its tributary, Diamond Y Spring, in Pecos County, Texas (Williams et al. 1985; Page and Burr 1991); the Comanche Springs pupfish (*C. elegans*), which originally inhabited just two isolated spring systems in southwestern Texas; the Pecos River pupfish (*C. pecosensis*), endemic to the Pecos River and associated springs and gypsum sinkholes; the Pecos gambusia (*Gambusia nobilis*), whose native habitat consisted of four small spring-fed systems, two in New Mexico and two in Texas; the greenthroat darter (*Etheostoma lepidum*), found in the Pecos system and also on the Edwards Plateau of Texas (in ecoregions [31] and [32]); and the Rio Grande cichlid (*Cichlasoma cyanoguttatum*), shared between this ecoregion and the Lower Rio Grande [20] ecoregion. Additionally, the Pecos River is home to the Pecos bluntnose shiner (*Notropis simus pecosensis*), the only extant form of *N. simus*.

At the species level, there are no endemic unionid mussels, crayfish, amphibians, or aquatic reptiles in this ecoregion, though there are a number of aquatic snails and amphipods found only in the ecoregion's spring-fed systems. Some of these, notable for their limited distributions, are the Pecos assiminea (*Assiminea* sp.), native to the Diamond Y Spring complex; the Phantom Spring cochliopa (*Cochliopa texana*); the Roswell springsnail (*Fontelicella* sp.); the Blue Spring springsnail (*Fontelicella* sp.); the Phantom Lake tryonia (*Tryonia cheatumi*); the Roswell tryonia (*Tryonia* sp.); the Diamond Y tryonia (*Tryonia* sp.); the Phantom Spring amphipod (*Gammarus hyalelloides*); the Pecos amphipod (*G. pecos*); the San Solomon amphipod (*G.* sp.); and Noel's amphipod (*G. desperatus*), endemic to springs along the Pecos River (Williams et al. 1985).

Conservation Status

As with other ecoregions in the desert Southwest, the Pecos [18] is fraught with threats stemming from competition for scarce water between humans and other organisms. As an indication of the level of disturbance, three of the ecoregion's seven endemic fish species, as well as the Pecos bluntnose shiner, are federally listed as endangered or threatened.

On the mainstem Pecos, the construction of impoundments has flooded upstream habitats and dewatered those below. For instance, Amistad Reservoir inundated much of the ranges of the proserpine shiner and the Rio Grande darter, and the federally threatened Pecos bluntnose shiner's range in the Pecos River has been greatly reduced largely because of reservoir construction (Williams et al. 1985). Extensive irrigation diversion canals throughout the ecoregion, particularly in the Artesia-Carlsbad area of New Mexico, have had impacts on the federally endangered Comanche Springs pupfish, the Phantom spring cochliopa, the Phantom Spring amphipod, and the San Solomon amphipod. Excessive pumping of groundwater, and resultant failure of springs, has been implicated in the declines of the Comanche Springs pupfish, the Pecos River pupfish, the Pecos gambusia, the Pecos assimea, the Roswell springsnail, the Diamond Y pondsnail, the Roswell tryonia, the Diamond Y tryonia, the Phantom Spring cochliopa and amphipod, and the San Solomon amphipod. These springs and spring-fed habitats are also particularly susceptible to point source pollution; for instance, pollution from an oil refinery may have contributed to the decline of the federally endangered Leon Springs pupfish (Williams et al. 1985).

An equally important threat to Pecos [18] ecoregion fish fauna is interaction with exotic species. The Pecos River was once home to the federally endangered Rio Grande silvery minnow (*Hybognathus amarus*), but the species was extirpated within the course of a decade by competition and hybridization with the introduced plains minnow (*H. placitus*), which fared better in the modified habitats of the river (New Mexico Ecological Services 1997). Closely related exotics have also affected the Leon Springs pupfish, the Comanche Springs pupfish, and the Pecos gambusia (Williams et al. 1985).

Suite of Priority Activities to Enhance Biodiversity Conservation

- Maintenance of spring flows is critical to many of this ecoregion's endemic species. It will require mandates to reduce groundwater pumping, or the purchase of sensitive groundwater areas.
- Guard against the introduction of new exotic species.
- Restore streams by removing irrigation diversion canals where possible.
- Enforce strict antipollution measures, and monitor potential point source polluters.

Conservation Partners

For contact information, please see appendix H.

- Canyon Preservation Trust
- Forest Guardians
- The Nature Conservancy (New Mexico and Texas Field Offices)
- Rio Grande Alliance
- Southwest Center for Biodiversity

Ecoregion Number:	**19**
Ecoregion Name:	**Mapimí**
Major Habitat Type:	**Endorheic Rivers, Lakes, and Springs**
Ecoregion Size:	**187,773 km²**
Biological Distinctiveness:	**Continentally Outstanding**
Conservation Status:	**Snapshot—Endangered Final—Critical**

Introduction

This large ecoregion within the Rio Grande complex stretches from southeastern Chihuahua through eastern Durango and southwestern Coahuila into northern Zacatecas, northern San Luis Potosí, and southern Nuevo León. The watersheds of the Río Oro, Río Nazas, and the Río Aguanaval define the boundaries of this ecoregion. Additionally, the Continental Divide marks the southwestern boundary.

Biological Distinctiveness

The Mapimí [19] ecoregion is a very large endorheic basin located in north-central Mexico. Its freshwater biodiversity is characteristic of endorheic basins, with a relative paucity of species but high endemism. Thirteen of the region's twenty-six fish are found nowhere else. One endemic crayfish and one endemic species of aquatic herpetofauna (the salamander *Pseudoeurycea galeanae*) occur here.

This region is characterized by closed-basin streams and spring environments, all associated with high levels of endemism. Unique and isolated environments encourage the development of highly specialized biota and speciation by extant species to fill ecological niches. Species in the Mapimí [19] ecoregion are likely derived from the Rio Grande drainage to the east. The fish of the area are similar to those of the Rio Grande but have adapted to the different ecological pressures of an endorheic system. Among these historic endemics are the stumptooth minnow (*Stypodon signifer*) and the Parras pupfish (*Cyprinodon latifasciatus*), both of which are extinct (Williams et al. 1985), and a regional morph of the chub *Gila nigrescens*. All have been found only in the isolated Parras spring system.

Conservation Status

One of the most serious problems in the Mapimí [19] ecoregion is overuse of its scarce water resources for agricultural and municipal use. The springs of the Parras region are being lowered by pumping of its source aquifer and by diversion of water from natural channels for field irrigation. This overpumping has led to the extinction of the majority of the region's native fish fauna. The few populations of endemic fish left are marginal, restricted to minimum habitat,

and in immediate danger of extinction (Williams et al. 1985). Similar problems exist in the other isolated habitats of the region, such as La Concha (a large thermal spring) and the watershed of the Río Nazas.

Water quality has been degraded throughout this area by overgrazing, excessive use of pesticides and other agrochemicals, and animal and human waste. Clear-cutting and uncontrolled burns of the forests in the upper Nazas watershed have also contributed to sedimentation of the waters of the Mapimí [19] ecoregion.

Suite of Priority Activities to Enhance Biodiversity Conservation

- Encourage responsible agricultural practices, such as terracing and drip irrigation, to control degradation of the Mapimí [19] ecoregion's water quality and to minimize withdrawal of surface and groundwater.
- Enforce efficient municipal water use, such as through a requirement for flow-control devices.
- Ban clear-cutting in the upper Río Nazas watershed.
- Enforce forestry practices that include a wide riparian zone to minimize sedimentation of receiving waters.

Conservation Partners

For contact information, please see appendix H.

- Centro Interdisciplinario de Investigación para el Desarrollo Integral Regional (CIIDIR)
- La Comisión Nacional para el Conocimiento y uso de la Biodiversidad (CONABIO)
- Fondo Mexicano para la Conservacion de la Naturaleza (FMCN)
- Institito Nacional de Ecología
- The Nature Conservancy
- Universidad Autónoma de Nuevo León (UANL)
- World Wildlife Fund México

Ecoregion Number:	**20**
Ecoregion Name:	**Lower Rio Grande/Bravo**
Major Habitat Type:	**Large Temperate Rivers**
Ecoregion Size:	**162,576 km²**
Biological Distinctiveness:	**Continentally Outstanding**
Conservation Status:	**Snapshot—Critical**
	Final—Critical

Introduction

This ecoregion is defined primarily by the Rio Grande, or Río Bravo del Norte, from its confluence with the Río Conchos to the Gulf of Mexico. It comprises the drainage area of the Lower Rio Grande, with the exception of the basins of the Pecos [18], Río Salado [21], and Río San Juan [23] ecoregions, each of which supports distinct freshwater faunas. Also included in this ecoregion are the watersheds of the San Fernando River, located south of the Rio Grande in Mexico and draining to the Gulf of Mexico, and the Devil's River, a tributary to the Rio Grande directly east of the Pecos River. The ecoregion covers a small part of southwestern Texas, and in Mexico parts of eastern Chihuahua, northern Coahuila, the extreme northern section of Nuevo León, and northern Tamaulipas.

Biological Distinctiveness

Of the eighty-four fish species native to the Rio Grande, sixteen are endemics, such as the Rio Grande silvery minnow (*Hybognathus amarus*), phantom shiner (*Notropis orca*), bluntnose shiner (*N. simus*), and Devil's River minnow (*Dionda diaboli*). Also associated with this ecoregion are one endemic unionid mussel species and one endemic crayfish species.

One outstanding feature of the Lower Rio Grande [20] ecoregion is the area known as Zona Carbonifera. This is a subterranean aquifer with an extensive system of springs and caves. A distinctive cave fauna occurs here, which is imperiled by overuse of the aquifer's water.

Conservation Status

The lower reach of the Rio Grande is facing major, immediate ecological problems. Because the Rio Grande flows through a xeric region, its water is in high demand by the human populations of Texas and Mexico. Overuse of water for municipalities and irrigation has reduced water

levels to the point of drying the entire watercourse at times. *Gambusia amistadensis*, a local endemic of the region, went extinct when the Amistad Reservoir was completed (Smith and Miller 1986).

The Rio Grande has been heavily affected by pollution of all types: unregulated urban discharge, agricultural chemicals, nutrient loading, pasture runoff (fecal coliform), hazardous waste dumps, and industrial pollution. The most persistent water quality problem is high salinity. In some cases, the fresh water from the Rio Grande must be desalinated before it can be used.

Exotic species introduction is also a severe problem. Such voracious exotics as carp, bass, and tilapia have been introduced and compete with native species. Exotic plants and mussels (such as the Asiatic clam, *Corbicula fluminea*) have also been introduced into this area.

Groundwater of the region is also in peril. In Zona Carbonifera, a subterranean aquifer with cave and spring habitats, water resources are under severe pressure because there is no regulation of withdrawals.

Suite of Priority Activities to Enhance Biodiversity Conservation

- Regulate the volume of water used by irrigators and municipalities.
- Enforce efficiency measures for water use, such as flow regulators and drip irrigation.
- Enforce responsible agricultural practices, such as terracing, to minimize nutrient loading and sedimentation.
- Enforce better municipal and industrial effluent treatment.
- Compile a complete biological inventory of the aquatic species of the Rio Grande, including exotics.
- Establish refuge areas for migratory species.
- Regulate water use and discharge in Zona Carbonifera.
- Restore riparian vegetation and channelized river reaches, where possible.

Conservation Partners

For contact information, please see appendix H.

- Bioconservación, A.C.
- La Comisión Nacional para el Conocimiento y uso de la Biodiversidad (CONABIO)

- Instituto Tecnologico y de Estudios Superiores de Monterrey (ITESM)
- National Audubon Society
- The Nature Conservancy (Headquarters and Texas Field Office)
- Universidad Autónoma de Nuevo León (UANL)
- World Wildlife Fund México

Ecoregion Number: **21**
Ecoregion Name: **Río Salado**
Major Habitat Type: **Xeric-Region Rivers, Lakes, and Springs**
Ecoregion Size: **48,675 km²**
Biological Distinctiveness: **Continentally Outstanding**
Conservation Status: **Snapshot—Vulnerable**
Final—Endangered

Introduction

Largely defined by the watersheds of the Río Sabinas and Río Salado, this ecoregion includes central and eastern Coahuila, northern Nuevo León, and a small portion of Tamaulipas. The Cuatro Ciénegas [22] ecoregion spring complex drains eastward into the headwaters the Río Salado.

Biological Distinctiveness

Of the forty-one species of fish native to the Río Salado [21] ecoregion, six are endemic. They are the minnow *Notropis saladonis*, one species of *Cyprinella*, one species of *Dionda*, one darter (*Etheostoma* sp.), a species of mosquitofish (*Gambusia marshi*), and a species of platyfish (*Xiphophorus meyeri*). The Río Salado [21] ecoregion possesses no known endemic mussels, crayfish, or aquatic herpetofauna.

Conservation Status

Because of imminent population pressures and the relative scarcity of water resources in the Río Salado [21] ecoregion, the freshwater biota is endangered. Agricultural and industrial development threaten Río Salado with both point and nonpoint source pollution. Population expansion also puts pressure on the region's water resources. Urban pollution in the form of solid

wastes and residual runoff could imperil its specialized and highly endemic fauna. *Cyprinella* sp., *Notropis braytoni*, and *N. saladonis*, fish species known to be indicators of clean water, have already disappeared from the Río Salado. Population expansion within the watershed also means increased water withdrawals. Water depletion and salinization are ominous consequences of expansion within the Río Salado watershed.

Exotic species introduction is another serious problem. At least six non-native species have been introduced to the basin: blue tilapia (*Oreochromis aureus*), warmouth (*Chaenobryttus gulosus*), common carp (*Cyprinus carpio*), threadfin shad (*Dorosoma petenense*), the crayfish *Procambarus clarkii*, and water hyacinth (*Eichornia* sp.) These species threaten to replace native species through competetion, predation, and habitat modification.

Suite of Priority Activities to Enhance Biodiversity Conservation

- Control water use within the Río Salado Basin, including withdrawals from wells, to prevent desiccation and salinization.
- Control land clearing for agricultural, urban, and industrial development.
- Carry out a complete biological inventory for the freshwater biota of the Río Salado watershed, including native and introduced species.
- Monitor the effects of physical and chemical landscape changes on freshwater habitats.

Conservation Partners

For contact information, please see appendix H.

- Bioconservación, A.C.
- La Comisión Nacional para el Conocimiento y uso de la Biodiversidad (CONABIO)
- PROFAUNA
- The Nature Conservancy
- Universidad Autónoma de Nuevo León (UANL)
- Universidad Autónoma Agraria Antonio Narro (UAAAN)
- World Wildlife Fund México

Ecoregion Number: **22**
Ecoregion Name: **Cuatro Ciénegas**
Major Habitat Type: **Xeric-Region Rivers, Lakes, and Springs**
Ecoregion Size: **492 km²**
Biological Distinctiveness: **Globally Outstanding**
Conservation Status: **Snapshot—Relatively Stable**
Final—Vulnerable

NOTE: See essay 13 for additional ecoregion details.

Introduction

This small ecoregion, whose Spanish name means "four marshes," is found entirely within the Mexican state of Coahuila, on the eastern edge of the Sierra Madre Oriental, and includes a series of basins with pools fed by freshwater springs. Its climate is extremely arid, as it receives less than 200 mm of rain a year. This intermontane valley supports diverse habitats, including springs, marshes, rivers, lakes, playas, and canals (Marsh 1984).

Biological Distinctiveness

The Cuatro Ciénegas [22] ecoregion harbors an incredible assemblage of species, characterized by its richness and globally outstanding endemism. It boasts sixty-six species of amphibians and reptiles (including an endemic freshwater box turtle, *Terrapene coahuila*, among the six endemic species in the herpetofauna). Half of the twelve crustacean species are endemic, as are twenty-three of the thirty-four mollusks and nine of the twelve snail species.

Eight of Cuatro Ciénegas' [22] native fish species are endemic. The Cuatro Ciénegas cichlid (*Cichlasoma minckleyi*), is particularly interesting because it exhibits two distinct morphs. Other endemic fish include *Lucania interioris*, *Xiphophorus gordoni*, and *Gambusia longispinis*.

Conservation Status

The Mexican government declared 150,000 hectares of the Cuatro Ciénegas Valley protected in 1994. Much of the valley's land, however, is still held by private landowners. With few restrictions, landowners are able to exploit the valley's resources. Development has proven to be Cuatro Ciénegas's [22] biggest problem. Water is diverted for uses such as alfalfa agriculture

and pivot irrigation. The delicate ecology has also been affected by tourists, who bathe and swim in the pools.

Population growth in the region and associated water demands threaten the delicate balance of life that has developed in the isolated springs of Cuatro Ciénegas [22] over millions of years. Because the biota of the springs is highly specialized, any change in water levels could severely disturb one or more of these species. Indeed, the basin has already been heavily modified. Before water diversions within Cuatro Ciénegas [22], the area had a large system of lakes and marshes. These have now largely been drained. Gypsum mining on nearby dunes, where vegetative cover is sparse, has contaminated many of the pools, or *pozas*.

Exotic species also threaten the native biota of Cuatro Ciénegas [22]. Aggressive invaders such as the water lily *Thiara tuberculata*, the crayfish *Procambarus clarkii*, and the cichlid *Hemichromis guttatus* threaten to force native species from their habitats. Added to this are the threats of illegal hunting and fishing, and overcollection, within the boundaries of the protected Cuatro Ciénegas Valley.

Suite of Priority Activities to Enhance Biodiversity Conservation

- Regulate and minimize the extraction of gypsum from the basin.
- Increase the total amount of land under direct protection.
- Close unused artificial channels.
- Regulate the withdrawal of water from the *pozos*, and consider reconstructing wetlands.
- Regulate tourist visitation of the site to minimize pollution.
- Regulate agricultural chemical use to prevent pollution of the springs.
- Complete a biological survey of the biota of the springs and the hypothesized biota of the underground connecting channels.
- Limit settlement and agricultural expansion into the basin.
- Enforce hunting and fishing regulations within the protected areas of Cuatro Ciénegas.
- Control populations of established exotic species and work to prevent the introduction of new species.

Conservation Partners

For contact information, please see appendix H.

- La Comisión Nacional para el Conocimiento y uso de la Biodiversidad (CONABIO)
- Desert Fishes Council
- Instituto Nacional de Ecología
- PROFAUNA
- The Nature Conservancy
- Universidad Autónoma de Nuevo León (UANL)
- World Wildlife Fund México

Ecoregion Number:	**23**
Ecoregion Name:	**Río San Juan**
Major Habitat Type:	**Xeric-Region Rivers, Lakes, and Springs**
Ecoregion Size:	**46,677 km²**
Biological Distinctiveness:	**Globally Outstanding**
Conservation Status:	**Snapshot—Endangered** **Final—Endangered**

Introduction

Part of the Rio Grande complex, this ecoregion stretches from southeastern Coahuila through central Nuevo León, and northwestern Tamaulipas to the Rio Grande (Río Bravo) River. The ecoregion is defined by the watershed of the Río San Juan and its tributary the Pilón. The Río San Juan originates in the mountains of the Sierra Madre Oriental and the Altiplanicie Mexicana and flows eastward onto the coastal plains of eastern Mexico, where it converges with the Rio Grande (Río Bravo) on the U.S.–Mexico border.

Biological Distinctiveness

Like many xeric regions, the Río San Juan [23] exhibits high endemism among its freshwater fauna. Of its forty-four native fish species, six are found nowhere else. These endemic fish include the spotted minnow (*Dionda melanops*), Saltillo chub (*Gila modesta*), Mexican red shiner (*Cyprinella rutila*), Monterrey platyfish (*Xiphophorus couchianus*), and Rio Grande darter (*Etheostoma grahami*). Other endemic species include one crayfish (*Procambarus regiomontanus*) and two isopods (*Sphaerolana* spp.).

Conservation Status

Extensive changes have been made to the landscape of the region in the form of dams and canals. The El Cuchillo Dam on the Río San Juan has significantly reduced flow levels below the dam, causing a legal dispute over water rights between two states in Mexico and most likely negatively affecting the fishes in this 45-mile stretch of river (International Rivers Network 1996). The water quality is degraded as a result of human population increase, agricultural expansion, and industrial pollution. The water resources of the region are being overexploited by municipalities, industry, and agriculture, leading to water scarcity and salinization of groundwater reserves. Unregulated fishing activities also represent a persistent problem. Use of explosives in fishing activities is widespread.

Another of the pressing ecological problems of the Río San Juan [23] ecoregion stems from introductions of exotic species such as the plant *Hydrilla*, the water hyacinth (*Eichornia crassipes*), and the grass *Zosterella dubia*. Many types of exotic aquatic fauna also have become established. Among these are the blue tilapia (*Oreochromis aurea*), Mozambique tilapia (*O. mossambicus*), smallmouth bass (*Micropterus dolomieu*), white bass (*Morone chrysops*), common carp (*Cyprinus carpio*), threadfin shad (*Dorosoma petenense*), gizzard shad (*Dorosoma cepedianum*), several types of silversides, including *Menidia beryllina* and *Chirostoma* spp., warmouth (*Chaenobryttus gulosus*), a crayfish (*Procambarus clarkii*), the lampmussel *Lampsilis* sp., and the Asiatic clam (*Corbicula fluminea*).

Suite of Priority Activities to Enhance Biodiversity Conservation

- Control industrial, urban, and agricultural runoff and discharge.
- Establish more and better water treatment plants.
- Regulate surface and groundwater extraction.
- Perform a thorough biological inventory of the San Juan [23] ecoregion, cataloging native and exotic species.
- Promote the basin's status as an area of refuge and feeding for migratory species.

Conservation Partners

- La Comisión Nacional para el Conocimiento y uso de la Biodiversidad (CONABIO)
- Instituto Tecnologico y de Estudios Superiores de Monterrey (ITESM)
- The Nature Conservancy
- Universidad Autónoma de Nuevo León (UANL)
- World Wildlife Fund México

Ecoregion Number: **24**
Ecoregion Name: **Mississippi**
Major Habitat Type: **Large Temperate Rivers**
Ecoregion Size: **478,549 km²**
Biological Distinctiveness: **Bioregionally Outstanding**
Conservation Status: **Snapshot—Endangered**
Final—Endangered

NOTE: See essay 2 for a description of the entire Mississippi River Basin.

Introduction

A part of the Mississippi complex, this ecoregion includes most of Minnesota, most of western Wisconsin, part of northeastern South Dakota, most of Iowa, parts of northeastern and southeastern Missouri, most of Illinois, and a small extension into northwestern Indiana. The ecoregion is defined by the drainage area of the Mississippi from its headwaters to its confluence with the Ohio River, as well as its many tributaries along this reach, except the Missouri. Among these are the Chippewa and Wisconsin Rivers in Wisconsin, the Minnesota River in Minnesota, the Des Moines and Iowa Rivers in Iowa, and the Illinois, Wabash, and White Rivers in Illinois. There are numerous lakes, with the heaviest concentration in the northern half of the ecoregion, which was covered by a continental glacier as recently as ten to fifteen thousand years ago. This ecoregion also contains the Driftless Area, located in the northwest corner of Illinois, southern Wisconsin, and eastern Iowa; this area is completely surrounded by once-glaciated land, but was itself never glaciated.

Biological Distinctiveness

With only one endemic fish (*Percina aurora*), one endemic crayfish, and one endemic unionid mussel, the Mississippi [24] ecoregion is distinguished more by the presence of several remarkable species whose distributions include other ecoregions. Rare or near-endemic species of fish within the bounds of the region include the greater redhorse (*Moxostoma valenciennesi*), silver lamprey (*Ichthyomyzon unicuspis*), and pugnose shiner (*Notropis anogenus*) (The Nature Conservancy 1994). It is the northernmost habitat for the largest of the gars, the alligator gar (*Atractosteus spatula*), and the paddlefish (*Polyodon spathula*). Also common in these waters are predatory fishes such as northern pike (*Esox lucius*) and muskellunge (*Esox masquinongy*).

Some of the herpetofauna that inhabit the mainstem of the Mississippi River farther south also extend as far north as this ecoregion and are possibly invading northward from the Mississippi refugium. Salamanders such as the lesser siren (*Siren intermedia*) have ranges that extend north of Tennessee only along the Mississippi.

Common in this region (as well as in ecoregions [26], [27], and [52]) are isolated wetlands called prairie potholes (see essay 9), which harbor a variety of aquatic species. These wetlands are so numerous that a full catalog of species inhabiting them has never been completed.

Conservation Status

Probably the most enduring and onerous problem for the freshwater ecology of this region is gross modification of the entire landscape for pasturage and agriculture (see essay 2). These activities lead to pollution of streams by animal waste, pesticides, herbicides, fertilizers, and other agricultural chemicals. This ecoregion has among the highest-recorded sedimentation levels as well (Burr and Taylor pers. comm.). In some parts of the Upper Mississippi River, sediments contain large enough concentrations of adsorbed lead to be considered moderately polluted (40 μg of lead per gram of sediment) to highly polluted (60 μg of lead per gram of sediment). The effects of agriculture and grazing are exacerbated by the associated loss of riparian-zone vegetation, which normally acts as a natural buffer. Direct modification of in-stream habitat as a result of gravel extraction, channelization, and impoundments is also widespread. Side-channel and backwater habitat losses are a major concern because they serve as critical habitats for a broad array of flora and fauna, including several species of economically important fish. Additionally, fire suppression and agriculture have eliminated most of the floodplain prairies, which were once a widespread landscape feature (USGS 1999). Sprawl and related pollution threaten drainages that are near urban areas.

Exotic species have caused measurable change in the region. For example, the Chinese carp (*Cyprinus carpio*), introduced into this area, now accounts for some 90 percent of the biomass in mainstem habitats of the Mississippi River (Burr and Taylor pers. comm.). The rusty crayfish (*Orconectes rusticus*) is another aggressive exotic. Also, the ubiquitous zebra mussel (*Dreissena polymorpha*) has made significant inroads in the ecoregion, posing a severe threat to native freshwater mussels (USGS 1999).

Overharvesting of freshwater mussels, particularly pearly mussels for the seed pearl industry, is a growing concern. Although the extent of the problem is not fully known, mussel biologists believe that current levels of exploitation cannot be sustained, and efforts are underway to end the harvest. It is estimated that more than 15 million pounds of mussel shells were harvested in the Illinois and Mississippi Rivers in 1989 and 1990. Poaching of endangered mussel species by collectors and subsistence fishermen is an additional potential threat (Mueller 1993).

Suite of Priority Activities to Enhance Biodiversity Conservation

- Promote responsible agricultural practices, such as terracing, to minimize runoff and sedimentation.
- Maintain riparian zones along waterways to minimize pollution from runoff.
- Prohibit in-stream gravel removal.
- Enforce conservation of populations and habitats of endangered and rare pearly mussel stocks.
- Develop and enforce protocols for the management of exotic species.
- Develop partnerships between private landowners and public agencies.
- Work to incorporate a more natural flow regime in navigation practices.

- Implement "smart-growth" land use planning and work to reduce stormwater discharges and other nonpoint source runoff in urban areas (American Rivers 1999).

Conservation Partners

For contact information, please see appendix H.

- Columbia Environmental Research Center
- Illinois Department of Natural Resources
- Indiana Department of Natural Resources
- Minnesota Department of Natural Resources
- Missouri Department of Conservation
- MoRAP
- The Nature Conservancy (Illinois, Indiana, Minnesota, Missouri, and Wisconsin Field Offices)
- Wisconsin Department of Natural Resources

Focus on the Southeast

When the first settlers arrived in the southeastern United States, they found it blessed with an abundance of rivers and creeks with high water quality. Because this resource was so abundant, it was generally taken for granted. Unlike the Southwestern United States with its water rights and water allocation struggles, water in the Southeast has generally been considered a free commodity. This abundance is in part responsible for the abuse and outright destruction of many biologically productive rivers and creeks, magnificent in their pristine state, that have been altered to the point that in many places they can hardly be called rivers. Not surprisingly . . . many fishes, freshwater mussels, and snails that were common in these rivers are now among our endangered and threatened species.
—*Ono et al. 1983*

Ecoregion Number:	**25**
Ecoregion Name:	**Mississippi Embayment**
Major Habitat Type:	**Large Temperate Rivers**
Ecoregion Size:	**258,675 km²**
Biological Distinctiveness:	**Globally Outstanding**
Conservation Status:	**Snapshot—Endangered**
	Final—Critical

Introduction

The Mississippi Embayment [25] ecoregion is defined roughly by the mainstem Mississippi River below its confluence with the Ohio River, excluding the upper Red and upper Ouachita Rivers. It also includes the lowermost reaches of the Arkansas and White Rivers. The entire ecoregion lies within the lowland Gulf Coastal Plain, and the Fall Line constitutes its northern boundary (Robison 1986). Structurally, the embayment is a trough between the Appalachians to the east and the Ozark and Ouachita Mountains to the west, and this, combined with the Mississippi's turbid character, has served as a barrier to dispersal of upland fish species between river systems on either side (Robison 1986).

The ecoregion ranges across parts of southwestern Kentucky, southeastern Missouri, western Tennessee, eastern Arkansas, most of Mississippi, and the eastern half of Louisiana. Other major rivers include the lower reaches of the Arkansas and White Rivers in Arkansas, the Big Black River in Mississippi, and the Tensas River in Louisiana. The entire original extent of the Mississippi alluvial plain, stretching 1,120 km from the northern portion of the ecoregion to the confluence with the Gulf of Mexico, occurs here (U.S. EPA 1998). Swamps, marshes, and other wetland areas, including bottomland forests, were once dominant features throughout this ecoregion. Although these areas still exist in many places, they are not as extensive as in presettlement times (USGS 1996).

Biological Distinctiveness

The Mississippi Embayment [25] is distinguished by its extraordinary species richness, particularly in fish. The entire Mississippi Basin has served as a center for fish distribution as well as a glacial refugium, and as such it is home to many of the species found in surrounding drainages (see essay 2). In total, 206 species of fish are found in this ecoregion, making it (and the Teay-Old Ohio [34] ecoregion, with an equal number) the second-richest ecoregion in North America, after the Tennessee-Cumberland [35] ecoregion. Only 11 (5 percent) of these fish species are endemic, and these are found in tributary drainages rather than in the Mississippi mainstem. These endemics are two minnows (*Notropis rafinesquei, N. roseipinnis*), one catfish (*Noturus hildebrandi*), one cavefish (*Forbesichthys agassizi*), two killifish (*Fundulus euryzonus, F. notti*), and five darters (*Etheostoma chienense, E. pyrrhogaster, E. raneyi, E. rubrum, E. scotti*). The ecoregion is more noted for its assemblages of large-river fish, which include five lamprey species, four sturgeon, the only North American paddlefish, four gar, and the bowfin. Additionally, numerous marine species have commonly been recorded in the Mississippi's lower reaches.

This ecoregion supports a moderate number of unionid mussel and crayfish species (sixty-three and fifty-seven species, respectively) compared with the Tennessee-Cumberland [35] and Teays-Old Ohio [34] ecoregions to the north, but an impressive 58 percent of its crayfish species are endemic. With its warm, humid climate, it is also home to sixty-eight species of amphibians and aquatic reptiles, including the American alligator (*Alligator mississippiensis*). Two reptiles are endemic: the ringed map turtle (*Graptemys oculifera*) is restricted to the Pearl River system and the yellow-blotched map turtle (*G. flavimaculata*) is found only in the Pascagoula River system (Conant and Collins 1991).

Conservation Status

Historically this area was heavily forested, but today most of the forest has been converted for agriculture, constituting one of the most serious ecological disturbances in the ecoregion (USGS 1996). Bottomland forests in the floodplains of the major rivers, especially the Mississippi, have been lost, and nonpoint source pollution from agriculture results in sedimentation and pesticide contamination of aquatic habitats (Ricketts et al. 1999). Water quality is also affected by wastewater discharges in urban areas (larger cities include Memphis, Vicksburg, Jackson, and Baton Rouge). Research is currently under way to assess the effects of crayfish and catfish farming on water quality (USGS 1996).

Equally if not more important, extensive hydrological modifications have destroyed in-stream habitat for native fishes and other species, including native mussels (U.S. EPA 1998). The U.S. Army Corps of Engineers sought to control flooding from the Mississippi, and it continues to pursue new flood-control projects. Recently, the Corps proposed a backwater pumping project for the Yazoo River and a channelization project for the Big Sunflower River, with the intention of providing protection to farms that lie in the floodplain. Besides eliminating rich mussel beds, the project would destroy more than 350 acres of remaining bottomland forests and 10,000 acres of wetlands. American Rivers has launched a campaign to encourage implementation of newer, less ecologically destructive alternatives, including but not limited to land acquisition in flood-prone areas (American Rivers 1997).

Research suggests that upstream damming and levee construction also increase flood levels farther downstream, a recent problem as evidenced by the Mississippi River floods of 1993 and 1997 (Bayley 1995; Lower Mississippi River Conservation Committee 1998). Importantly, these artificial structures serve to destroy the phenomenon of flood pulses that is normally associated with large temperate rivers. Flood pulses generally help maintain species diversity by triggering the migration of species from main channel habitats to previously inaccessible habitat. Combined with destruction of floodplain forests and other wetland areas, disappearance or reduction of the flood pulse threatens aquatic species reliant upon this phenomenon for access to new habitats. Furthermore, elimination of this phenomenon threatens to lower the biological productivity of the system, thereby potentially affecting overall species numbers, including commercially important species (Bayley 1995).

Although the extent of the problem is not fully known, overharvesting of freshwater mussels, particularly pearly mussels, may be a problem in this ecoregion. Mussels in the United States are primarily harvested for use as seeds for the Japanese pearl industry; current levels of exploitation probably cannot be sustained, and the permitted limits for harvesting may need to be adjusted. Poaching of endangered mussel species by collectors and subsistence fisherman may also be a smaller-scale threat (Mueller 1993).

Suite of Priority Activities to Enhance Biodiversity Conservation

- Protect remaining blocks of forest and wetlands. Identify strategic areas where these systems should be restored.
- Remove physical barriers such as dams that obstruct fish movements.
- Pursue land acquisition strategies in flood-prone areas, instead of older channelization and levee approaches (American Rivers 1997).
- Direct new development away from streamside areas, especially floodplains and other biologically important areas.
- Continue to monitor and assess the problem of mussel poaching and potential overharvesting. Additional funding is necessary for research into the effects of this problem (Williams et al. 1993).

Conservation Partners

For contact information, please see appendix H.

- Alabama Natural Heritage Program
- American Rivers
- Arkansas Natural Heritage Commission
- Illinois Natural Heritage Division
- Kentucky Natural Heritage Program
- Louisiana Natural Heritage Program
- Lower Mississippi River Conservation Committee
- Mississippi Natural Heritage Program
- Mississippi Wildlife Federation
- Missouri Natural Heritage Database
- MoRAP
- National Wildlife Federation
- The Nature Conservancy (Alabama, Arkansas, Illinois, Kentucky, Louisiana, Mississippi, Missouri, and Tennessee Field Offices, and Southeast Regional Office)
- Tennessee Division of Natural Heritage

Ecoregion Number:	**26**
Ecoregion Name:	**Upper Missouri**
Major Habitat Type:	**Large Temperate Rivers**
Ecoregion Size:	**678,741 km²**
Biological Distinctiveness:	**Nationally Important**
Conservation Status:	**Snapshot—Vulnerable**
	Final—Vulnerable

Introduction

This large ecoregion covers all of Montana east of the Continental Divide, most of northern Wyoming, the northwestern corner of Nebraska, western South Dakota, southwestern North Dakota, and a small portion of extreme southeastern Alberta and southern Saskatchewan. The ecoregion is largely defined by the watershed of the Upper Missouri River. Other rivers include the Yellowstone in Montana, the Bighorn River in Wyoming, the Little Missouri in all four states, the Grand River, the Moreau River, and the Cheyenne River in South Dakota.

This ecoregion represents the uppermost drainages of the Mississippi Basin, the largest watershed on the North American continent. The headwaters of this drainage are on the arid eastern slope of the Rocky Mountains. The land gradually slopes downward to the east. The streams change from high-gradient mountain streams to slower-moving, larger rivers on the plains.

Biological Distinctiveness

The northern portions of the Upper Missouri [26] ecoregion were subjected to heavy glaciation as recently as ten to fifteen thousand years ago. Therefore, the landscape of this ecoregion is much younger than that of areas farther south. As a result of this glaciation, the Upper Missouri [26] has no known endemic fish, mussel, crayfish, or aquatic herpetofauna species. The ecoregion is, however, important for its large-river habitat, which supports remarkable species such as the pallid sturgeon (*Scaphirhyncus albus*). This large, ancient fish is completely restricted to the main channels of the Mississippi and Missouri Rivers from Montana all the way to Louisiana. Quite uncommon, it requires the turbid, sediment-filled waters of large rivers for its habitat. Also present in these streams is the shovelnose sturgeon (*Scaphirhyncus platorynchus*), the pallid sturgeon's smaller cousin.

This ecoregion forms the southernmost extent west of the Mississippi for some northern fishes as well. These fish include the brook stickleback (*Culaea inconstans*), the burbot (*Lota lota*), and the spoonhead sculpin (*Cottus ricei*).

A distinctive feature of this ecoregion (along with ecoregions [24], [27], and [52]) is the presence of isolated wetlands called prairie potholes (see essay 9). These wetlands, which are highly endangered, may harbor endemic species of aquatic invertebrates and plants.

Conservation Status

This ecoregion is sparsely settled but is vulnerable to ecological degradation from extensive agriculture and ranching. Agricultural chemicals, sediment, and animal waste can easily find their way into its waterways. One of the more onerous problems associated with this ecoregion is contamination of the groundwater by pesticides, as in the Squaw Creek/Baldwin Creek watershed in central Wyoming (U.S. EPA 1995).

Overappropriation of water is another problem. Because of its situation on the eastern side of the Rockies, the Upper Missouri [26] ecoregion is considered arid to semiarid. In this area of limited water, there are now dry stream reaches, and reaches with elevated water temperatures and low dissolved oxygen (U.S. EPA 1995).

Ironically, in this arid region flooding and erosion are a concern for streamside landowners, and in some cases their efforts to stabilize river banks have seriously impacted the river habitats. The Yellowstone River has received attention because excessive "rip-rapping" of banks with rock and the construction of levees have increased river velocity, eliminated pools and backwater habitats, reduced natural turbidity, and eliminated alluvial deposition on the floodplain (American Rivers 1999; The Wilderness Society 1999). These activities threaten productive trout fisheries and rarer species such as the sauger (*Stizostedion canadense*) and pallid sturgeon. The Yellowstone River has also been highlighted as in danger from increased logging and road building within its catchment's national forests (The Wilderness Society 1999).

Suite of Priority Activities to Enhance Biodiversity Conservation

- Place moratoria on the issuance of new water rights.
- Enforce responsible agricultural practices, such as terracing, to minimize sedimentation and pollution of streams.
- Enforce strict limits on application of agricultural chemicals, to minimize the risk of groundwater contamination.
- Prevent drainage and filling of prairie potholes.
- Place moratoria on non-emergency streambank stabilization projects (American Rivers 1999).
- Restrict new logging and road building permits on federal lands (The Wilderness Society 1999).

Conservation Partners

For contact information, please see appendix H.

- Alberta Wilderness Association
- Canadian Nature Federation
- Canadian Parks and Wilderness Society
- Ducks Unlimited Canada
- Federation of Alberta Naturalists
- Greater Yellowstone Coalition
- Missouri River Coalition
- The Nature Conservancy (Montana, North Dakota, South Dakota, and Wyoming Field Offices)
- Natural Resources Conservation Service
- Prairie Conservation Forum
- Society of Grassland Naturalists
- World Wildlife Fund Canada

Ecoregion Number:	**27**
Ecoregion Name:	**Middle Missouri**
Major Habitat Type:	**Large Temperate Rivers**
Ecoregion Size:	**594,095 km²**
Biological Distinctiveness:	**Nationally Important**
Conservation Status:	**Snapshot—Vulnerable**
	Final—Vulnerable

Introduction

This ecoregion extends from central North Dakota to northwestern Colorado, including eastern South Dakota, a small portion of southwestern Minnesota, western Iowa, northwestern Missouri, virtually all of Nebraska, the northern half of Kansas, and southeastern Wyoming. The ecoregion is defined by the watersheds that constitute the middle portion of the Missouri River. Other rivers include the James River in North and South Dakota; the Platte River in Nebraska, Colorado, and Wyoming; the Niobrara River in Nebraska; the Kansas River in Kansas; and the Republican River in Kansas and Nebraska.

Biological Distinctiveness

The far northern portion of the Middle Missouri [27] ecoregion was glaciated as recently as ten to fifteen thousand years ago. The portion that sits on the eastern slope of the Rocky Mountains is semiarid, with limited freshwater habitats other than ephemeral prairie potholes. The Middle Missouri [27] ecoregion harbors no known endemic fish, mussels, crayfish, or aquatic herpetofauna, though prairie potholes may harbor unknown endemic invertebrate or plant species. Parts of the ecoregion also provide winter habitat for the bald eagle and crucial staging grounds for more than 100 species of migratory birds (American Rivers 1998).

Several fish that are adapted to large, mainstem environments are found as far up the Missouri River as the Middle Missouri [27] ecoregion. These fish include the pallid sturgeon (*Scaphirhynchus albus*) and the shovelnose sturgeon (*Scaphirhynchus platorhyncus*). This ecoregion forms the northernmost and westernmost extent of the range of at least one other species of large fish, the shortnose gar (*Lepisosteus platostomus*).

Conservation Status

The status of the Middle Missouri [27] ecoregion is vulnerable. Even as sparsely settled as it is, the ecoregion has historically faced many problems and may be increasingly imperiled in the near future. Levees, dams, bank armoring, channelization, and weirs have been used extensively throughout this region. Natural habitats, from pool and riffle sequences to snags and deadfalls, have been eliminated by these practices.

Also problematic is agriculture. This entire region is extremely productive agricultural land, despite its aridity. Overuse of water for irrigation and heavy application of agricultural chemicals have endangered habitats and aquatic biota.

Other major threats are associated with resource extraction. Logging in roadless areas threatens water quality in National Forest land, such as that along the Wyoming–Colorado border (The Wilderness Society 1999). Commercial sand and gravel mining in rivers such as the Kansas have increased erosion and destabilized river banks; in the Kansas River these operations threaten to expand into a potential Wild and Scenic River segment (American Rivers 1998).

Suite of Priority Activities to Enhance Biodiversity Conservation

- Enforce application of agricultural chemicals.
- Enforce limits on water drawn from groundwater and surface sources.
- Enforce responsible agricultural practices, such as terracing, to minimize sedimentation and runoff.
- Prevent drainage and filling of prairie potholes.
- Acquire flood-prone land from willing landowners and restore riparian and upland vegetation (American Rivers1998).
- Re-operate dams to resemble natural flow conditions more closely.

Conservation Partners

For contact information, please see appendix H.

- Kansas Natural Heritage Inventory
- MoRAP
- The Nature Conservancy (Colorado, Iowa, Kansas, Missouri, Nebraska, and Wyoming Field Offices)
- Nebraska Natural Heritage Program
- Sierra Club
- West Central Research and Extension Station, University of Nebraska

Ecoregion Number:	**28**
Ecoregion Name:	**Central Prairie**
Major Habitat Type:	**Temperate Headwaters and Lakes**
Ecoregion Size:	**158, 635 km²**
Biological Distinctiveness:	**Bioregionally Outstanding**
Conservation Status:	**Snapshot—Relatively Stable**
	Final—Relatively Stable

Introduction

This ecoregion includes central and southwestern Missouri, southeastern Kansas, northwestern Oklahoma, and a portion of northwestern Arkansas. It is defined by the watersheds of several rivers, including the middle portion of the Arkansas River and its tributary the Neosho River, and the lower Missouri River to Kansas City and its tributary the Osage River.

Biological Distinctiveness

The Central Prairie [28] ecoregion is characterized by relatively low endemism of aquatic species. It contains eight endemic fish species, including the Niangua darter (*Etheostoma nianguae*), bluestripe darter (*Percina cymatotaenia*), and Missouri saddled darter (*Etheostoma tetrazonum*), all endemic to the Osage and Gasconade watersheds. Also endemic are the Neosho madtom (*Noturus placidus*) and orangethroat darter (*Etheostoma spectabile squamosum*) in the middle Arkansas River (Hocutt and Wiley 1986). This ecoregion also includes one endemic mussel, one endemic aquatic herpetofauna species (the salamander *Eurycea tynerensis*), and a remarkable thirteen endemic crayfish, including the prairie crayfish (*Procambarus gracilis*), often found considerable distances from surface water in the grasslands (McDonald 1996).

One species of interest is the Ozark cavefish (*Amblyopsis rosae*), a federally endangered species. The range of the fish is limited to a small area where Arkansas and Oklahoma meet. It prefers caves housing bat colonies, such as Cave Springs Cave, where the endangered gray bat (*Myotis grisescens*) breeds (The Nature Conservancy 1996a).

Conservation Status

Despite the extensive agriculture and ranching that take place in the Central Prairie [28] ecoregion, it is considered relatively stable. Extensive areas of intact habitat remain, but local species declines and disruptions of ecological processes have occurred.

Three areas of note are the Upper Niangua, the Meramec River, and Crooked Creek. The Upper Niangua, home of the threatened Niangua darter, suffers from nutrient loading and fecal coliform contamination of both its surface water and groundwater. Agriculture and pasturing immediately adjacent to the watercourse are to blame for the pollution (U.S. EPA 1995). The Meramec River, located near the border of Missouri and Illinois, is critically imperiled. It has suffered extensive habitat losses due to wetland loss and channel modification, including dredging operations and levee construction. The water is also heavily contaminated by fecal coliform and nutrients from agricultural runoff (U.S. EPA 1995). In Crooked Creek, a tributary of the White River that is home to over forty species of fish, commercial in-stream gravel mining is degrading water quality and fish habitat (American Rivers 1998).

Suite of Priority Activities to Enhance Biodiversity Conservation

- Enforce the designation of riparian zones forbidden to livestock.
- Enforce responsible agricultural practices, such as terracing, to minimize nutrient loading.
- End dredging and levee construction operations to reestablish natural flow regimes.
- Establish stricter monitoring and management of water quality.

Conservation Partners

For contact information, please see appendix H.

- American Rivers
- Columbia Environmental Research Center
- Kansas Natural Heritage Inventory
- Missouri Department of Conservation
- MoRAP
- The Nature Conservancy (Arkansas, Kansas, Missouri, and Oklahoma Field Offices)
- U.S. Fish and Wildlife Service

Ecoregion Number:	**29**
Ecoregion Name:	**Ozark Highlands**
Major Habitat Type:	**Temperate Headwaters and Lakes**
Ecoregion Size:	**85,600 km²**
Biological Distinctiveness:	**Bioregionally Outstanding**
Conservation Status:	**Snapshot—Relatively Intact**
	Final—Relatively Intact

Introduction

This ecoregion extends from southern Missouri into northern Arkansas. It is largely defined by the watersheds of the middle portion of the Arkansas River and the White River, including its tributary the Black River. The Ozark Highlands [29] ecoregions contains abundant springs, which feed the headwaters of larger, free-flowing streams.

Biological Distinctiveness

The Ozark Highlands [29] ecoregion is part of the western Mississippi drainage but is distinctive because of its relative biogeographical isolation. It is a region of high-gradient headwater streams surrounded by coastal plains and prairie. This ecoregion and its neighbor, the Ouachita Highlands [30], boast a highly distinctive aquatic fauna, especially in crayfish. The Ozarks contain ten endemic fish species, three endemic mussels, and fifteen endemic crayfish.

The Ozark Highlands [29] is remarkable for its populations of aquatic herpetofauna, many of which are restricted to the confines of the Ozark Plateau and the adjacent Ouachita Highlands [30]. In all, there are forty-four species of salamanders, frogs, and aquatic snakes in this ecoregion. Though none is found exclusively in the Highlands, several are near-endemics, such as Strecker's chorus frog (*Pseudacris streckeri*) and the ringed salamander (*Ambystoma annulatum*).

Both the Ozark Highlands [29] and the Ouachita Highlands [30] ecoregions are remarkable for their rich crayfish fauna. Particularly distinctive are the populations of hypogean crayfish present in the extensive caves of the region, such as the Salem Cave crayfish (*Cambarus hubrichti*) and the bristly cave crayfish (*Cambarus setosus*). Both these animals are slimmer and smaller than their epigean cousins and almost completely lack pigment (Laws 1998).

Conservation Status

The Ozark Highlands [29] ecoregion is considered relatively intact. In this sparsely populated ecoregion, the environmental problems typical of the eastern United States are not as severe. However a number of of threats may imperil the biodiversity of this relatively pristine region in the near future.

Most pressing is the problem of population growth. Industrial pollution and exotic species invasions are likely to follow quickly on the heels of settlement of the area. Settlement also leads to residential and commercial development, and to point and nonpoint source pollution in the streams.

Mining companies have expressed interest in the mineral resources of the Highlands but have not been able to extensively exploit the region. The prospect of large scale heavy-metal pollution and sedimentation, such as that plaguing areas of the Appalachian Mountains, raises fears of ecological destruction in this region.

Suite of Priority Activities to Enhance Biodiversity Conservation

- Establish protected, publicly owned riparian zones, which cannot be logged or developed and function as pollution buffers.
- Restrict and strictly regulate current mining operations.
- Prevent further industrial development, especially by mining interests.
- Manage currently established exotic species and prevent introductions of other exotics.

Conservation Partners

For contact information, please see appendix H.

- American Rivers
- Arkansas Natural Heritage Commission
- Missouri Department of Conservation
- MoRAP
- The Nature Conservancy (Arkansas and Missouri Field Offices)
- The Ozark Society

Ecoregion Number:	**30**
Ecoregion Name:	**Ouachita Highlands**
Major Habitat Type:	**Temperate Headwaters and Lakes**
Ecoregion Size:	**115,370 km²**
Biological Distinctiveness:	**Bioregionally Outstanding**
Conservation Status:	**Snapshot—Relatively Stable**
	Final—Relatively Stable

Introduction

This ecoregion covers southeastern Oklahoma, northeastern Texas, southern Arkansas, and northwestern Louisiana. The rivers that drain the Ouachita Highlands [30] largely define this ecoregion. They include the middle portion of the Red River in the southwestern part of the ecoregion and the Ouachita River in the northeastern part. A small portion of the Arkansas River below Little Rock is also in this ecoregion.

The Ouachita Highlands [30] ecoregion is separated from the Ozark Highlands ecoregion [29] by the Arkansas River. Like the Ozark Highlands [29], this ecoregion is distinguished by its relative biogeographic isolation. As a source area for several larger streams and an area of high-gradient and spring-fed springs, it can almost be considered an island surrounded by the Great Plains, coastal plains, and prairie.

Biological Distinctiveness

The Ouachita Highlands [30] ecoregion has nine endemic fish species, three endemic mussels, and one endemic aquatic herpetofauna species. Most remarkable however, is that twenty-one of the thirty-four native species of crayfish are endemic. The springs that form part of the headwaters of the region also support highly specialized assemblages of fish and crayfish species.

The relatively untouched Kiamichi River is the only locality of the Ouachita rock-pocketbook (*Arkansia wheeleri*), the only known species of the mussel genus *Arkansia*. The Kiamichi is also home to twenty-eight other mussel species and over a hundred species of native fish. This river remains relatively intact thanks to the efforts of local landowners to preserve riparian zones and in-stream habitat (Master et al. 1998).

Conservation Status

The ecologic state of this region is fairly good, but introductions of exotic species have hurt populations of native fish. The primary source of these non-native species is Arkansas' aquaculture industry. Escapes from aquaculture pens are fairly commonplace, which lead to feral breeding populations.

Agriculture has also had a negative impact on the biodiversity of the Ouachita Highlands [30]. Runoff from sediment and agricultural chemicals (pesticides, herbicides, and fertilizers) ends up in the streams and harms native populations. Other sources of pollution are industrial interests such as manufacturing and mining.

Suite of Priority Activities to Enhance Biodiversity Conservation

- Establish protected, publicly owned riparian zones, which cannot be logged or developed and function as pollution buffers.
- Enforce strict management of aquaculture species, including pollution associated with concentrated breeding operations.
- Strictly regulate industrial development, especially by mining interests.

Conservation Partners

For contact information, please see appendix H.

- Arkansas Natural Heritage Commission
- Louisiana Natural Heritage Association
- The Nature Conservancy (Arkansas, Louisiana, Oklahoma, and Texas Field Offices)
- Oklahoma Natural Heritage Inventory
- Texas Conservation Data Center

Ecoregion Number:	**31**
Ecoregion Name:	**Southern Plains**
Major Habitat Type:	**Temperate Headwaters and Lakes**
Ecoregion Size:	**417,780 km²**
Biological Distinctiveness:	**Nationally Important**
Conservation Status:	**Snapshot—Vulnerable Final—Vulnerable**

Introduction

This ecoregion covers southeastern Colorado, northeastern New Mexico, most of southern Kansas, western Oklahoma, and the panhandle of Texas. It is largely defined by the watersheds of three rivers—the upper portions of the Arkansas, the South Canadian, and the upper half of the Red.

Biological Distinctiveness

This ecoregion is not distinguished by particularly high endemism, with only two endemic crayfish and three endemic fish. The fish are the Arkansas River shiner (*Notropis girardi*), Red River shiner (*Notropis bairdi*), and Arkansas darter (*Etheostoma cragini*). In addition, several fish and herpetofauna that are near-endemics, such as the bluntface shiner (*Cyprinella camura*), Topeka shiner (*Notropis tristis*), and crawfish frog (*Rana areolata*), are shared with only one or two other areas.

One distinctive feature of this semiarid region is the presence of heavily used aquifers, such as the famous Ogalalla Aquifer and the Arbuckle-Simpson Aquifer. The latter is relatively undeveloped, but critical subterranean and spring ecosystems in the area are threatened by pumping. The Oklahoma cave amphipod (*Allocrangonyx pellucidus*) is native to this region. This rare, endemic cave dweller is found only in the Arbuckle Mountains, and it depends on high water quality (The Nature Conservancy 1996a).

Another important freshwater ecosystem is the Cheyenne Bottoms wetland, the largest system of permanent and ephemeral wetlands left in the state of Kansas. It serves not only as a critical habitat for aquatic species, but also as a stopover point for more than half the population of northward-migrating shorebirds of North America (U.S. EPA 1995; The Nature Conservancy 1996e).

Conservation Status

Continued development of land and water resources for agriculture and municipal use is likely to have a negative impact on sensitive fauna, especially in cave and spring systems. The region faces not only overuse of surface water and groundwater, but the contamination of that water as well. Agrochemicals and animal wastes have found their way into the rivers from runoff, and into the groundwater through percolation. Another source of contamination has been mining in the western part of the ecoregion, especially on the upper Arkansas River (U.S. EPA 1995).

TNC has purchased 6,800 acres of Cheyenne Bottoms to be managed as a wetland instead of as marginal farmland (The Nature Conservancy 1996e). The total area of the wetland is 41,000 acres.

Suite of Priority Activities to Enhance Biodiversity Conservation

- Encourage responsible agricultural practices, such as terracing and drip irrigation, to prevent pollution and minimize aquifer drawdown.
- Designate Cheyenne Bottoms as a protected area.
- Remediate the effects of past mining activities through elimination of contaminated sediments.
- Enforce strict water-use limits.

Conservation Partners

For contact information, please see appendix H.

- Columbia Environmental Research Center
- Kansas Natural Heritage Inventory
- National Cattlemen's Beef Association
- The Nature Conservancy (Colorado, Nebraska, New Mexico, Oklahoma, and Texas Field Offices)
- Society for Range Management

Ecoregion Number:	**32**
Ecoregion Name:	**East Texas Gulf**
Major Habitat Type:	**Temperate Coastal Rivers, Lakes, and Springs**
Ecoregion Size:	**408,405 km²**
Biological Distinctiveness:	**Continentally Outstanding**
Conservation Status:	**Snapshot—Vulnerable; Final—Vulnerable**

Introduction

This ecoregion stretches from eastern New Mexico to southwestern Louisiana, covering most of central Texas. It is defined by the watersheds of the Neches, Trinity, Brazos, and Texas' Colorado Rivers and their numerous tributaries. In the southwestern portion of this ecoregion, and split with the West Texas Gulf [33] and the Lower Rio Grande/Río Bravo del Norte [20] ecoregions, lies the Edwards Plateau. This is a karst area characterized by the Edwards Aquifer, a body of groundwater that has a distinct biota associated with its caverns and springs (see essay 5).

Biological Distinctiveness

There are 268 aquatic species in the analyzed taxa in the East Texas Gulf [32] ecoregion, of which thirty-six species (13 percent) are considered endemic. This region harbors six endemic fish species, five endemic mussels, twelve endemic crayfish, and thirteen endemic aquatic herpetofauna.

The Edwards Plateau alone contains twelve endemic hydrobiid snails and two endemic unionid mussels, *Lampsilis bracteata*, and *Quincuncina guadalupensis* (Bowles and Arsuffi 1993). It is also the home of the endangered Texas blind salamander (*Typhlomolge rathbuni*) and the Edwards Plateau shiner (*Cyprinella lepida*). The aquatic invertebrate fauna of the region are, as one might expect, poorly known compared with its vertebrates (Edwards et al. 1989).

One particularly important river running through the East Texas Gulf [32] ecoregion is the Guadalupe River. The upper part of the watercourse cuts through the karst bedrock of the Edwards Plateau. This one river is responsible for four of the East Texas Gulf's [32] six endemic fish species: the Guadalupe bass (*Micropterus treculi*), the endangered fountain darter (*Etheostoma fonticola*), the greenthroat darter (*Etheostoma lepidum*), and the gray redhorse (*Moxostoma congestum*). In addition, the lower reach of the river harbors one of the region's unique herpetofauna, Cagle's map turtle (*Graptemys caglei*) (Master et al. 1998).

Conservation Status

Both the vertebrate and the invertebrate faunas of central Texas are vulnerable to rapidly increasing anthropogenic pressures. This ecoregion is arid to semiarid, and large stores of water are held underground. Agriculture and municipal growth are placing more and more demands on water that is already scarce. In all but the easternmost part of the state, existing water supplies are insufficient to meet the demand (Edwards et al. 1989). Not only are spring and aquifer communities threatened with extinction, but so are communities in large-river or permanent stream habitats. Flow regimes and vital habitats are being destroyed by channel modification and excessive water withdrawal. Riparian vegetation is being cleared to facilitate channelization and agriculture. This vegetation provides important habitat for aquatic species (in the form of rootmass, overhangs, and snags) and also serves a pollution-buffering function.

Pollution also contributes to the vulnerability of the East Texas Gulf [32] ecoregion. Extensive agriculture and pasturage in the region introduce large amounts of sediment, nutrients, fecal coliform bacteria, and agrochemicals into the surface water and groundwater.

Suite of Priority Activities to Enhance Biodiversity Conservation

- Enforce responsible agricultural practices, such as terracing and drip irrigation, to minimize pollution and aquifer drawdown.
- Enforce establishment and maintenance of riparian buffer zones.
- Enforce strict water limits for industrial and municipal uses.
- Designate protected areas based on freshwater habitats, including springs and aquifers.
- Complete a full biotic survey of the aquatic species living in this ecoregion, especially invertebrates.

Conservation Partners

For contact information, please see appendix H.

- Hill Country Federation
- The Nature Conservancy (Louisiana, New Mexico, and Texas Field Offices)
- Save Barton Creek Association
- Sierra Club
- Texas Parks and Wildlife

Ecoregion Number:	**33**
Ecoregion Name:	**West Texas Gulf**
Major Habitat Type:	**Subtropical Coastal Rivers, Lakes, and Springs**
Ecoregion Size:	**75,596 km²**
Biological Distinctiveness:	**Continentally Outstanding**
Conservation Status:	**Snapshot—Vulnerable**
	Final—Vulnerable

Introduction

This ecoregion is primarily defined by the watershed of the Nueces River and is contained entirely within southern Texas. Other watersheds of this ecoregion include those of the Mission River, the Aransas River, and Petronila Creek.

Biological Distinctiveness

The West Texas Gulf [33] ecoregion contains eight endemic fish species, no known endemic mussels or aquatic herpetofauna, and five endemic species of crayfish.

The northernmost portion of this ecoregion reaches into the Edwards Plateau, a zone of spring and distinctive subterranean habitats and fauna (see essay 5). Among the endemic fish of this ecoregion is the Edwards Plateau shiner (*Cyprinella lepida*). The Edwards Plateau is shared among the West Texas Gulf [33] ecoregion, the East Texas Gulf [32] ecoregion, and the Lower Rio Grande/Río Bravo del Norte [20] ecoregion, and is noted for its subterranean habitats and highly endemic fauna.

The arid coastal plains region that forms the West Texas Gulf [33] ecoregion shares many faunal assemblages with ecoregions farther south, on Mexico's relatively dry Gulf Coastal Plain. Examples of these shared species are the black spotted newt (*Notophthalmus meridionalis*) and the Mexican burrowing toad (*Rhinophrynus dorsalis*).

Conservation Status

This ecoregion is sparsely settled: the total population of the Nueces River watershed is slightly less than 200,000 (Nueces River Authority 1997). Because of the region's aridity, however, its water resources have been modified extensively to provide drinking water and irrigation for farmland. Two impoundments, Choke Canyon Reservoir and Corpus Christi Reservoir, provide water for municipal, industrial, and agricultural needs, but not enough to meet the demands of the region, so inhabitants must rely on groundwater as well. The deep aquifers that provide water for irrigation and municipal use are not easily recharged, however, and are being pumped at a rate that will eventually leave them dry. This imperils both human pursuits within this ecoregion and the future of its freshwater biodiversity. The Edwards Aquifer, shared among the West Texas Gulf [33], the East Texas Gulf [32], and the Lower Rio Grande (Bravo del Norte) [20] ecoregions, is in particular danger. The 1934–97 estimated annual discharge from wells ranges from 101,900 acre-ft in 1934 to 542,400 acre-ft in 1989. The average and median estimated annual recharges to the aquifer for 1934–97 are 676,000 and 547,100 acre-ft, respectively (Slattery et al. 1997). The obvious conclusion is that, in the very near future, drawdown rates in the area will exceed the aquifer's recharge rate.

Agricultural, industrial, and municipal pollution have impaired this region's water quality. Nutrient loading and fecal coliform contamination from pastured animals are common. Oil and natural gas exploration are dominant industries in this region, as is petrochemical manufacturing. Concerns have been raised regarding elevated fecal coliform levels on many of the West Texas Gulf [33] ecoregion's waterways and heavy-metal contamination on the Mission River (Texas Natural Resource Conservation Commission 1997b).

Suite of Priority Activities to Enhance Biodiversity Conservation

- Promote responsible agricultural practices, such as drip irrigation and terracing, to minimize evaporative water loss from irrigation and to prevent nutrient and agrochemical runoff.
- Enforce strict pollution management for industrial and municipal sources.
- Enforce riparian zones forbidden to livestock and pasturage.

Conservation Partners

For contact information, please see appendix H.

- The Nature Conservancy (Texas Field Office)
- Sierra Club
- Texas Center for Policy Studies
- Texas Parks and Wildlife

Focus on the Appalachian Mountains

The Appalachian region is one of the oldest and least-disturbed landscapes on earth, having undergone slow erosion as a once-mighty range as high as the Himalayas. When ancient supercontinents broke up around 200 million years ago, remnants of this vast mountain system drifted to present-day Greenland, Ireland, Great Britain, Norway, and North America. This last portion, by far the largest, today stretches from northern Alabama to Newfoundland.

An extraordinarily long period of relatively stable conditions, coupled with a highly complex terrain, have promoted a great diversification of biotas in the Appalachian region. High species richness, endemism, and beta-diversity within both freshwater and terrestrial habitats distinguish much of this region at a global scale. The Appalachians have also acted as a refuge for representatives of ancient taxa such as the hellbender (*Cryptobranchus alleganiensis*), a primitive salamander with relatives that now occur only in Japan and central China. Indeed, the equally ancient freshwater habitats of central China, and particularly of the Yangtze headwaters, share numerous biogeographic features with the Appalachian region.

The northern extent of the Appalachians, from New York to Newfoundland, has been subjected to extensive glaciation in the recent geologic past. The ubiquitous presence of ice prevented the survival of many ancient forms and the development of rich and endemic biotas. In contrast to the 375-million-year-old landscape of the southern Appalachians, the northern Appalachians have a biological landscape of only about ten to fifteen thousand years old. The southern freshwater ecoregions referred to as the Teays-Old Ohio [34], Tennessee-Cumberland [35], Mobile Bay [36], Apalachicola [37], South Atlantic [40], and Chesapeake Bay [41] escaped glaciation and exhibit the highest levels of freshwater biodiversity.

Other ancient temperate landscapes exist that have never been glaciated, but the Appalachians are different. When the mountains formed, the rock was folded in such a way that a series of long, nearly continuous ridges and valleys was created. At the beginning of the Appalachians' lifetime, the mountains were very high (possibly 27,000 to 29,000 ft); consequently, the individual valleys between ridges were almost completely isolated. Even as they wore down, the valleys were still relatively isolated, and the ridges themselves formed niches that could be colonized. In addition, the southern Appalachians are within a large karst (limestone-based bedrock) zone; the result is a diverse array of freshwater habitats, with lengthy isolation of freshwater biotas restricted to separate watersheds.

Ecoregion Number:	**34**
Ecoregion Name:	**Teays-Old Ohio**
Major Habitat Type:	**Temperate Headwaters and Lakes**
Ecoregion Size:	**373,887 km²**
Biological Distinctiveness:	**Globally Outstanding**
Conservation Status:	**Snapshot—Vulnerable**
	Final—Vulnerable

Introduction

Predominantly within the physiographic provinces of the Appalachian Plateau in the east, the Central Lowlands, and the Interior Low Plateau in the southwest, this ecoregion is defined largely by the watershed of the present-day Ohio River. Three other provinces, the Ridge and Valley, Blue Ridge, and a small part of the Gulf Coastal Plain, occur here as well. Other major rivers in this ecoregion include the Wabash in Indiana, the Green River and Kentucky River in Kentucky, the Scioto and Muskingum Rivers in Ohio, the New River in West Virginia, and the Monongahela, Youghiogheny, and Allegheny Rivers in Pennsylvania. In total, the ecoregion covers parts of eleven states: Illinois, Indiana, Ohio, Kentucky, West Virginia, Tennessee, North Carolina, Virginia, Maryland, Pennsylvania, and New York.

The ecoregion's name is derived from the historic Teays River and the ancient course of the Ohio River before the last ice age. Before advancing glaciers blocked their flows, many of the rivers in the eastern part of the region, including the Allegheny and Monongahela, flowed northward into the Laurentian system that today is composed of the St. Lawrence River and its tributaries. Consequently, fishes that had been confined to the Hudson Bay and

Laurentian system were displaced into the Old Ohio during glaciation (Burr and Page 1986).

Historically, much of this ecoregion was forested, including areas where rich soils were deposited by the last glaciers. Much of the lower, downstream portion of the ecoregion, which was not glaciated, includes an extension of the Mississippi alluvial plain, where bottom-land hardwood forests and swamps were once common (USFWS 1995).

Biological Distinctiveness

The Teays-Old Ohio [34] ecoregion is considered globally outstanding because of the sheer numbers of aquatic species found within it. With its 206 native fish species, 122 unionid mussels, 49 crayfish, and 60 native amphibians and aquatic reptiles, this ecoregion is second only to the Tennessee-Cumberland [35] ecoregion in total number of species. This high level of richness is derived principally from the diversity of upland and lowland habitats and the presence of both glaciated and unglaciated areas (Burr and Page 1986).

Endemism is moderately high in the ecoregion: 12 percent of fish, 14 percent of mussels, 47 percent of crayfish, and 5 percent of herpetofauna are found nowhere else. Certain basins within this ecoregion have markedly higher endemism than others. For instance, the upper Green River drainage has an endemic sucker and four endemic darters (*Moxostoma atripinne*, *Etheostoma barbouri*, *E. barrenense*, *E. bellum*, and *E. rafinesquei*), while the Wabash River has no endemics (Burr and Page 1986). Within the entire ecoregion, the endemic fish fauna consists of six minnows, two catfish, two cave springfish, and fourteen darters. Several of these endemics are also found in the Tennessee-Cumberland [35] to the south but have such limited distributions that they can be considered endemic to both. Endemism in herpetofauna is limited to three species of salamanders: the Black Mountain salamander (*Desmognathus welteri*), West Virginia spring salamander (*Gyrinophilus subterraneus*), and streamside salamander (*Ambystoma barbouri*). Like some of the endemic fishes, the Black Mountain salamander has a restricted range that falls within the southeastern portion of this ecoregion and the northeastern part of the Tennessee-Cumberland [35]. In general, the fauna of this ecoregion is more cosmopolitan than that of the Tennessee-Cumberland [35].

A native Ohio River crayfish species known as the rusty crayfish (*Orconectes rusticus*) is of special interest. Much like the flathead catfish, this voracious predator has been introduced into numerous other rivers and streams across the United States, primarily by bait dealers. Evidence suggests that this crayfish generally threatens to eliminate native crayfishes wherever it is introduced (Clancy 1997; Taylor pers. comm.).

Conservation Status

All threats that face flowing-water systems—impoundments, pollution (industrial, agricultural, urban runoff, acid rain), excessive sedimentation, mining and associated acid-mine runoff, general poor land-use practices, direct alteration and destruction of wetland habitats, and rapid urbanization—are prevalent within this ecoregion. According to Burr and Page (1986), this ecoregion shares with the Mississippi [24] the dubious distinction of having experienced the most dramatic changes in fish fauna of any area within the eastern United States. Furthermore, the authors suggest that changes in species distribution and abundance are "continuing at ever-increasing rates." Like most of the eastern United States, much of this ecoregion's pre-settlement landscape was forested and has been logged at least once since the arrival of Europeans. Although some areas have regrown, agriculture continues to be a predominant land use. It has been estimated that well over 50 percent of the land cover has been altered.

Today, portions of the ecoregion are heavily populated, particularly in larger industrial cities such as Pittsburgh, Cincinnati, and Indianapolis. Growing urban areas and their populations demand goods, more space, and sewage facilities. As a result, swamps and other types of wetlands have been drained and rivers have been altered to ensure navigable channels. With the widening, dredging, and straightening of streams, important habitats for aquatic species are lost, further intensifying stress on rivers and their biotas.

Degraded water quality poses a serious threat to this ecoregion's biota. Agricultural pollution, including pesticides, as well as chemicals and other hazardous materials from industrial and urban development, are prevalent in virtually all parts of the ecoregion. Surface mining in large areas of eastern Kentucky, Illinois, and parts of West Virginia has had a tremendous effect on water quality. Acid-mine runoff can cause dramatic fish kills, and mine siltation adds even more particles to already cloudy waters. Areas where mining has been extensive are often devoid of fish. Examples include the Saline River in southeastern Illinois and the Pond River in west-central Kentucky (Burr and Page 1986).

Areas like Cincinnati, Ohio, are not only heavily populated but also heavily industrialized. Many rivers and streams support the economic growth of the towns and cities that surround them. Landfills, factories, toxic waste sites, and residential sewage treatment facilities are located near rivers. An example is Mill Creek, which drains roughly 60 percent of the city of Cincinnati. The problems caused by these point sources of pollution, as well as by uncontrolled urban and suburban stormwater runoff, exemplify the ways that cities across the ecoregion tend to affect the waters on which they rely. The problems at Mill Creek prompted American Rivers to designate it as one of the nation's ten most-threatened rivers in 1997 (American Rivers 1997).

Sedimentation, whether from agriculture, mining, logging, or urban development, is of particular concern as a major threat to the highly distinctive mussel fauna of this ecoregion. Additionally, off-road vehicles in some areas can create or add to already heavy sediment loads. Areas of the Daniel Boone National Forest (DBNF), which spans this ecoregion and the neighboring Tennessee-Cumberland [35], are threatened by the growing popularity of this recreational activity (Kentucky State Nature Preserves Commission 1996).

Within the last ten years a practice known as mountaintop removal coal mining has been increasingly employed in this ecoregion, particularly in West Virginia. This method of mining low-sulfur coal involves the literal removal of mountaintops and the dumping of the resultant soil and rocks in stream valleys. The environmental costs are enormous, including the burial of entire stream reaches, normally in headwater areas. According to the U.S. Fish and Wildlife Service, more than 470 miles of West Virginia's streams in the Logan mining region were destroyed between 1986 and 1998 (American Rivers 1999).

An emerging threat to native mussels is the introduced zebra mussel (*Dreissena polymorpha*). First sighted in the lower reaches of the Ohio River in 1991 (U.S. Army Corps of Engineers 1998), this small but prolific exotic often colonizes native species, suffocating them in the process.

Overharvesting of freshwater mussels, particularly pearly mussels, primarily for the use of shells as seed pearls, may be a problem in this ecoregion. Although the extent of the problem is not fully known, mussel researchers believe that current levels of exploitation cannot be sustained. The USFWS and state agencies are largely responsible for enforcement of existing regulations. As part of a strategic plan to address the entire Ohio River Valley ecosystem, the USFWS proposed developing baseline data and monitoring existing populations in the region through Fiscal Year 2000 to assess the degree of threat. Potentially, the permitted limits for harvesting will have to be adjusted. The poaching of endangered mussel species by collectors and subsistence fisherman may be a smaller-scale threat (Mueller 1993).

Suite of Priority Activities to Enhance Biodiversity Conservation

- Remove physical barriers obstructing fish movements in selected sites, specifically the Green River Lake Dam on the Green River in Kentucky.
- Work to form cooperative partnerships among local residents, government, local businesses, and industry to protect habitat of imperiled species.
- Develop enforceable regulations to minimize nonpoint source pollution from agriculture, which has resulted in major fish kills.
- Develop innovative ways to stop the spread of zebra mussels and control established populations. Additional resources should be allocated to fund research in this area.
- Initiate programs that work with local farmers to mitigate the effects of cattle grazing on soil erosion. An example is the program within the Tennessee-Cumberland's [35] Clinch River area, through which a consortium of government agencies and TNC have worked to "create financial incentives to help farmers keep cattle out of rivers voluntarily" (The Nature Conservancy 1996b).

- Develop a national urban stream restoration program that would work cooperatively with local citizens and grassroots groups to reclaim degraded urban rivers (American Rivers 1997).
- Revise the forest management plan of the DBNF to ensure that biologically important areas are preserved; special attention should be given to restricting off-road vehicle use in and near these sites. Additionally, more resources should be allocated to strictly enforce these restrictions (Kentucky State Nature Preserves Commission 1996).
- Continue to monitor and assess the problems of mussel poaching and potential overharvesting. Secure funding to continue work by the USFWS to conduct this research. Loopholes in legislation that allow for falsification of harvesting records should be closed. Additional funding is necessary for research into the effects of these problems.
- Eliminate mountaintop removal coal mining by forcing state and federal agencies to comply with the Clean Water Act and Surface Mine Control and Reclamation Act (American Rivers 1999).

Conservation Partners

For contact information, please see appendix H.

- American Rivers
- Illinois Natural Heritage Division
- Indiana Natural Heritage Data Center
- Kentucky Natural Heritage Program
- The Nature Conservancy (Illinois, Indiana, Kentucky, Maryland, New York, Ohio, Pennsylvania, and West Virginia Field Offices, and Midwest and Southeast Regional Offices)
- New York Natural Heritage Program
- Ohio Natural Heritage Database
- Ohio Valley Environmental Coalition
- Pennsylvania Natural Diversity Inventory–West
- Sierra Club
- Tennessee Division of Natural Heritage
- U.S. Fish and Wildlife Service
- U.S. Forest Service
- Virginia Division of Natural Heritage
- West Virginia Highlands Conservancy
- West Virginia Natural Heritage Program
- Western Pennsylvania Conservancy
- Wildlands Project

Ecoregion Number:	**35**
Ecoregion Name:	**Tennessee-Cumberland**
Major Habitat Type:	**Temperate Headwaters and Lakes**
Ecoregion Size:	**152,292 km²**
Biological Distinctiveness:	**Globally Outstanding**
Conservation Status:	**Snapshot—Vulnerable**
	Final—Endangered

Introduction

The watersheds of the Tennessee and Cumberland Rivers, which drain to the larger Mississippi Basin, define this ecoregion. Originating in the Appalachian Highlands of Virginia, the Tennessee River drainage covers more than 103,600 km² (Ono et al. 1983). The majority of this area is centered in Tennessee; the river also drains parts of southeastern Kentucky, western North Carolina, two disjunct areas in northern Georgia, northern Alabama, and the extreme northeastern corner of Mississippi. In the northern portion of the ecoregion, the mainstem Cumberland River originates at the confluence of the Poor and Clover forks; in total, the Cumberland drains more than 46,000 km² before joining the Ohio River at Smithland, Kentucky (Ono et al. 1983). Major tributaries to the Tennessee include the Clinch, Powell, Holston, French Broad, Duck, Elk, Buffalo, Bear Creek, Paint-Rock, Sequatchie, Little Tennessee, and Hiwassee Rivers. Tributaries to the Cumberland include the Big South Fork, Rockcastle, and Little Rivers.

Although the Tennessee and Cumberland Rivers flow quite close to each other near their confluence with the Ohio, they were not physically linked historically. Today, dam construction on both rivers has changed this situation. The construction of Barkley Dam impounded the Cumberland, forming Lake Barkley, while just a few miles away the Kentucky Dam was built to impound the Tennessee River, thereby creating Kentucky Lake. This alone was not enough to link the two reservoirs, so a channel was cut not far from the head of each lake to link them together. Other main-stem and tributary reservoirs constructed by the Tennessee Valley Authority (TVA) for flood storage and power generation are also major surface water features.

From east to west, the Tennessee-Cumberland [35] ecoregion traverses a number of physiographic provinces, creating a broad diversity of freshwater habitats. The lower Tennessee River Basin drains a small portion of the coastal plain, and moderate- to lower-gradient streams occur in this area. Swamps occur in the Big Sandy system, a major lower tributary to the Tennessee in the northwest corner of the ecoregion. Cave and spring habitats are abundant in the Highland Rim province, which covers most of the western half of the ecoregion. To the east of the Highland Rim is the Cumberland Plateau, and to the east of that is the Ridge and Valley province. Finally, the southeastern headwaters of the Tennessee drainage are found in the Blue Ridge province, where streams are typically high gradient and cold (Starnes and Etnier 1986).

Much of the ecoregion is covered in forest, particularly in the Jefferson, Pisgah, Cherokee, Nantahala, and Chattahoochee National Forests. Agriculture also accounts for a major land use; most agricultural land is used for pasturage (Hampson 1995).

Biological Distinctiveness

This ecoregion contains the highest level of freshwater diversity in North America and is possibly the most diverse temperate freshwater ecoregion in the world (Starnes and Etnier 1986; Olson and Dinerstein 1998). It is the most species-rich ecoregion in fish, mussel, and crayfish species, and has the highest number of endemics of all North American ecoregions. This high diversity is derived largely from the range of habitat types represented in the ecoregion, as well as the ecoregion's location adjacent to Atlantic Slope, eastern Gulf Slope, lower Mississippi River, and Ohio River drainages, all with distinctive faunas (Starnes and Etnier 1986).

The ecoregion is perhaps best known for its fish fauna, which numbers an extraordinary 231 species, of which 67 (29 percent) are endemic. These endemics are made up of sixteen minnows, five suckers, two cave springfish, one killifish, one pygmy sunfish, one sculpin, and an incredible forty-one darters. Of the many physiographic provinces in this ecoregion, the Highland Rim and Ridge and Valley tend to support the largest numbers of fish species. Many species with the most restricted ranges are found where provinces meet and overlap; for instance, the palezone shiner (*Notropis* sp.), smoky madtom (*Noturus baileyi*), and duskytail darter (*Etheostoma* sp.) all apparently require habitat created by the combination of features in two provinces (Etnier and Starnes 1993). New species of fish continue to be discovered and described in this ecoregion, despite fairly extensive historical study of the region's fauna (Etnier and Starnes 1993; Burr pers. comm.).

The conditions that have engendered this region's distinct fish fauna have likewise produced a globally outstanding unionid mussel and crayfish fauna. Historically, the ecoregion has supported 125 mussel and 65 crayfish species, of which 20 (16 percent) and 40 (62 percent) are endemic, respectively. These numbers exceed those for any other ecoregion in North America. Certain areas within this ecoregion are particularly noteworthy. For example, the combined drainages of the Clinch, Powell, and Holston Rivers in southwest Virginia are the most outstanding in terms of mussel richness, with sixty species (The Nature Conservancy 1996a, 1996b).

Recognized as a center of diversity for salamanders, the Appalachian Mountains are home to numerous species, many of which are restricted to the Tennessee-Cumberland [35] (Conant and Collins 1991; Constanz 1994). In particular, this ecoregion harbors eight endemic lungless salamanders of the family Plethodontidae. These are the Santeetlah dusky salamander (*Desmognathus santeetlah*), Black Mountain salamander (*Desmognathus welteri*), pigmy salamander (*Desmognathus wrighti*), imitator salamander (*Desmognathus imitator*), shovelnose salamander (*Leurognathus marmoratus*), Junaluska salamander (*Eurycea junaluska*), Blue Ridge two-lined salamander (*Eurycea wilderae*), and subterranean Tennessee cave salamander (*Gyrinophilus palleucus*) (Conant and Collins 1991).

Conservation Status

According to Master et al. (1998), more than fifty-seven species of fish and forty-seven species of mussels are at risk in the Tennessee-Cumberland [35] ecoregion. Among the numerous endangered fishes are the boulder darter (*Etheostoma wapiti*), smoky madtom (*Noturus baileyi*), and pygmy madtom (*Noturus stanauli*), all endemic to the Tennessee River drainage. In the Cumberland, the endemic bluemask (or jewel) darter (*Etheostoma (doration) sp.*) is endangered. Endangered mussels species include the Appalachian elktoe (*Alasmidonta raveneliana*), dromedary pearlymussel

(*Dromus dromas*), shiny pigtoe mussel (*Fusconaia cor*), birdwing pearlymussel (*Lemiox rimosus*), Cumberland monkeyface mussel (*Quadrula intermedia*), and Cumberland bean mussel (*Villosa trabalis*). Anthony's river snail (*Athearnia anthonyi*), an endemic aquatic gastropod, is another endangered invertebrate whose habitat has been largely eliminated by impoundments on the Tennessee River. The endangered and endemic Nashville crayfish (*Orconectes shoupi*), located in only a few streams in middle Tennessee, including Mill Creek, is one of four federally listed crayfish in the United States (Clancy 1997; State Goals and Indicators Project 1998)

This rich ecoregion has been disturbed by every category of threat facing flowing-water systems: impoundment, channelization, pollution (industrial, agricultural, urban runoff, acid rain, acid-mine runoff), excessive sedimentation, and rapid urbanization (around Knoxville, Nashville, and the paired tourist meccas of Gatlinberg and Pigeon Forge). Specific threats to water quality include sedimentation from logging on steeply sloping lands and clearing of riparian zones for planting; nutrient enrichment from agricultural nonpoint source pollution; and contamination by industrial organic chemicals and trace elements (PCBs, dioxins, and mercury, as well as radionuclides in the area around the Department of Energy's Oak Ridge Reservation; Hampson 1995). The combination of a high level of threat and species with highly localized distributions has led to the listing of a number of this ecoregion's species as imperiled at the federal and state levels.

An example of the multiple impacts of human activity on the ecoregion's biota is the boulder darter, which was once found in three tributaries of the Elk River but today is restricted to only two of these, Richland and Indian Creeks. According to the USFWS (1990), the species' decline has resulted from construction of reservoirs, phosphate mining in the Elk River watershed, land-use changes, pesticides, toxic chemical spills, and possible elimination of habitat from the Elk River by cold-water releases from the Tims Ford Reservoir. Extirpation of the boulder darter from the Tennessee River as a result of flooding from the Wheeler and Wilson Dams is also a possibility. Pollution in Shoal Creek, a tributary of the Tennessee, is another threat (USFWS 1990).

The rich mussel fauna of this ecoregion is clearly under attack as well. Species such as the Appalachian elktoe are threatened by sedimentation and other pollutants (fertilizers, pesticides, heavy metals, oil, salts, organic wastes) from nonpoint sources (USFWS 1990). Sedimentation has also been implicated in the decline of the imperiled birdwing pearlymussel, fanshell mussel (*Cyprogania stegaria*), and dromedary pearly mussel. Local farmers use nearby streams as water sources for cattle, a practice that causes streambank erosion. Additionally, environmentally unsound forestry practices, including the construction and use of erosion-prone logging roads (The Nature Conservancy 1996a), cause increased sediment loads in streams, further threatening these sensitive mussel species.

Another major cause of soil erosion is the growing use of off-road vehicles. Perhaps more than in any other part of the ecoregion, this threat has major implications for the Daniel Boone National Forest (DBNF), an important macrosite for the conservation of many freshwater species, particularly mussels (Kentucky State Nature Preserves Commission 1996). One affected mussel species is the little-wing pearly mussel (*Pegias fabula*), found in Horse Lick Creek. The issue of off-road vehicle damage is addressed in detail in Kentucky's management plan for the forest.

This ecoregion has been extensively dammed, largely since the 1933 creation of the TVA in response to population growth and poor living conditions. As the TVA efficiently converted the free-flowing river system to a series of impoundments, the varied habitats of the Tennessee and its tributaries were lost and extensive mussel shoals were destroyed (Ono et al. 1983). Today, the Tennessee River alone has more than fifty major dams. Similarly, the Cumberland River has been dammed to a large extent since the early part of the twentieth century.

Channelization of streams and rivers in the ecoregion is extensive and has touched nearly all flowing-water habitats. Channelization has historically been undertaken to benefit navigation but today is also used to control flow in areas of encroaching development. For example, habitat for the slackwater darter (*Etheostoma boschungi*), known from only a few creeks in north-central Alabama and south-central Tennessee, is threatened by channelization activities related to suburban sprawl from nearby Huntsville, Alabama. Because the fish breeds in flooded fields or pastures, preservation of these open spaces is also critical for its survival. The sprawl-related destruction of this habitat is only one pressure; these breeding areas are also threatened by possible conversion to fish-farm ponds (USFWS 1998b).

Another example of a potential future threat is provided by the Cumberland bean mussel, known to exist in only a half dozen stretches of the Cumberland and Tennessee Rivers and their tributaries. A recent proposal to dam a portion of the Laurel Fork of the Rockcastle River in Jackson County, Kentucky, threatens an entire population of the species. The dam is intended to provide a recreational lake that could create tourism jobs and help secure an ample supply of water for homes and factories that might result from the new lake (Tagami 1998).

The introduction of non-native species also poses a problem, with the zebra mussel (*Dreissena polymorpha*) presenting perhaps the most urgent case. This aggressive exotic is present in both the Cumberland and Tennessee Rivers, though it is less prevalent in the Tennessee where barge traffic is lighter (U.S. Army Corps of Engineers 1998). Species native to the Mobile Bay [36] ecoregion to the south have also made their way into the Tennessee-Cumberland [35] via the Tennessee-Tombigbee Waterway (see Mobile Bay ecoregion description); these include the blacktail shiner (*Cyprinella venusta*) and weed shiner (*Notropis texanus*). Intentionally introduced fish include the alewife (*Alosa pseudoharengus*), common carp (*Carpiodes carpio*), brown trout (*Salmo trutta*), and striped bass (*Morone saxatilis*).

Overharvesting of pearly mussels, collected and sold to the Japanese for use by the cultured pearl industry, is another potential problem (Williams et al. 1993). Although the extent of the problem is not known, levels of exploitation are probably not sustainable. Currently no federal restrictions exist on the harvesting of mussel species that are not endangered or threatened. Efforts to assess the level of this threat should be undertaken. Poaching of endangered mussel species also poses a threat to the mussel fauna of this ecoregion (Mueller 1993).

Suite of Priority Activities to Enhance Biodiversity Conservation

- Continue to develop public education campaigns, in coordination with groups such as TNC and the Tennessee Aquarium.
- Focus protection and restoration efforts on selected stream reaches with high biodiversity value (e.g., the Clinch River biopreserve).
- Work to form cooperative partnerships among local residents, government, local businesses, and industry to protect habitat of imperiled species, including breeding habitats.
- Keep the Laurel Fork of the Rockcastle River free flowing, working to ensure that it is not dammed for recreational purposes (Tagami 1998).
- Expand programs that work with local farmers to mitigate the effects of cattle grazing on soil erosion. For example, TNC has formed a partnership with USFWS, the TVA, the Natural Resource Conservation Service, and the Virginia Division of Soil and Water Conservation to create financial incentives that will help farmers keep cattle out of the rivers voluntarily (The Nature Conservancy 1996b).
- Continue efforts that focus on streambank restoration, including fencing off streambanks to cattle and constructing cattle stream crossings (The Nature Conservancy 1996b).
- Develop zoning guidelines that direct urban development into areas away from biologically sensitive stream reaches. Avoid channelization of these areas and maintain riparian buffers.
- Revise the forest management plan of the Daniel Boone National Forest to ensure that biologically important areas are preserved; special attention should be given to restricting off-road vehicle use in and near these sites. Additionally, more resources should be allocated for strict enforcement of these restrictions (Kentucky State Nature Preserves Commission 1996).
- Conduct research to assess the impact on freshwater biota of three high-capacity chip mills on the Tennessee River, and associated clear-cut logging. Take action to halt impacts if necessary.
- Develop enforceable regulations to minimize nonpoint source pollution from agriculture, which has resulted in major fish kills.
- In lieu of dissolving all TVA nonpower programs (including its environmental programs), as currently proposed, work to reorganize the TVA's programs to meet environmental needs (American Rivers 1997).
- Develop innovative ways to stop the spread of zebra mussels and control established populations. Additional resources should be allocated to fund research in this area.

Conservation Partners

For contact information, please see appendix H.

- Alabama Natural Heritage Program
- Georgia Natural Heritage Program
- Kentucky Natural Heritage Program
- Mississippi Natural Heritage Program
- The Nature Conservancy (Alabama, Kentucky, North Carolina, Tennessee, and Virginia Field Offices, and Midwest and Southeast Regional Offices)
- North Carolina Heritage Program
- The Sierra Club
- Tennessee Aquarium
- Tennessee Division of Natural Heritage
- U.S. Forest Service
- Virginia Division of Natural Heritage
- Wildlands Project
- The Wilderness Society, Southeast Region

Ecoregion Number:	**36**
Ecoregion Name:	**Mobile Bay**
Major Habitat Type:	**Temperate Coastal Rivers, Lakes, and Springs**
Ecoregion Size:	**115,514 km²**
Biological Distinctiveness:	**Globally Outstanding**
Conservation Status:	**Snapshot—Endangered Final—Critical**

Introduction

The Mobile Bay [36] ecoregion encompasses the Mobile, Tombigbee-Black Warrior, and Alabama-Coosa-Tallapoosa Basins, along with the drainages of numerous smaller streams and lakes. These connected systems form the largest drainage basin in the East Gulf Coastal Plain. This ecoregion is centered in central Alabama and includes eastern Mississippi, western Georgia, and a small area in southern Tennessee. Above the Fall Line, the ecoregion's rivers flow through the Valley and Ridge Province, the Piedmont Upland, and the Appalachian Plateau, then descend into the East Gulf Coastal Plain. Historically, rivers and streams in this ecoregion stretched over 1,000 miles; today, flow in the Mobile River is regulated by a series of upstream reservoirs on the Etowah, Coosa, and Tallapoosa Rivers as well, and to a lesser extent by the locks and dams of the Tombigbee River (Livingston 1992; USFWS 1993; Stolzenburg 1997).

Biological Distinctiveness

Due largely to the variety of physiographic provinces occurring in this ecoregion, its size, and its escape from Pleistocene-era glaciation, the Mobile Bay [36] ecoregion has the highest level of aquatic diversity and endemism in the eastern Gulf (USFWS 1993). The basin once supported 120 species of aquatic snails, comprising the most diverse such assemblage in the world (Stolzenburg 1997), and the combined system of the Etowah, Oostanaula, and Coosa Rivers once supported the largest diversity of all freshwater mollusks in the world (American Rivers 1997). The Cahaba River, a tributary to the Alabama, itself is home to 131 fish species, more than any other river of its size in North America, as well as numerous mussels, crayfish, and snails (Stolzenburg 1997).

The Mobile Bay [36] ecoregion gains its extraordinary fish richness from its combination of lowland and upland species. In addition to supporting forty-seven endemic species, it also shares a number with only one other ecoregion. Ninety-nine of 187 species are from the Cyprinidae (minnows), Catostomidae (suckers), and Percidae (darters, perches, walleye, sauger) families. Of particular interest are three species known only from a few isolated upland springs in this ecoregion—the coldwater darter (*Etheostoma ditrema*) and pygmy sculpin (*Cottus pygmaeus*) of the Coosa drainage, and the endangered watercress darter (*Etheostoma nuchale*) of the Black Warrior drainage (Swift et al. 1986).

Conservation Status

This ecoregion's freshwater fauna has been heavily affected by human activities, with fifty-four documented extinctions since European settlement. In-stream threats to habitats and species of the Mobile Basin include channel modification for navigation and flood control, impoundment, pollution, and gravel dredging. Within the basin, thirty-three major dams have been built in the past century for navigation, flood control, water supply, and hydroelectric production (USFWS 1993), with observable impacts. Dams on the Coosa River, for example, eliminated important habitat for aquatic snails, with the resultant loss of at least twenty-seven species (Stolzenburg 1997).

In 1993, the USFWS listed eight Mobile Bay [36] unionid mussel species as endangered and three as threatened, and described the general threats to the species as "habitat modification, sedimentation, and water quality degradation." In its Final Rule on the eleven species, the USFWS (1993) wrote that "none of the species are known to tolerate impoundments . . . . Impoundments adversely affect riverine mussels by: killing them during construction and dredging; suffocation by accumulating sediments; lowered food and oxygen availability by the reduction of water flow; and the local extirpation of host fish. Other forms of habitat modification such as channelization, channel clearing and desnagging, and gravel mining result in stream bed scour and erosion, increased turbidity, reduction of groundwater levels, sedimentation, and changes in the aquatic community structure."

Modifications of the Mobile Bay's [36] hydrology and channel structure are numerous, but the Tennessee-Tombigbee Waterway deserves special mention. Completed in 1984, the waterway is a 234-mile navigation link that connects the Tennessee River (of ecoregion [35]) to the Black Warrior/Tombigbee River system with a series of locks and dams. According to the USFWS (1990), "Most of the Tombigbee River was modified by construction of the Tennessee-Tombigbee Waterway. This resulted in the loss of river habitat by impoundment, channelization, and flow diversion. Habitat that should continue to support mussel populations has been destroyed by heavy accumulations of sediment. The entire Tombigbee River has been modified for navigation by impoundment; channelization and frequent dredging are required to maintain the navigation channel." Interbasin transfer also results in altered flow and possible net loss of water from the basin. A future threat is the proposed transfer of sewage water from the Apalachicola [37] to the Etowah River.

Water withdrawals to satisfy the growing needs of the Atlanta metropolitan area also pose a major threat to this ecoregion. According to American Rivers (1999), "Water managers have proposed construction, re-operation, and modification of dams in the headwaters of the Coosa and Tallapoosa Rivers. Most notably, the proposal includes a plan to build a major dam and impoundment on one of the last remaining healthy, free-flowing stretches of the Tallapoosa. Construction and poor operation of dams in the basin threaten to block and destroy habitat . . . and dramatically alter the timing and magnitude of water flows in the Alabama-Coosa-Tallapoosa River Basin." A water-allocation plan known as the ACT Compact, to be completed by the states of Alabama, Georgia, and Florida by December 1999, has the potential to either prevent or encourage further degradation of the basin (American Rivers 1999).

Terrestrial land uses pose threats to freshwater biota that are potentially as grave. Natural vegetative cover has been extensively cleared in the Mobile Basin for agriculture, and mixed hardwood forests have been converted to pine plantations. These land uses, in addition to construction and general development, have led to erosion and sedimentation of the ecoregion's freshwater habitats, as well as additional impacts associated with loss of riparian cover. The Etowah River is a prime example of the effects of these land uses, with the highest level of imperiled aquatic organisms for any similarly sized river in the United States (American Rivers 1997).

Point and nonpoint source pollution have also degraded freshwater habitats and their biota. Point sources include municipal and industrial effluents and coalbed methane wells; sources of nonpoint source pollution are agriculture, feed lots and poultry houses, pastures, coal-mine sites, and highway and road drainage. Pollution is particularly severe in the Black Warrior River Basin, especially around the Birmingham-Jefferson County area, because of the presence of feed lots and abandoned and working coal mines. The Cahaba River, though unimpounded, also suffers from poor water quality, a result of discharges and runoff from ten municipal wastewater treatment plants, thirty-five surface mining areas, one coalbed methane operation, and sixty-seven other permitted discharges in the Cahaba River Basin (USFWS 1993).

Introduced species, particularly the Asiatic

clam (*Corbicula fluminea*) and the zebra mussel (*Dreissena polymorpha*), pose threats to native species, especially mollusks. The Asiatic clam is already well established and may compete with native mussels for space and nutrients, as well as disrupt the predator-prey balance between muskrats and mussels. The zebra mussel has not yet been reported from the Mobile Bay Basin; however, its movement into the basin via the Tennessee-Tombigbee Waterway is anticipated, as it has been reported from the Tennessee River (USFWS 1993).

Suite of Priority Activities to Enhance Biodiversity Conservation

- Protect the Cahaba River, one of the most species-rich freshwater rivers of its size, which remains unimpounded.
- Prevent the city of Atlanta from constructing a pipeline to pump sewage discharge from the Chattahoochee Basin to the Etowah Basin (American Rivers 1997).
- Eliminate point source pollution by developing and enforcing stricter regulations, and by cleaning up abandoned mine sites.
- Develop enforceable regulations to reduce nonpoint source pollution.
- Build no new impoundments, and investigate possible impoundments for removal.
- Develop an education program on the protection of aquatic biodiversity and aquatic habitats, combined with action aimed at protection of riparian communities.
- Work to control the spread of the introduced Asiatic clam, *Corbicula fluminea*, and the likely introduction of the zebra mussel, *Dreissena polymorpha*, via the Tennessee-Tombigbee Waterway.
- Encourage state and federal management authorities to work more actively toward watershed protection.

Conservation Partners

For contact information, please see appendix H.

- Alabama Natural Heritage Program
- Alabama Rivers Alliance
- Alabama Wilderness Association
- American Rivers
- Cahaba River Society
- Coosa River Basin Initiative

- Coosa River Society
- Georgia Natural Heritage Program
- Mississippi Natural Heritage Program
- The Nature Conservancy (Alabama, Georgia, and Mississippi Field Offices, and Southeast Regional Office)
- The Sierra Club

Ecoregion Number:	**37**
Ecoregion Name:	**Apalachicola**
Major Habitat Type:	**Temperate Coastal Rivers, Lakes, and Springs**
Ecoregion Size:	**54,403 km²**
Biological Distinctiveness:	**Continentally Outstanding**
Conservation Status:	**Snapshot—Endangered Final—Critical**

Introduction

The Apalachicola ecoregion [37] encompasses the drainages of the Apalachicola and Econfina Rivers, and stretches from northern Georgia along the western border with Alabama to the Gulf Coast through the central part of the Florida panhandle. With the exception of the Mobile Bay drainage, the Apalachicola is the only northeastern Gulf drainage that extends above the Fall Line. The Apalachicola's two main tributaries, the Chattahoochee and the Flint, arise in the Blue Ridge physiographic province of northern Georgia and the Piedmont Plateau near Atlanta, respectively (Livingston 1992; Carr 1994). The Chattahoochee then flows through the Red Hills of the Piedmont province before entering the upper coastal plain, while the Flint crosses a fall area before reaching the coastal plain and eventually meeting the Chattahoochee to form the Apalachicola. Although the Apalachicola lies entirely within the coastal plain, the variety of habitats found in its two tributary rivers provides the foundation for a diverse freshwater fauna (Livingston 1992). Because reaches of the Apalachicola flow through shaded ravines with cool spring inputs, immigrant species that have found their way south via the Chattahoochee and Flint Rivers find habitat resembling that of more northerly regions (Livingston 1992; Carr 1994).

Biological Distinctiveness

While this ecoregion contains only slightly more than half as many fish as the Mobile Bay [36] ecoregion (104 species versus 187), it supports more species than do adjacent lowland ecoregions (Swift et al. 1986). Endemic fish species in this ecoregion are the bluefin stoneroller (*Campostoma pauciradii*), broadstripe shiner (*Pteronotropis euryzonus*), highscale shiner (*Notropis hypsilepis*), bluestripe shiner (*Cyprinella callitaenia*), bandfin shiner (*Luxilus zonistius*), greater jumprock (*Moxostoma lachneri*), grayfin redhorse (*Moxostoma* sp.), and shoal bass (*Micropterus* sp.) (Swift et al. 1986; Page and Burr 1991). Euryhaline marine, diadromous, and secondary freshwater fish are found in the Apalachicola in addition to the more dominant primary freshwater species. Anadromous fish once abundant in the river are the Alabama shad (*Alosa alabamae*), striped bass (*Morone saxatilis*), and sturgeon (*Acipenser oxyrhynchus*); construction of the Jim Woodruff Lock and Dam has truncated their migration routes (Livingston 1992).

This ecoregion is distinguished more for its unionid mussel fauna than for its fish: it is home to thirty-eight mussel species, of which seventeen (45 percent) are endemic. The Apalachicola not only contains the largest number of freshwater gastropods and bivalves of all of the drainages from the Escambia west to the Suwannee River (encompassing ecoregions [37], [38], and [39]), but studies have found that it is also the center of endemism for these mollusks (Livingston 1992).

Due largely to the diversity of physical habitats, the upper Apalachicola Basin is also rich in herpetofauna, with a greater density of reptiles and amphibians than any other region in North America north of Mexico (Livingston 1992).

Conservation Status

Taken as a whole, the Apalachicola is one of the least-disturbed ecoregions in the Southeast. Dams have perhaps caused the greatest amount of damage to the ecoregion's biodiversity; in total there are sixteen impoundments on the Apalachicola, Flint, and Chattahoochee Rivers, thirteen of them on the Chattahoochee (Livingston 1992). As a result of these impoundments, much of the habitat on the lower and middle Chattahoochee has been eliminated (Gilbert 1992). The only dam on the

Apalachicola is the Jim Woodruff Lock and Dam, constructed in the mid-1950s for navigation and hydropower (Livingston 1992; Gilbert 1992). With only one impoundment, the Apalachicola is one of the most free-flowing rivers in the Gulf Coastal Plain.

A second major disturbance to this ecoregion is pollution from agriculture, urbanization, and industrialization. The Flint River is largely undisturbed in its upper reaches but suffers from pollution from the Atlanta area and from agriculture downstream. The upper Chattahoochee flows directly through Atlanta, which gets 70 percent of its drinking water from the river, and the lower half flows through a heavily agricultural area (Livingston 1992; American Rivers 1998). Rapid population growth in Atlanta has placed enormous stresses on the Chattahoochee, seen particularly in the discharge into the river of large amounts of undertreated and untreated sewage (American Rivers 1998). As of 1997, more than 330 miles of the upper Chattahoochee River basin did not support designated fishing, recreation, and drinking water uses (American Rivers 1998).

The Apalachicola River begins with the polluted reservoir of Lake Seminole, flowing first through an area of light industrialization and then through rural areas dominated by wetlands (Livingston 1992). Fish species living below the Jim Woodruff Lock and Dam, such as the grayfin redhorse and bluestripe shiner, are protected by the dam from upstream silt and pollution but are unable to escape potentially more substantial pollution inputs below the dam (Gilbert 1992). A battery plant on the Chipola River near Marianna, Florida, for years released acids and heavy metals into the water, but this point source pollution has largely abated since the plant was made a Superfund site (Livingston 1992; Deyrup and Franz 1994). The Apalachicola's mussel fauna is especially threatened by silt from agriculture, road construction, and other development, though local mussel extinctions have been associated with the introduced Asiatic clam, *Corbicula fluminea*, as well (Livingston 1992; Deyrup and Franz 1994). In general, the coastal plain portion of the basin is plagued by the agriculture-driven problems of sediment runoff, land clearing, and pumping of groundwater.

A third major threat to this ecoregion's biota is the continued dredging and channelization of the Apalachicola and Chattahoochee Rivers by the U.S. Army Corps of Engineers to aid barge navigation. Despite these disturbances, the freshwater habitats of this ecoregion remain in relatively good condition, particularly in the upper Flint and the lower reaches of the Apalachicola. This is in part a result of a tradition of informed management in this river system. To protect the system and its biota, which supports many of the ecoregion's human inhabitants, the Apalachicola has received an almost unprecedented level of protection. The Apalachicola River has been designated an Outstanding Florida Water; Apalachicola Bay is a state aquatic preserve; the lower river and bay have been designated the largest national estuarine reserve in the United States; and the lower valley is part of the Experimental Ecological Reserve System of the National Science Foundation (Livingston 1992). With only a single impoundment on the Apalachicola proper and continued protection, this ecoregion could retain its distinct biological features into the future.

Suite of Priority Activities to Enhance Biodiversity Conservation

- Remove the Jim Woodruff Dam on the Apalachicola River at Chattahoochee. Construction of this dam in the mid-1950s cut off access to the Flint and Chattahoochee Rivers, severing the migration route of spawning Atlantic sturgeon (*Acipenser oxyrhynchus*) and possibly striped bass (*Morone saxatilis*) and several species of shad (*Alosa* spp.) (Gilbert 1992; Carr 1994). (The 1987 removal of the dam on Dead Lake on the lower Chipola River restored natural flow, but it may take two decades or more for the accumulated silt to flush out of the old lake [Deyrup and Franz 1994].)
- End barge channel maintenance on the Apalachicola, which destroys crucial mussel habitat.
- Maintain good water quality in the Chipola River to protect sensitive mussel populations.
- Protect cave sites in this ecoregion from changes in water quality and volume.
- Prevent the city of Atlanta from constructing a pipeline to pump sewage discharge from the Chattahoochee Basin to the Etowah Basin (American Rivers 1997).

- Protect bottomland hardwoods and associated riparian habitat.
- Expand existing watershed protection programs to include additional streams.
- Take action to reduce sediment load in all watersheds of the ecoregion.

Conservation Partners

For contact information, please see appendix H.

- Alabama Rivers Alliance
- American Rivers
- Chattahoochee Riverkeeper
- Florida Audubon Society
- The Nature Conservancy (Alabama, Georgia, Florida Field Offices, and Southeast Regional Office)
- Upper Chattahoochee Riverkeeper

Ecoregion Number:	**38**
Ecoregion Name:	**Florida Gulf**
Major Habitat Type:	**Temperate Coastal Rivers, Lakes, and Springs**
Ecoregion Size:	**35,514 km²**
Biological Distinctiveness:	**Continentally Outstanding**
Conservation Status:	**Snapshot—Relatively Stable**
	Final—Relatively Stable

Introduction

This ecoregion drains a relatively small area along the Gulf Coast and covers southern Alabama and the western portion of the Florida panhandle. It is defined by the lowland drainages of the Perdido, Escambia, Blackwater, Yellow, and Choctawhatchee Rivers. The Perdido River, which forms the northwest boundary of Florida, flows through forests in its upper reaches and marshes and swamps below. The Blackwater, Yellow, and Shoal Rivers provide relatively cool freshwater habitat, and the Choctawhatchee flows through extensive floodplain forests. The largest of this ecoregion's rivers, the Escambia, has its headwaters in Alabama (Livingston 1992).

Biological Distinctiveness

Confined to the Gulf Coastal Plain, the Florida Gulf [38] ecoregion does not boast the same fish richness or endemism as the neighboring Mobile Bay [36] ecoregion, but with 106 species and 7 endemics it rivals the larger Apalachicola [37] to the east. Several of the endemic fish are relatively new to science, suggesting the need for additional taxonomic research in this region. The seven endemics are the blacktip shiner (*Lythrurus atrapiculus*), coastal darter (*Etheostoma colorosum*), Choctawhatchee darter (*E. davisoni*), southern logperch (*Percina autroperca*), Florida sand darter (*E. bifascia*), blackmouth shiner (*Notropis melanostomus*), and Okaloosa darter (*E. okaloosae*). More distinctive is this ecoregion's unionid mussel fauna, with seven of thirty endemic species. It also supports relatively large populations of the rare Alabama map turtle (*Graptemys pulchra*) in the Escambia and Yellow River drainages (Moler 1992).

Conservation Status

The conservation status of the Florida Gulf [38] ecoregion varies substantially from river to river. The Escambia has been recognized as quite polluted, especially in its lower reaches. This pollution comes from industrial discharge and agricultural runoff. Heavy siltation from the latter threatens mussel populations (Deyrup and Franz 1994). Multiple fish kills have also occurred in the Escambia because of hypoxic conditions (Livingston 1992).

Portions of the Choctawhatchee are heavily polluted from silt, but sections of the river also exhibit relatively good water quality, particularly those flowing through intact wetlands and floodplain forests (Livingston 1992). The Blackwater River is relatively unpolluted, protected in part by the heavily forested Blackwater State Forest in Florida and Conecuh National Forest in Alabama, though removal of in-stream woody debris may be responsible for the river's low productivity (Livingston 1992). The Yellow and Shoal Rivers are cool, relatively undisturbed streams with fairly high fish richness. The Perdido River watershed is dominated by forestry and agriculture, with the lower reaches flowing through largely undisturbed marshes and swamps (Livingston 1992).

Suite of Priority Activities to Enhance Biodiversity Conservation

- In order to protect remaining fish and particularly mussel populations, restoration of good water quality in the polluted Escambia and Choctawhatchee Rivers and protection of current water quality in unpolluted tributaries is essential. This will require establishing enforceable guidelines and cooperation between Alabama and Florida to reduce siltation especially.

- Protect important stream habitat in and around Eglin Air Force Base in northwest Florida. This would protect the habitat of the Okaloosa darter, Alabama map turtle, bog frogs, midland snakes, and several species of wetland plants. Currently, poor stewardship on the base has led to excessive siltation from erosion (Gilbert 1992).
- Conduct species surveys of this ecoregion's rivers to ascertain the current distribution of rare freshwater species.

Conservation Partners

For contact information, please see appendix H.

- Alabama Rivers Alliance
- American Rivers
- Florida Audubon Society
- The Nature Conservancy (Alabama and Florida Field Offices, and Southeast Regional Office)

Ecoregion Number:	**39**
Ecoregion Name:	**Florida**
Major Habitat Type:	**Subtropical Coastal Rivers, Lakes, and Springs**
Ecoregion Size:	**161,393 km²**
Biological Distinctiveness:	**Globally Outstanding**
Conservation Status:	**Snapshot—Endangered Final—Critical**

Introduction

Restricted to the Gulf-Atlantic Coastal physiographic province, this ecoregion covers southeastern Georgia and all of peninsular Florida. The northern boundary of the ecoregion is defined largely by the watershed of the Suwannee River, flowing from Georgia through Florida to the Gulf, and several smaller rivers that flow into the Gulf of Mexico or the Atlantic Ocean. Among these are the Ochlockonee and the Saltilla. In northern Florida, other major rivers

include the St. Johns and the Ocklawaha. Southern Florida is dominated by the Everglades freshwater system. Another area of interest is Okefenokee Swamp, located in the northern part of the ecoregion along the Georgia-Florida border. Several major lakes occur throughout this ecoregion, including Lake Okeechobee.

The Florida [39] ecoregion is characterized by a variety of aquatic habitats that is virtually unequaled elsewhere in North America. Though intermittent streams are few, there are abundant marshes, swamps, mangrove swamps, ponds and lakes, springs, and large rivers. In fact, Florida has more artesian springs than any other region of the world, with 100 major springs occurring beside or within the channels of the Suwannee River or its main tributaries alone (Carr 1994). The lower reaches of all the rivers that flow into the Atlantic or Gulf of Mexico experience tidal effects and are often inhabited by both euryhaline (salt-tolerant) freshwater and marine species, including tropical peripheral species (Gilbert 1992).

Biological Distinctiveness

Unlike adjacent ecoregions, the Florida [39] ecoregion is not distinguished by a particularly rich freshwater fish or mussel fauna, but instead gains its globally outstanding status from its highly rich and endemic crayfish fauna, as well as its diversity of habitat types. Cambarid crayfishes are found in nearly every kind of freshwater habitat in this ecoregion, and cave crayfishes dominate the list of endemic species. This ecoregion supports thirty-six species of crayfish, twenty-nine (81 percent) of which are endemic. Interestingly *Procambarus alleni*, one of these endemics, is the only crayfish known from the Everglades (Lodge 1994). Because of its abundant springs and cave systems, the Florida [39] ecoregion is also home to a high diversity of amphipod and isopod crustaceans as well as freshwater snails; though poorly investigated to date, many of these species apparently are restricted to single spring or cave localities (Gilbert 1992). One such species is the Seminole rams-horn (*Planorbella duryi*), which is restricted to the Everglades. More widely distributed snails include a tropical species known as the Florida applesnail (*Pomacea paludosa*), found throughout the Everglades and in warm waters farther north (Lodge 1994).

The ecoregion supports 110 species of freshwater fish, ten of which are endemic. Coastal freshwater habitats support some of the more distinctive fish communities, among which many marine forms are represented. For instance, several diadromous fish species, such as the rare opossum pipefish (*Microphis brachyurus lineatus*), migrate up a few tributaries of the Indian River lagoon. Into this same lagoon migrate millions of larval gobioid fish of Caribbean origin. Within the mangrove swamps of southwest Florida live fish specially adapted to harsh conditions, such as rivulus (*Rivulus marmoratus*), the only species of its genus found in the United States. Seagrass meadows common in most of Florida's coastal estuaries support a rich fish community that includes spottail goby (*Gobionellus stigmaturus*). In the Loxahatchee and St. Lucie Rivers and in Sebastian Creek occur such tropical peripheral fishes such as the river goby (*Awaous tajasica*), bigmouth sleeper (*Gobiomorus dormitor*), and slackcheek goby (*Gobionellus pseudofasciatus*) (Gilbert 1992). Other euryhaline species found fairly often in Florida's freshwater habitats are bull shark (*Carcharhinus leucas*), Atlantic stingray (*Dasyatis sabina*), tarpon (*Megalops atlanticus*), striped anchovy (*Anchoa mitchelli*), fat snook (*Centropomus parallelus*), common snook (*C. undecimalis*), fat sleeper (*Dormitator maculatus*), spinycheek sleeper (*Eleotris pisonis*), darter goby (*Gobionellus boleosoma*), freshwater goby (*G. shufeldti*), violet goby (*Gobioides broussoneti*), naked goby (*Gobiosoma bosci*), and clown goby (*Microgobius gulosus*) (Swift et al. 1986).

Conservation Status

The freshwater habitats of the Florida [39] ecoregion have in many cases been extensively modified as a result of development, interbasin water transfer, staggering human population growth, agricultural pollution, phosphorus mining, and impoundments. Most natural stream channels have been replaced by box-cut canals, in which herbicides are applied to impede the growth of aquatic vegetation. In these canals, water quality can be quite degraded by agricultural and urban runoff. All of the large rivers of this ecoregion except the Suwannee have been dammed, and many have also been affected by channelization and extreme pollution. The rivers that flow through the cities of Miami, Tampa, and Jacksonville have virtually no intact habitat remaining. Tidal creeks, which are abundant along Florida's southeast coast, have suffered

from a variety of impacts, and exotic fish species have taken hold. Mangrove forests along Florida's central east coast have been flooded as a result of mosquito-control impoundment projects, and human population growth has damaged or destroyed an even larger proportion of the coast's mangroves (Gilbert 1992).

Despite this extensive list of habitat alterations, freshwater fishes of the Florida [39] ecoregion have probably been less seriously affected than those of adjacent ecoregions, where impoundments, channelization, and pollution are even more extensive (Gilbert 1992). No freshwater fish is known to have been extirpated from the ecoregion, although Gilbert (1992) identifies thirteen freshwater and euryhaline species as rare or endangered.

The situation for other freshwater fauna in this ecoregion is grimmer, especially for species inhabiting springs. Deyrup and Franz (1994) identify eight freshwater snail species, two groundwater amphipods, two groundwater isopods, and thirteen cavernicolous crayfish species inhabiting restricted sites and thus vulnerable to extinction in this ecoregion. All of these species are threatened by local land-use practices and physical disturbance of habitat, both of which can result in changes in water quality and volume. Cave crayfish are particularly threatened by the disruption of flow into caves of organic debris, their food source (Deyrup and Franz 1994).

Although widely studied and described, threats to the Everglades warrant special attention here. Chief among these threats are the extensive hydrological modifications, including channelization and pumping to drain large portions of the native sawgrass prairies, that have altered the historical timing and flow of water in the system (Lodge 1994). Pollution from the resulting agricultural land use has placed further stresses on the species found here, as excessive amounts of phosphorous and mercury have been introduced into the environment. Today, continued population growth in Miami-Dade County has intensified pressure on the availability of water for the ecosystem. Taken together, these two stresses have had a devastating effect on the quality and overall availability of water for aquatic species as well as the numerous terrestrial species dependent on them. For example, populations of wading birds that feed on fishes and aquatic invertebrates have been severely affected.

Currently, the U.S. Army Corps of Engineers is undertaking a comprehensive review of water management in the Everglades system. The Corps has been given the task of developing a long-term restoration plan for the ecosystem. Care must be taken to ensure that water flow is restored in a way that replicates the natural hydrology as closely as possible, including amount, flow, timing, and distribution of water through the system. Efforts are also being made to ensure that water quality in the system is addressed, including creating storage and buffer zones around agricultural and urban areas (WWF and National Audubon Society 1996).

Suite of Priority Activities to Enhance Biodiversity Conservation

- Remove Rodman Dam on the Ocklawaha River, a vestige of the Cross-Florida Barge Canal that currently floods 9,000 acres of floodplain forest and has likely affected populations of anadromous shortnose sturgeon (*Acipenser brevirostrum*) and the southern tesselated darter (*Etheostoma olmstedi maculaticeps*) (Gilbert 1992).
- Support funding for an agreement that would prevent DuPont Corporation from building a titanium-mining operation along 38,000 acres of the eastern border of the Okefenokee Swamp, a disturbance that would affect both the swamp and the St. Mary's River (Sierra Club 1998; The Wilderness Society 1998).
- Protect the series of small freshwater streams that empties into the Indian River lagoon, from which the only U.S. populations of *Gobionellus pseudofasciatus*, *Gobiomorus dormitor*, *Awaous tajasica*, and *Microphis bachyurus lineatus* are known. Protection of these tributaries requires the elimination of point and nonpoint source pollution, especially the spraying of herbicides along stream edges (Gilbert 1992).
- Protect the upper Santa Fe River drainage, which serves as a relatively pristine refugium for the Suwannee moccasinshell (*Medionidus walkeri*) and Florida pigtoe (*Pleurobema reclusum*). Water quality and riparian-zone cover must be protected and development restricted (Deyrup and Franz 1994).
- The Suwannee River is the only unimpounded large river in this ecoregion and should be

protected as such; also, phosphate mining in the Suwannee Basin should be halted.
- Spring habitats, including those in caves, must be protected wherever possible, as many harbor unique invertebrate species with highly restricted ranges.
- Remove the Jackson Bluff Dam on the Ochlockonee River. This dam, built in the late 1920s to create Lake Talquin, destroyed the riverine habitat of the Ocklockonee arc-mussel (*Alasmidonta wrightiana*), which may be extinct, and blocked the upstream movement of anadromous fishes that may have been hosts to the mussel's glochidia (Deyrup and Franz 1994).
- Continue efforts to restore the Everglades. Replicate the essential features of the natural hydrology of the Everglades, including the amount, flow, depth, timing, and distribution of water that once flowed through the system. Conserve water entering the system and increase self-sufficiency of urban and agricultural water supplies. Restore water quality to pristine conditions.
- Work to control established populations of the Australian tree *Melaleuca quinquenervia* and the Brazilian pepper, *Schinus terebinthifolius*.

Conservation Partners

For contact information, please see appendix H.

- Everglades Coalition
- Florida Audubon Society
- Florida Defenders of the Environment
- Florida Natural Areas Inventory
- Georgia Natural Heritage Program
- National Audubon Society
- The Nature Conservancy (Florida and Georgia Field Offices, and Southeast Regional Office)
- Sierra Club

Ecoregion Number: **40**

Ecoregion Name: **South Atlantic**
Major Habitat Type: **Temperate Coastal Rivers, Lakes, and Springs**
Ecoregion Size: **295,608 km²**
Biological Distinctiveness: **Globally Outstanding**
Conservation Status: **Snapshot—Endangered Final—Critical**

Introduction

This Atlantic coastal ecoregion ranges from eastern Georgia to southern Virginia, covering all of South Carolina and most of North Carolina. Many of the rivers begin their journeys to the Atlantic as small, fast-flowing mountain streams in the eastern slopes of the Blue Ridge physiographic province of the Appalachian Mountains. From the hills and mountains they flow across the Piedmont Plateau until they reach the Fall Line, descending and flowing through the southern portion of the Atlantic Coastal Plain. Major rivers include the Altamaha and its two tributaries, the Oconee and Ocmulgee, in Georgia; the Savannah River, which forms the border between South Carolina and Georgia; the Cooper-Santee River system and Pee Dee River in South Carolina; the Cape Fear River in North Carolina; and the Roanoke River in North Carolina and Virginia.

As a result of the broad flat coastal plain and a high water table, this ecoregion contains an abundance of wetlands (McNab and Avers 1994). Approximately 9,816 km² of coastal marsh exist on the Atlantic Coast (Alexander et al. 1986), roughly three-fourths of which occurs predominantly within this ecoregion in the states of North Carolina, South Carolina, and Georgia (Chabreck 1988). The ecoregion also includes swamps, bogs, freshwater marshes, and shallow lakes (McNab and Avers 1994). A subset of these lakes concentrated primarily along the coast from southern North Carolina to eastern Georgia, including Lake Waccamaw, is known as the Carolina Bays. These features were formed by the impact of extraterrestrial bodies. The unusual chemical makeup of Lake Waccamaw may be attributable to the lake's origins and may have played a part in the evolution of the lake's distinctive fauna as well as its high productivity (Eyton and Parkhurst 1975; Stager and Cahoon 1987).

Biological Distinctiveness

For a temperate ecoregion, the South Atlantic [40] contains noteworthy levels of richness and endemism, but its outstanding biodiversity is particularly evident when compared with that of other temperate coastal ecoregions. The ecoregion historically supported more than 177 fish species, of which forty-eight (27 percent) were endemic. Equally impressive, nineteen of its fifty-nine unionid mussel species (32 percent) are endemic, as are an extraordinary thirty-nine of its fifty-six crayfish species (70 percent). Like the other ecoregions radiating from the Appalachian Mountains, age, favorable climate, and geologic stability have provided a wealth of varied habitats, allowing a diverse aquatic fauna to evolve and survive (Rohde et al. 1994).

With 177 species of native fish, the South Atlantic [40] is the fifth-richest ecoregion for fish in North America and is second only to the Mobile Bay [36] ecoregion in the temperate coastal rivers, lakes, and springs MHT. Among the forty-eight endemic species are the federally endangered Cape Fear shiner (*Notropis mekistocholas*), restricted to a small section of the upstream portion of the Cape Fear River; the Waccamaw silverside (*Menidia extensa*), restricted solely to Lake Waccamaw; the Waccamaw killifish (*Fundulus waccamensis*), known only from Lake Waccamaw and Lake Phelps; and the Waccamaw darter (*Etheostoma perlongum*), found in Lake Waccamaw and the headwaters of the Waccamaw River. This concentration of endemics in and around Lake Waccamaw gives further distinction to the South Atlantic [40] ecoregion, as does the large degree of endemism encountered in the Roanoke River drainage near the ecoregion's northern boundary.

Other endemic fish include two of the six species of pygmy sunfishes in the family Elassomatidae, which is restricted to the southeastern United States (Rohde et al. 1994). These species are the blue barred pygmy sunfish (*Elassoma okatie*) and the Carolina pygmy sunfish (*E. boehlkei*). The ecoregion is also home to a newly discovered species, a relative of the golden redhorse (*Moxostoma erythrurum*), tentatively known as the Carolina redhorse (*Moxostoma* sp.) (Southeastern Fishes Council 1997). In total, the endemic fish fauna comprises twenty minnow species, six suckers, three catfish, one spring cavefish, two killifish, one silverside, two pygmy sunfish, two bass, and thirteen darters. It should be noted that new species may yet be discovered, because of all of the ecoregions in the southeastern United States, this is perhaps the least studied biologically—"a veritable black hole of life history knowledge for fishes," according to one expert (Burkhead pers. comm.).

Several species of anadromous fish that are widely distributed along the East Coast, including the alewife (*Alosa pseudoharengus*), American shad (*Alosa sapidissima*), and blueback herring (*Alosa aestivalis*), return in the spring to the coastal rivers of this ecoregion where they were born.

The South Atlantic [40] ecoregion also harbors a number of endemic amphibians, five of which are salamanders. Two of these belong to the Plethodontidae family. They are the many-lined salamander (*Stereocheilus marginatus*) and the shovelnose salamander (*Leurognathus marmoratus*), whose restricted range also occupies the neighboring Tennessee-Cumberland ecoregion [35]. Mabee's salamander (*Ambystoma mabeei*), the dwarf waterdog (*Necturus punctatus*), and the Neuse River waterdog (*Necturus lewisi*) are the other three endemic salamanders found in the ecoregion. The other endemic amphibian is the pine barrens tree frog (*Hyla andersonii*). Although this tree frog is also found farther north in the pine barrens of southern New Jersey and the western panhandle of Florida (Conant and Collins 1991), its total range is so small that it is considered endemic to all three ecoregions.

Conservation Status

According to Master et al. (1998), at least forty-seven species of fish and mussels are at risk of extinction in this ecoregion. Major threats to the biota include urban development; channelization; agricultural runoff and other nonpoint source pollution; dam construction for navigation and hydropower; and introductions of non-native species. Both small and large impoundments are common, most rivers have at least one high dam, and many streams are multiply impounded. Furthermore, all or nearly all major drainages are impounded at the Fall Line in this ecoregion. Much of the original forest cover in the ecoregion has been lost, converted either to agriculture or to pine plantations. Areas such as Charlotte and the Raleigh-Durham corridor are

experiencing tremendous population growth and consequent suburban sprawl.

Runoff from largely unregulated hog farming is a relatively new threat to parts of this ecoregion, particularly in areas of North Carolina. Factory farms raise hogs in high densities, and each hog produces three to five times more waste than the average human. Much of this waste ends up in nearby streams and rivers, causing severe eutrophication that results in dramatic fish kills. In recent years, the problem has been raised in the North Carolina legislature, but beyond basic water-quality appropriations and limited regulations, the legislature has failed to seriously address it (American Rivers 1997). Another agriculture-related problem is sedimentation from farms in the Piedmont region, largely due to unconsolidated soils.

As in so many other ecoregions, introductions of non-native species into the South Atlantic have had a detrimental effect on native biotas. Here the worst culprit is the flathead catfish (*Pylodictus olivaris*), a native of the Mississippi River drainage [24 and 25] and the Mobile Bay [38] ecoregion. Introduced for sport fishing in 1966, this predatory catfish wiped out populations of the native bullhead (*Ictalurus* sp.) in the Cape Fear River before it began preying on American shad. Today it is found throughout the South Atlantic [40] ecoregion (The Nature Conservancy 1996d).

Overharvesting of pearly mussel species, a threat faced by other ecoregions rich in mussels, may also be a problem in the South Atlantic [40]. These species are collected and sold to Japan for the cultured pearl industry (Williams et al. 1993). Although the extent of the problem is not fully known, mussel researchers believe that current levels of exploitation cannot be sustained. Currently no federal restrictions exist on the harvesting of mussel species that are not endangered or threatened. Efforts to assess the level of this threat should be undertaken. Additionally, the poaching of endangered mussel species by collectors and subsistence fishers may pose a threat to the survival of these species (Mueller 1993).

Suite of Priority Activities to Enhance Biodiversity Conservation

- Direct new development away from remaining blocks of forested lands, particularly streamsides and other biologically important areas.

- Undertake significant restoration efforts in selected areas, including removal of physical barriers obstructing fish movements, and allow forest regeneration (Ricketts et al. 1999).
- Protect blackwater river systems in the northeast Cape Fear area and the northeast Black River (Ricketts et al. 1999).
- Enforce existing regulations for hog farms. If waterways and their biotas are to be protected from the effects of accidental discharge, it is critical that stricter regulations for hog-farm waste lagooning and processing be enacted.
- Create economic incentives for hog farmers to reduce current densities. Additionally, develop new facilities that reduce the risk of accidental waste discharge.
- Increase protection for Lake Waccamaw and the Lake Waccamaw River (Ricketts et al. 1999).
- Identify remaining intact stream habitats and protect these areas.
- Enhance protection of areas held by the USFS and move away from a policy of cutting in the last remaining blocks of forest.
- Initiate cooperative programs with farmers to develop practices that reduce soil erosion and sedimentation in streams.
- Undertake more research on the freshwater fauna of this ecoregion.

Conservation Partners

For contact information, please see appendix H.

- American Rivers
- Coastal Plains Institute and Land Conservancy
- Georgia Natural Heritage Program
- The Nature Conservancy (Georgia, North Carolina, South Carolina, and Virginia Field Offices, and Southeast Regional Office)
- North Carolina Coastal Federation
- North Carolina Heritage Program
- The Sierra Club
- South Carolina Heritage Trust
- Virginia Division of Natural Heritage

Ecoregion Number:	**41**
Ecoregion Name:	**Chesapeake Bay**
Major Habitat Type:	**Temperate Coastal Rivers, Lakes, and Springs**
Ecoregion Size:	**179,243 km²**
Biological Distinctiveness:	**Continentally Outstanding**
Conservation Status:	**Snapshot—Vulnerable**
	Final—Endangered

Introduction

The extent of this ecoregion is defined by the river drainages of the Chesapeake Bay. The ecoregion covers most of northern Virginia, the eastern extension of West Virginia, most of Maryland, part of southwestern Delaware, roughly the central third of Pennsylvania, and part of western New York. The Appalachian Plateau and the Ridge and Valley physiographic provinces are found in the western and northern portions of the ecoregion, the Piedmont Plateau in the south-central portion, and the Atlantic Coastal Plain in the southeastern portion. Major rivers in the southern portion of the ecoregion include the Potomac River, the Rappahannock River, and the James River on the western shore of the Chesapeake. Rivers originating on the Delmarva Peninsula include the Sassafras, Chester, Choptank, and Nanticoke. The largest tributary to the Chesapeake is the Susquehanna River, contributing 50 percent of the fresh water in the bay. The headwaters of the Susquehanna originate on the Appalachian Plateau. The Susquehanna and its tributaries cut through select mountain ridges of the Ridge and Valley province on their way to the Piedmont Plateau. Unlike the other major rivers in this ecoregion, the Susquehanna does not reach the Coastal Plain until just before its confluence with the Chesapeake itself.

Biological Distinctiveness

This ecoregion supports ninety-five species of native freshwater fishes, of which seven are endemic. Endemics include the Maryland darter (*Etheostoma sellare*), restricted to one small section of a stream in central Maryland, and the roughhead shiner (*Notropis semperasper*), found only in the headwaters of the James River.

Like other coastal ecoregions, the Chesapeake [41] is host to several species of widely distributed anadromous fishes. Among these are the American shad (*Alosa sapidissima*), alewife (*A. pseudoharengus*), blueback herring (*A. aestivalis*), white perch (*Morone americana*), and striped bass (*M. saxatilis*). Known locally as rockfish, the striped bass has historically been an important commercial fish. After experiencing serious declines, due largely to overfishing, populations are beginning to respond to stricter conservation measures.

This ecoregion supports fourteen species of native crayfish and twenty-two species of unionid mussels, four of which are endemic.

Conservation Status

Major threats include rapid urban development, pollution from agriculture, and hydroelectric dams. It has been estimated that well over 50 percent of the catchment area has been altered in one form or another. Much of the original forest in this ecoregion has been logged at one time, and many second- or third-growth areas are now being cleared again for new housing developments.

The number of large-scale poultry farms is growing in parts of Maryland and Virginia. Preliminary evidence suggests that outbreaks of *Pfisteiria*, a tiny protozoan that can cause fish kills and is potentially dangerous to humans, are related to an excess of nutrients in the water. Crop farmers' overapplication of chicken manure to fields is probably one cause of this eutrophication. Agricultural runoff from dairy farms in Pennsylvania has also traditionally been a problem in the ecoregion.

Dams, including unused relicts, block the migrations of fish like the American shad. One such impoundment is Conowingo Dam, located ten miles from the mouth of the Susquehanna. Despite a growing recognition of the negative effects of dams on aquatic ecosystems, new projects continue to be pursued. For example, a proposed water development project on the Mattaponi River in Virginia's tidewater region threatens the ecological integrity of what The Nature Conservancy describes as "the heart of the most pristine freshwater complex on the Atlantic Coast" (American Rivers 1999).

Other problems are related to nuclear power plants. In addition to the potential for radioactive leaks, water temperature is elevated by discharges from the cooling systems of these power plants. The Susquehanna is home to Three Mile Island and the Peach Bottom nuclear power plants. A total of seven reactors can be found along the banks of the Susquehanna alone. Only the Mississippi River has more (Stranahan 1995).

Suite of Priority Activities to Enhance Biodiversity Conservation

- Promote innovative development initiatives such as Maryland's recent SmartGrowth, designed to curb runaway suburban sprawl by focusing growth in already-developed corridors.
- Undertake significant restoration efforts in selected areas, including removal of physical barriers obstructing fish movements and restoration of blocks of forest.
- Direct new development away from remaining blocks of forested lands, particularly streamsides and other biologically important areas.
- Continue cooperative initiatives with farmers to implement more ecologically sound agricultural practices, such as no-till farming, that reduce soil erosion.
- Work with poultry farmers to address the issue of runoff from large-scale industrial farms, including creation of economic incentives. Where appropriate, work to ensure that this issue is sufficiently regulated to protect the long-term health of the rivers and their biotas.

Conservation Partners

For contact information, please see appendix H.

- Anacostia Watershed Society
- Chesapeake Bay Foundation
- Natural Resources Defense Council
- The Nature Conservancy (Maryland, New York, Pennsylvania, Virginia, and West Virginia Field Offices)
- Save Our Streams
- Sierra Club

Ecoregion Number: **42**
Ecoregion Name: **North Atlantic**
Major Habitat Type: **Temperate Coastal Rivers, Lakes, and Springs**
Ecoregion Size: **335,412 km²**
Biological Distinctiveness: **Nationally Important**
Conservation Status: **Snapshot—Vulnerable Final—Endangered**

Introduction

Occupying the northeastern seaboard, this ecoregion stretches from eastern Delaware to southern Nova Scotia. In between it covers eastern Pennsylvania, southern New York, southeastern Vermont, part of southern Quebec, and southern New Brunswick. The states of New Jersey, Connecticut, Rhode Island, Massachusetts, New Hampshire, and Maine lie entirely within the ecoregion. The southern half is largely defined by the watersheds of the Delaware, Hudson, and Connecticut Rivers. The northern portion includes the Merrimack River, the Kennebec River, the Penobscot River, the St. Croix River, and the St. John River. Numerous smaller coastal rivers drain directly into the Atlantic. There are numerous lakes in this ecoregion.

Biological Distinctiveness

Most of this ecoregion has been glaciated as recently as ten to fifteen thousand years ago. The ubiquitous presence of glaciers in the north prevented development of endemic freshwater fauna. However, the southern extent of the area was not glaciated. As a result, two endemic species of fish and one endemic amphibian occur here. These are the Atlantic whitefish (*Coregonus huntsmani*), pygmy smelt (*Osmerus spectrum*), and the pine barrens treefrog (*Hyla andersonii*), represented by a disjunct population.

The North Atlantic [42] ecoregion is distinguished by runs of anadromous fish, such as Atlantic salmon (*Salmo salar*) and shad (*Alosa* sp.). There are also populations of Atlantic sturgeon (*Acipenser oxyrhynchus*) and shortnose sturgeon (*A. brevirostrum*) in the North Atlantic [42] ecoregion. The aquatic fauna also includes the endangered dwarf wedge mussel (*Alasmidonta heterodon*) in New York, Vermont, and New Hampshire, the brook floater mussel (*A. varicosa*), and the endangered ringed boghaunter dragonfly (*Williamsonia lintneri*).

Conservation Status

Some of the largest urban centers in the United States are within the North Atlantic [42] ecoregion. These include New York, Boston, and Philadelphia. Further growth of urban centers and their suburbs threaten water quality and volume, and therefore the aquatic biodiversity, of this ecoregion. With more people and increased urban development come more point and nonpoint source pollution. The energy needs of these highly populated areas have prompted the development of many hydroelectric dams, which act as insurmountable barriers to anadromous fish that need to swim upstream to spawn.

Also threatening the region are extensive logging and clear-cutting in the north. A large percentage of the forests of Maine are industrial stands. These tracts are owned, logged, and reseeded by private corporations. Such practices usually maintain forest cover in the area, but it is usually a monotypic population, disrupting normal distribution patterns and biodiversity.

North American Atlantic salmon populations reached an all-time low in 1998, and a number of organizations devoted to recovery of Atlantic salmon populations are active in the ecoregion. Conservation efforts have been focused on both freshwater habitats and overharvesting of salmon in the ocean. On July 1, 1999, Edwards Dam was removed from Maine's Kennebec River to promote restoration of fish populations, marking the first federally mandated dam removal in the United States under the FERC relicensing process (see essay 16).

Suite of Priority Activities to Enhance Biodiversity Conservation

- Make concerted efforts to legislate control over pollution, especially nonpoint source pollution.
- Restore natural flows below dams, especially on the Connecticut River, which has particularly rich runs of anadromous fish.
- Protect bogs and rivers (especially the Connecticut and Neversink Rivers) that are habitats for federally listed endangered species.
- Legislate river protection for pollution, discharge, and natural flow regimes.
- Continue to work for Atlantic salmon recovery by restoring river habitat and limiting salmon harvesting in the ocean.

Conservation Partners

For contact information, please see appendix H.

- Atlantic Salmon Federation
- Connecticut River Watershed Council
- Conservation Council of New Brunswick
- Ecology Action Centre
- Federation of Nova Scotia Naturalists
- New Brunswick Federation of Naturalists
- Maine Natural Areas Program
- Natural Resources Council of Maine
- The Nature Conservancy (Connecticut, Delaware, Maine, Maryland, Massachusetts, New Hampshire, New Jersey, New York, Pennsylvania, Rhode Island, and Vermont Field Offices)
- New Hampshire Natural Heritage Inventory
- Northern Appalachian Restoration Project
- Restore the Northwoods
- Society for the Protection of New Hampshire Forests
- Trout Unlimited
- Vermont Nongame and Natural Heritage Program
- World Wildlife Fund Canada

Focus on the St. Lawrence Complex and the Great Lakes

Although often considered one ecosystem, this complex is made up of five ecoregions. Positioned between the Arctic drainages to the north and the Mississippi and Atlantic drainages in eastern North America, the entire system drains into the Atlantic Ocean—the majority of water by way of the Gulf of St. Lawrence. In all, this area contains approximately one-fifth of the earth's fresh water.

Until ten to fifteen thousand years ago, the entire region was covered by glaciers associated with the Wisconsin Age, and the basins of the Great Lakes were created by the movements and the erosional forces of these glaciers. From a biogeographic perspective, this area is quite young (Underhill 1986). Although these ecoregions generally support a rich diversity of fishes, they harbor relatively few endemics. They do, however, contain unique forms of widely distributed species. This region's freshwater species, particularly its fish, tend to be adapted to one of two habitat types: lacustrine (lake) or lotic (river and stream) (Underhill 1986).

Among threats to the Great Lakes ecoregions, pollution has historically received the most attention. Nonpoint source runoff from agricultural lands and urban centers, point source discharges from municipalities and industry, and aerial deposition of pollutants from distant sources are problems, to a greater or lesser extent, within all five ecoregions. Because the lakes are essentially closed systems, pollutants entering them through various pathways tend to become trapped and concentrated over time. The resuspension of pollutants sequestered in lake sediments is an additional problem (Government of Canada and U.S. EPA 1995). As part of the Great Lakes Water Quality Agreement, the U.S. and Canadian governments have identified areas of concern where environmental criteria are exceeded or beneficial uses are impaired. Most of the areas of concern are located on the lakes near the mouths of tributaries where cities and industries are found or along connected channels between the lakes, and nearly all of the areas have contaminated sediments (Government of Canada and U.S. EPA 1995; U.S. EPA 1999).

Before European settlement, the streams draining into each of the Great Lakes were clear year-round. The widespread clearing of the naturally forested basins for agriculture and logging has led to increased erosion and runoff, and streams now carry higher sediment and nutrient loads, which are deposited in the lakes and at tributary mouths. The Great Lakes themselves were historically cool and clear, but today they are warmer and eutrophic as a result of nutrient and organic matter inputs. With excess nutrients, especially phosphorus from urban sources, aquatic plant growth in the lakes increases, often resulting in oxygen depletion (Government of Canada and U.S. EPA 1995).

An additional change to the landscape has been the loss of wetlands, which were once widespread along the coasts of the Great Lakes. Presently, more than two-thirds of the region's wetlands have been destroyed through development, draining, and pollution (Government of Canada and U.S. EPA 1995).

As closed systems, the lakes are highly vulnerable to impacts from introduced species. Sea lamprey (*Petromyzon marinus*), carp (*Cyprinus carpio*), Eurasian ruffe (*Gymnocephalus cernuus*), alewife (*Alosa pseudoharengus*), and zebra mussel (*Dreissena polymorpha*) are among the more notorious aquatic invaders.

Ecoregion Number:	**43**
Ecoregion Name:	**Superior**
Major Habitat Type:	**Large Temperate Lakes**
Ecoregion Size:	**128,464 km²**
Biological Distinctiveness:	**Continentally Outstanding**
Conservation Status:	**Snapshot—Vulnerable**
	Final—Vulnerable

Introduction

Comprising the drainage area of Lake Superior, this ecoregion encompasses the southern portion of Ontario, eastern Minnesota, northern Wisconsin, and northern Michigan. Numerous small rivers and streams, which often are segmented by barrier falls, flow into Lake Superior—the largest, deepest, and coldest of the five Great Lakes, and the largest temperate freshwater lake (in surface area) in the world. Lake Superior's level is controlled by gates on the St. Marys River at Sault Ste. Marie. The retention time for water in the lake is 191 years. In addition to Lake Superior, several much smaller lakes can be found within this ecoregion. A large portion of the Superior [43] ecoregion remains forested and sparsely populated because a cool climate and poor soils discourage agriculture (Government of Canada and U.S. EPA 1995).

Biological Distinctiveness

Superior [43] supports fewer species among the analyzed taxa than do the other St. Lawrence complex ecoregions. It is distinguished both by its size and its support (along with Michigan-Huron [44]) of the remainder of the recently evolved complex of deepwater fishes in the Great Lakes. These include the cisco or lake herring (*Coregonus artedi*), bloater (*C. hoyi*), kiyi (*C. kiyi*), shortjaw cisco (*C. zenithicus*), pygmy whitefish (*Prosopium coulteri*), and round whitefish (*P. cylindraceum*).

In total, the ecoregion supports sixty species of native fishes—about half as many fish species as the rest of the Great Lakes ecoregions. This ecoregion has no endemics, but it does harbor disjunct populations of fishes generally found farther to the north in cold waters. Among these are the pygmy whitefish, a member of the family Salmonidae that is found more commonly in western Canada and southern Alaska. Normally found in shallower waters in the West, the pygmy whitefish inhabits water deeper than

18 m in Lake Superior (Lee et al. 1980). Another deepwater fish is the siscowet (*Salvelinus namaycush*), a much fatter form of lake trout found in Superior [43] that is found at depths of 270 m or more (Underhill 1986). Additional deepwater fishes of Lake Superior include the deepwater sculpin (*Myoxocephalus thompsoni*), spoonhead sculpin (*Cottus ricei*), brook stickleback (*Culaea inconstans*), and ninespine stickleback (*Pungitius pungitius*). Like the pygmy whitefish and siscowet, these species' ranges are predominantly north of Lake Superior.

Several fishes are largely restricted to Superior's [43] tributaries. They include the widespread johnny darter (*Etheostoma nigrum*), least darter (*E. microperca*), logperch (*Percina caprodes*), and smallmouth bass (*Micropterus dolomieu*) (Page and Burr 1991).

Only seven species of unionid mussels, and three species of crayfish, are found in the ecoregion. None of these is endemic.

Conservation Status

According to TNC (Natural Heritage Central Databases 1997), five fish species in this ecoregion are at risk. One is the lake sturgeon (*Acipenser fulvescens*), whose numbers have been diminished throughout its range from the effects of overfishing, dams, and pollution (Page and Burr 1991).

A primary threat in this ecoregion is introduced species. The invasion of zebra mussels (*Dreissena polymorpha*), efficient filter feeders, in Lake St. Clair has resulted in a large decline in the population of native walleye (*Stizostedion vitreum*), a species adapted to more turbid water conditions (MacIsaac 1996). In fact, most habitats in the Great Lakes have shifted from borderline eutrophic, because of agricultural nutrient loading and runoff, to oligotrophic, because of the abatement of phosphorus in the system and the introduction of the voracious zebra mussel (Corkum pers. comm.).

Pollution in one form or another is the primary problem for the entire Great Lakes area. Acid rain, shipping pollution, and runoff into tributary streams and the lakes themselves present problems for the biota of the region.

Suite of Priority Activities to Enhance Biodiversity Conservation

- Improve water quality, particularly in smaller streams and tributaries, through reduction and control of acid precipitation created by the transportation sector, electric power generation, and smelters (U.S. EPA 1995).
- Prohibit major water diversions, such as those previously proposed for redistribution of Great Lakes water to the southwestern United States, or ensure that they do not come to fruition (U.S. EPA 1995).
- Establish protected areas based on aquatic habitats, particularly deepwater habitats.
- Increase funding to study the ecological consequences of exotic species introductions.
- Support and enforce The 1987 Great Lakes Water Quality Agreement.

Conservation Partners

For contact information, please see appendix H.

- Environment North
- Federation of Ontario Naturalists
- Great Lakes Commission
- Michigan Environmental Council
- Michigan Natural Areas Council
- National Wildlife Federation
- The Nature Conservancy (Wisconsin and Minnesota Field Offices, and Great Lakes Program)
- The Nature Conservancy of Canada (Ontario Office)
- Northwatch
- University of Wisconsin Sea Grant Program
- The Wildlands League
- World Wildlife Fund Canada

Ecoregion Number:	**44**
Ecoregion Name:	**Michigan-Huron**
Major Habitat Type:	**Large Temperate Lakes**
Ecoregion Size:	**250,901 km²**
Biological Distinctiveness:	**Continentally Outstanding**
Conservation Status:	**Snapshot—Endangered**
	Final—Critical

Introduction

Defined by the watersheds of Lake Michigan and Huron, this ecoregion covers a portion of

northern Indiana, eastern Wisconsin, most of Michigan including the southern half of the peninsula, and the western half of southeastern Ontario. In addition, the coastline of Illinois along Lake Michigan is included in this ecoregion. The two lakes are joined hydrologically by the Straits of Mackinac. Numerous small rivers, streams, and lakes occur throughout the ecoregion. Among the larger rivers are the Grand River and Flint River in Michigan and the French River in Ontario.

The surrounding land uses in the basins of the two lakes vary, largely as a function of climate and soil. The northern portion of the Lake Michigan basin is sparsely populated except for the Fox River Valley, which drains into the highly industrialized Green Bay. The southern, warmer part of the Lake Michigan basin is much more urban, containing Milwaukee and Chicago and their associated suburbs. Lake Huron's basin contains a greater degree of agriculture, especially in the Saginaw Bay basin on the southwestern shore of Huron. Both Green Bay and Saginaw Bay are known for their productive fisheries (Government of Canada and U.S. EPA 1995). The world's largest freshwater dunes line Lake Michigan's shore, bringing millions of tourists annually (Great Lakes Information Network 1999).

Biological Distinctiveness

Together with Superior [43], this ecoregion supports what is left of the recently evolved cisco complex of deepwater fishes. This family of fish played a dominant ecological role within the ecoregion before its species were overharvested. They include the longjaw cisco (*Coregonus alpenae*), deepwater cisco (*C. johannae*), and blackfin cisco (*C. nigripinnus*). Today the blackfin cisco is believed to be extant only in Lake Nipigon in southern Ontario. Among the other extant species of this group are the shortnose cisco (*C. reighardi*), the rare kiyi (*C. kiyi*), and the shortjaw cisco (*C. zenithicus*) (The Nature Conservancy 1994). A total of 123 native fish are found in the ecoregion. Populations of globally rare species include the lake sturgeon (*Acipenser fulvescens*), pugnose shiner (*Notropis anogenus*), and greater redhorse (*Moxostoma valenciennesi*) (The Nature Conservancy 1994).

Unlike Lake Superior [43] farther to the north and west, the Michigan-Huron [44] ecoregion supports a relatively rich mussel fauna. With a total of thirty-five species, none of which is endemic, the ecoregion has the highest number of mussel species of any of the Great Lakes ecoregions.

Conservation Status

Like the rest of the Great Lakes ecoregions, Michigan-Huron [44] is highly degraded. The region faces severe problems from pollution, urbanization, overfishing, and exotic species introductions. The ecology has been severely damaged, and remedial strategies have been modest and marginalized.

Of primary concern is pollution. The Great Lakes serve as a particularly busy shipping lane, and spills and routine pollution are common. Also problematic is the pollution that results from agricultural runoff, nutrient loading, and logging runoff. The latter is particularly serious on the Spanish River, in Ontario. Because of the area's high level of industrialization and widespread mining, coal-burning, and smelting activity, acid precipitation is ubiquitous. Persistent pollution within the watershed has severly contaminated the sediments of these lakes.

Another major problem for the Michigan-Huron [44] ecoregion is the introduction of many types of exotic species. Most notorious is the zebra mussel (*Dreissena polymorpha*), which was released into the waters of the Great Lakes by a foreign ship during ballast-water transfer. This mussel is extremely voracious and fecund, reproducing by the millions and outcompeting native species. A few fish have learned to eat zebra mussels, but this is thought to have little control on the exploding population. Other problem species include alewife (*Alosa pseudoharengus*), smelt (*Osmerus* spp.), and ruffe (*Gymnocephalus cernuus*).

Suite of Priority Activities to Enhance Biodiversity Conservation

- Improve water quality, particularly in smaller streams and tributaries, through reduction and control of acid precipitation created by the transportation sector, electric power generation, and smelters (U.S. EPA 1995).
- Ensure that major water diversions, such as those previously proposed for redistribution of Great Lakes water to the southwestern United

States, do not come to fruition (U.S. EPA 1995).
- Support and enforce the 1987 Great Lakes Water Quality Agreement.
- Remediate areas of particular concern and establish protected areas based on aquatic habitats.
- Control ballast-water transfer in the Great Lakes.

Conservation Partners

For contact information, please see appendix H.

- Federation of Ontario Naturalists
- Great Lakes Commission
- Illinois Department of Natural Resources
- Michigan Land Use Institute
- National Wildlife Federation
- The Nature Conservancy (Indiana, Michigan, and Wisconsin Field Offices)
- The Nature Conservancy of Canada (Ontario Office)
- University of Wisconsin Sea Grant Program
- Wild Earth
- The Wildlands League
- Wisconsin Department of Natural Resources
- World Wildlife Fund Canada

Ecoregion Number:	**45**
Ecoregion Name:	**Erie**
Major Habitat Type:	**Large Temperate Lakes**
Ecoregion Size:	**79,527 km²**
Biological Distinctiveness:	**Nationally Important**
Conservation Status:	**Snapshot—Endangered**
	Final—Critical

Introduction

Defined by the watersheds of Lake Erie, this ecoregion includes northeastern Indiana, southeastern Michigan, the southern part of peninsular Ontario, part of western New York, extreme northwestern Pennsylvania, and northern Ohio. Numerous small rivers feed into Lake Erie. Among them are the Thames River in Ontario, the Detroit River between Michigan and Ontario, and the Portage and Grand Rivers in Ohio. Lake St. Clair, between Ontario and Michigan, also occurs within this unit.

Lake Erie is the smallest of the Great Lakes in terms of volume, and its basin is the most highly urbanized and agricultural. Intensive agriculture is practiced in southwestern Ontario and parts of Ohio, Indiana, and Michigan, and runoff from these areas drains to the lake (Government of Canada and U.S. EPA 1995). Large cities within the ecoregion include Cleveland, Toledo, and Detroit.

Biological Distinctiveness

Lake Erie is the warmest and most productive of the Great Lakes. It contains the richest freshwater biota of all the lakes and in Canada. The aquatic fauna present, however, has migrated from adjacent, unglaciated refugia. There are no known endemic fish, mussels, crayfish, or aquatic herpetofauna in the Lake Erie ecoregion. In total, Erie [45] supports 120 native fish species, at least 42 unionid mussels, 13 crayfish, and 32 aquatic herpetofauna species.

Erie [45] is home to several taxa that will require protection if they are to persist. The white cat's-paw pearlymussel (*Epioblasma obliquata perobliqua*), restricted to Fish Creek in Ohio, is considered endangered by the USFWS and critically imperiled by TNC (Natural Heritage Central Databases 1997). Fish Creek, on the Ohio-Indiana border, is home to thirty-one species of freshwater mussels (The Nature Conservancy 1996a). The Lake Erie water snake (*Nerodia sipedon insularum*), which has declined as a result of destruction of shoreline habitat, was proposed as a U.S. federally threatened species in 1994 and listed as endangered in Canada in 1991 (USFWS 1994; Committee on the Status of Endangered Wildlife in Canada 1998).

Conservation Status

Like the other Great Lakes, Lake Erie is in serious trouble. The biota of the ecoregion suffers from pollution of all types, from acid rain to agricultural runoff. Exotic species threaten to extirpate or eliminate native species. Commercial fishing depletes wild stocks. Protective legislation is neither particularly representative nor effective.

The entire Great Lakes/St. Lawrence waterway is an extremely busy shipping route. Industrial pollution has degraded benthic populations to a large extent and contaminated the sediments of the basin. Agricultural runoff and nutrient loading have historically caused eutrophication. Today, however, with the introduction of exotic filter feeders such as the zebra mussel, the trend is toward oligotrophication of the lakes. There are numerous pollution hot spots within the basin, especially at the western end of Lake Erie near Detroit, Cleveland, and Toledo.

With decreasing nutrient levels in recent years has come increased concern from both commercial and conservation interests about fish stocks. If the nutrient level falls too much, the basal groups of the food chain—aquatic plants and algae—dwindle, causing disruption of higher levels. Particular attention has been paid to declining yellow perch (*Perca flavescens*), which are native to the Great Lakes and an important commercial fishery. However, recent rebounds in yellow perch populations have allayed some concerns.

Suite of Priority Activities to Enhance Biodiversity Conservation

- Support and enforce the 1987 Great Lakes Water Quality Agreement.
- Increase funding for the study of the ecological effects of exotic species introductions.
- Establish protected areas based on aquatic habitats.
- Maintain conservation activities at Long Point Bird Observatory.

Conservation Partners

For contact information, please see appendix H.

- Federation of Ontario Naturalists
- Great Lakes Commission
- National Wildlife Federation
- The Nature Conservancy (Indiana, Michigan, New York, Ohio, and Pennsylvania Field Offices, and Great Lakes Program)
- The Nature Conservancy of Canada (Ontario Office)
- New York Natural Heritage Program
- University of Wisconsin Sea Grant Program
- Western Pennsylvania Conservancy
- The Wildlands League
- World Wildlife Fund Canada

Ecoregion Number:	**46**
Ecoregion Name:	**Ontario**
Major Habitat Type:	**Large Temperate Lakes**
Ecoregion Size:	**67,573 km²**
Biological Distinctiveness:	**Nationally Important**
Conservation Status:	**Snapshot—Endangered**
	Final—Critical

Introduction

This ecoregion is part of the St. Lawrence complex and is defined by the watershed of Lake Ontario, including parts of western New York and southern Ontario. Among the rivers draining into the lake are the Niagara, the Crowe, and the Oswego. The ecoregion includes the Finger Lakes of upstate New York and many other small lakes. Lake Ontario receives the entire outflow of the other four Great Lakes. Niagara Falls constituted a navigational barrier between Lake Ontario and the other lakes until the Trent-Severn Waterway as well as the Erie and Welland Canals were constructed. The smallest of the Great Lakes, Lake Ontario is second only to Superior in average depth.

Historically, the watershed of Lake Ontario was almost entirely covered by deciduous forest, but today the vast majority of this forest has been lost or altered, or is heavily fragmented (Ricketts et al. 1999). Canada's largest city, Toronto, and its surrounding metropolitan area, as well as Buffalo, New York, are found within this relatively small ecoregion. As a result, portions of the ecoregion have been heavily developed for urban and industrial purposes, especially the western part of the ecoregion in Canada. The remaining areas have been largely converted for agricultural purposes.

Biological Distinctiveness

The region has had little time to recover biologically from glaciation occurring as recently as ten to fifteen thousand years ago. Most of the freshwater habitats were restocked from more southerly glacial refugia (areas that were not glaciated, but are adjacent to glaciated areas); thus the low level of endemism for the Great Lakes region. The Ontario [46] ecoregion harbors no known endemic fish, mussel, crayfish, or aquatic herpetofauna.

Populations of globally rare species in this ecoregion include the lake sturgeon (*Acipenser*

fulvescens), pugnose shiner (*Notropis anogenus*), copper redhorse (*Moxostoma hubbsi*) and greater redhorse (*M. valenciennesi*) (The Nature Conservancy 1994). Also of concern are the endangered beluga whales (*Delphinapterus leucas*) of the St. Lawrence Estuary, which receives all the pollutants and other human-generated impacts from the entire Great Lakes Basin through Lake Ontario.

Conservation Status

Primary threats to biodiversity include point and nonpoint source pollution; pressure from suburban expansion; and alteration, degradation, and loss of aquatic habitat. Particularly troubling have been the historical problems caused by persistent substances such as DDT, dioxin, and PCBs, which had a direct impact on wildlife populations within the watershed. Today, many of the point sources for these chemicals have been eliminated and the loads of the toxics themselves diminished drastically (U.S. EPA 1998).

According to the U.S. EPA (1998), the "loss of fish and wildlife habitat is a lakewide impairment caused by artificial lake level controls; the introduction of exotic species; and the physical loss, modification, and destruction of habitat, such as deforestation and the damming of tributaries." Unlike the reduction of toxic pollution, these threats tend to be more difficult to reverse and have not been adequately addressed (The Nature Conservancy 1994).

Suite of Priority Activities to Enhance Biodiversity Conservation

- Emphasize enforcement of and compliance with the 1987 Great Lakes Water Quality Agreement between the United States and Canada.
- Establish funding from the United States and Canada for remediation of areas of concern within the Great Lakes Basin.
- Establish more protected areas based on aquatic habitats and ensure that those areas are representative of vital habitats of the region.
- Control ballast-water transfer within the Great Lakes.

Conservation Partners

For contact information, please see appendix H.

- Le Centre de Donnees sur le Patrimoine Naturel du Québec
- Federation of Ontario Naturalists
- Great Lakes Commission
- National Wildlife Federation
- The Nature Conservancy (New York Field Office and Great Lakes Program)
- The Nature Conservancy of Canada (Québec Office)
- New York Natural Heritage Program
- Union Québecoise pour la Conservation de la Nature (UQCN)
- University of Wisconsin Sea Grant Program
- Wild Earth
- The Wildlands League
- World Wildlife Fund Canada

Focus on Northern Ecoregions

Ecoregions 47–60, as cold-climate areas rich in surface water but relatively poor in freshwater species, share certain characteristics. Much of the area covered by these ecoregions is boreal, and Schindler (1998) provides an excellent summary of boreal region aquatic ecosystems and threats to them. Boreal fresh waters, draining recently glaciated areas, support species-poor communities that in their simplicity are highly vulnerable to perturbations. Lakes, which are common landscape features, may have few if any species of fish, and the numbers of individual animals are limited by low productivity. Moderate to high fishing pressure or competition with introduced species can therefore decimate native fish populations. Invertebrate diversity is similarly limited, and research suggests that if one species is lost there may not be others to serve as functional replacements. Many fish and invertebrate species have a low tolerance for warm temperatures or low oxygen, making them especially vulnerable to climate change or warming from other causes (Schindler 1998).

Until the mid-twentieth century, most of the area of the northern ecoregions was too remote and inhospitable for extensive development. Since then, technology has allowed easier access, and various industries have rapidly moved into this region. Poorly managed forestry practices have resulted in elevated runoff, silt loads, and chemical concentrations, and logging roads have increased access, impounded wetlands, fragmented rivers, and further increased erosion and siltation. Recreational and commercial fishing have stressed native populations and have contributed to the spread of non-natives, largely through the practice of dumping bait buckets. Exotic invertebrates, including the zebra mussel (*Dreissena polymorpha*) and the rusty crayfish (*Orconectes rusticus*), are also spreading northward. Acid deposition of contaminants from distant sources has resulted in highly elevated concentrations of PCBs, pesticides, dioxins, and mercury in aquatic fauna. Mercury has also been released from lands flooded as a result of hydropower reservoir construction. Pulp mills and associated chlor-alkali plants have historically discharged toxins, including mercury, and mills currently are the source of high nitrogen and phosphorus loads. Development of Canadian oil sands through open-pit mining is expanding rapidly, with nine to ten projects expected to be in operation by 2007. Land clearing for agriculture and recreational housing is also expanding. Finally, acid precipitation, climatic warming, and increasing exposure to ultraviolet radiation are threats to which boreal aquatic ecosystems are particularly vulnerable (Schindler 1998).

Schindler (1998) provides a list of actions that can be taken locally to ameliorate current threats and prevent future impacts. These are:

1. Immediate implementation of catch and release fishing policies.
2. More restrictions on the disruption of drainage patterns by roads and railways.
3. Discouragement of additional cottage development.
4. Severe restrictions on access.
5. More stringent reductions of sulfur oxide and nitrogen oxide emissions.
6. Greatly reduced forest cutting.
7. Termination of live-bait fisheries.
8. Further improvements to pulp mill effluents, with a long-term objective of no discharge to watercourses.
9. Restrictions on human populations in the Boreal.
10. No disruptions to natural flow patterns.
11. Elimination of local sources of mercury and other contaminants.
12. Protection of riparian areas.
13. Stringent management of nutrient inputs.

These apply more or less to all the northern ecoregions. Specific information on each ecoregion is provided in the following descriptions.

Ecoregion Number:	**47**
Ecoregion Name:	**Lower St. Lawrence**
Major Habitat Type:	**Temperate Coastal Rivers, Lakes, and Springs**
Ecoregion Size:	**855,097 km²**
Biological Distinctiveness:	**Continentally Outstanding**
Conservation Status:	**Snapshot—Vulnerable**
	Final—Endangered

Introduction

This ecoregion is defined largely by the St. Lawrence River from the point where it leaves Lake Ontario until it reaches the Atlantic Ocean. It includes Prince Edward Island, the Island of Newfoundland, Cape Breton Island in Nova Scotia, and the freshwater systems of Anticosti Island in the Gulf of St. Lawrence. It also covers part of northern New York, northern Vermont, southern Quebec, southern Ontario, southern mainland Newfoundland (Labrador), New Brunswick, and Nova Scotia. Among numerous rivers and lakes in the ecoregion are Lac Saint-Jean, Rivière Saint-Maurice, Rivière Manicouagan, Rivière Saint-Marguerite, Rivière Magpie, and the Rivière du Petit Mécatina.

Unlike most of the freshwater habitats in the Lower St. Lawrence [47] ecoregion and the larger Great Lakes region, which were formed as a result of glaciation, Lake Manicouagan, in Quebec, was formed by a different mechanism. A nearly circular ring around a central island, this lake is the crater of a huge bolide impact that occurred some 212 million years ago. The crater has been worn down by repeated glaciation (Strain and Engle 1993).

Biological Distinctiveness

This ecoregion is distinguished by the presence of sturgeon (*Acipenser* spp.) in the main stem of the St. Lawrence and runs of anadromous Atlantic salmon (*Salmo salar*). Although these populations are depleted, they still represent the healthiest surviving stocks of Atlantic salmon in the world (Master, pers. comm.). As one would

expect, endemism in this glaciated region is very low. Two fish species, the pygmy smelt (*Osmerus spectrum*) and the copper redhorse (*Moxostoma hubbsi*), are endemic to this ecoregion. In total, ninety-three native freshwater fish species, twenty unionid mussels, six crayfish, and twenty-eight species of aquatic herpetofauna occur here.

Conservation Status

A wide variety of threats imperil the Lower St. Lawrence [47] ecoregion. First, the region is a major shipping corridor. When foreign ships enter the St. Lawrence gateway and pass into the Great Lakes, they dump ballast water into the lakes. This practice has caused the invasion of several exotic species, including the zebra mussel (*Dreissena polymorpha*), one of the most notorious invasive species in North America.

Also problematic in the St. Lawrence is the pollution caused by ships, agriculture, logging, and municipal discharge. Particularly onerous in this ecoregion is acid precipitation, which sterilizes lake habitats, making them unsuitable for many years. The decline of the copper redhorse, considered threatened in Canada, has been attributed to acid pollution of the species' habitat and food (mollusks), as well as increased turbidity and siltation and the introduction of competitive species (Natural Resources Canada 1998a).

The area has numerous hydroelectric dams, which block access of anadromous fish to spawning grounds in the headwaters, and extensive logging activity contributes to sedimentation and change in water quality.

Beluga whales (*Delphinapterus leucas*), which inhabit the St. Lawrence River Estuary, are critically endangered, with fewer than 500 remaining. Like many other cetaceans, they are victims of shipping disturbances including collisions, noise pollution, habitat degradation, and toxic contaminants. The St. Lawrence is so contaminated by DDT and PCBs that dead whale carcasses have to be disposed of as toxic waste. For more than twenty years, WWF Canada has been supporting—against enormous odds—protection and recovery efforts for the St. Lawrence beluga and is cochairing a government recovery team for the whales (Kemf and Phillips 1998).

Suite of Priority Activities to Enhance Biodiversity Conservation

- Consider decommissioning and removing hydroelectric dams.
- Regulate coal burning and smelting in the ecoregion to minimize the occurrence of acid precipitation.
- Enforce the 1987 Great Lakes Water Quality Agreement.
- Maintain ballast-water inspections on ships about to enter the Great Lakes.
- Strengthen and enforce the regulation of nonpoint source pollution by agriculture and logging.

Conservation Partners

For contact information, please see appendix H.

- Canadian Parks and Wilderness Society
- Federation of Ontario Naturalists
- Le Centre de Donnees sur le Patrimoine Naturel du Québec
- The Nature Conservancy (New York Field Office, and Great Lakes Program)
- The Nature Conservancy of Canada (Atlantic Region, Ontario, and Québec Offices)
- New York Natural Heritage Program
- Strategies Saint-Laurent, Inc.
- Union Québecoise pour la Conservation de la Nature (UQCN)
- Wild Earth
- World Wildlife Fund Canada

Ecoregion Number:	**48**
Ecoregion Name:	**North Atlantic-Ungava**
Major Habitat Type:	**Temperate Coastal Rivers, Lakes, and Springs**
Ecoregion Size:	**626,555 km²**
Biological Distinctiveness:	**Nationally Important**
Conservation Status:	**Snapshot—Relatively Stable**
	Final—Vulnerable

Introduction

This ecoregion occupies northern Quebec and most of the northern portion of mainland Newfoundland. It is defined by several coastal rivers that empty into the Labrador Sea. These rivers include Rivière aux Feuilles, Rivière aux Mélèzes, Rivière Caniapiscau, Rivière a la Baleine, Rivière George, the Kanairiktok River, and the Churchill River.

Biological Distinctiveness

Along with most of the rest of Canada, the North Atlantic-Ungava [48] ecoregion was glaciated as recently as ten to fifteen thousand years ago. Consequently, the aquatic biota of this ecoregion is extremely depauperate. There are no known endemic fish, mussel, crayfish, or aquatic herpetofauna species. The region has only twenty native fish species, one native mussel, and four species of aquatic herpetofauna. According to Hocutt and Wiley (1986), geographers and hydrographers include the watershed of Ungava Bay as part of the Hudson Bay system. After glaciation, this ecoregion was repopulated from stocks in the Mississippian, Beringian, and Atlantic refugia.

Conservation Status

The North Atlantic-Ungava [48] ecoregion is vulnerable to ecological damage in the near future. Among the most serious threats to the aquatic ecology of the region is modification of water resources, such as by the giant Smallwood Reservoir. Other threats include aerial deposition of pollutants from industrial areas farther to the south and west, forestry, and mining. The Ungava Peninsula is ripe for development of mineral resources, at the peril of both terrestrial and freshwater biodiversity. The Raglan mine, located in the northwestern end of the Ungava Peninsula in Quebec, is one of the world's most significant areas of unexploited nickel deposits (Natural Resources Canada 1998b). Population growth and increasing tourism also imperil the freshwater habitats of the Ungava watershed.

Suite of Priority Activities to Enhance Biodiversity Conservation

- Enforce strict regulation of forestry in the area's boreal forests.
- Encourage development of protected areas based on freshwater habitats.
- Educate indigenous and nonindigenous peoples about the necessity of protecting fragile Arctic ecosystems.
- Eliminate unnecessary channel modifications.

Conservation Partners

For contact information, please see appendix H.

- Canadian Nature Federation
- Grand Council of the Crees
- Labrador Inuit Association
- The Nature Conservancy of Canada (Atlantic Region and Québec Offices)
- Union Québecoise pour la Conservation de la Nature (UQCN)
- World Wildlife Fund Canada

Ecoregion Number:	**49**
Ecoregion Name:	**Canadian Rockies**
Major Habitat Type:	**Temperate Headwaters and Lakes**
Ecoregion Size:	**36,928 km²**
Biological Distinctiveness:	**Continentally Outstanding**
Conservation Status:	**Snapshot—Relatively Stable**
	Final—Vulnerable

Introduction

Contained almost entirely within Alberta, Canada, this relatively small ecoregion also includes several disjunct areas in British Columbia and extends into northern Montana. The ecoregion is on the eastern side of the Continental Divide and its boundaries are largely defined by the upper reaches of several rivers. These include the Snake Indian River, the Saskatchewan River, the Clearwater River, the Highwood River, the Castle River, and the Waterton River. A good deal of this ecoregion lies within protected areas. Portions of Banff National Park, Jasper National Park, Waterton Lakes National Park, the Rocky Mountains Forest Preserve, and Glacier National Park are within its boundaries.

Biological Distinctiveness

The Canadian Rockies [49] ecoregion is distinguished from the rest of the Saskatchewan River headwaters region by endemism. During the Wisconsin Age glaciation, this region was completely surrounded by the continental ice sheet, but it remained unglaciated. Consequently, it served as a refugium for freshwater species. Endemic aquatic fauna include three gastropods (*Stagnico-*

la montanensis, *Physa jennessi athearni*, and *Physa johnsoni*), the Banff hot springs dace (*Rhinichthys cataractae smithi*), the Pyramid Lake sucker (*Catostomus catostomus lacustris*), a blind, unpigmented asellid isopod (*Salmasellus steganothrix*), and two blind, subterranean amphipods (*Stygobromus canadensis* and *S. secundus*). The Canadian Rockies [49] ecoregion is continentally outstanding because of its role as a glacial refugium, its unusual subterranean habitats, and its high level of endemism (Crossman and McAllister 1986).

Conservation Status

Because large portions of this area are already protected by the Banff and Jasper National Parks in Canada and Glacier National Park in the United States, much of its aquatic biodiversity has been preserved. However, this region faces several immediate threats, both within and outside the national parks.

Extractive industries such as logging and coal mining threaten the water quality of this area. In March 1996, Cardinal River Coals Ltd. announced plans to develop a huge open-pit coal mine in the Rocky Mountain foothills. The proposed Cheviot mine area is 23 km by 3.5 km, and is located just 2.8 km from Jasper National Park, a United Nations World Heritage Site (Alberta Wilderness Association 1998a). Under federal legislation, a careful assessment of the proposal's environmental effects is required prior to approval. Natural flow in the Cardinal River will have to be altered extensively to build and maintain the mine. Construction will involve bridges, culverts, and river and creek realignment; channel restoration; and installation of rock riprap armoring and retaining structures. Affected streams will include the McLeod River, Thornton Creek, Cheviot Creek, Prospect Creek, Whitehorse Creek, and two other creeks locally known as Cave Springs Creek and Leyland Creek (Alberta Wilderness Association 1998b). Pollution from the operating mine is expected to be severe. On October 2, 1997, the Canadian government released its decision to allow the Cheviot Project to proceed with no further environmental assessment (Alberta Wilderness Association 1998a).

This region is also threatened by development and tourism. Because of its unusual landscapes and breathtaking scenery, the Canadian Rockies [49] area is a popular tourist destination. Tourism itself can threaten sensitive habitats; more important, it leads to development of infrastructures and subsequent degradation of the landscape.

Suite of Priority Activities to Enhance Biodiversity Conservation

- Banff, Jasper, and Glacier National Parks provide some protection for this ecoregion's freshwater habitats and biotas, but the boundaries of the protected area need to be extended. Preferably, they would reach beyond the mountains and into the foothills.
- Set rigid controls on logging, mining, and development.
- Educate ecoregion inhabitants and tourists about the uniqueness of the freshwater habitats of the Canadian Rocky Mountains.

Conservation Partners

For contact information, please see appendix H.

- Alberta Wilderness Association
- Banff/Bow Valley Naturalists
- Canadian Arctic Resources Committee
- Canadian Parks and Wilderness Society
- The Nature Conservancy of Canada (Alberta Office)
- Sierra Club
- World Wildlife Fund Canada

Ecoregion Number:	**50**
Ecoregion Name:	**Upper Saskatchewan**
Major Habitat Type:	**Temperate Headwaters and Lakes**
Ecoregion Size:	**304,749 km²**
Biological Distinctiveness:	**Nationally Important**
Conservation Status:	**Snapshot—Vulnerable Final—Endangered**

Introduction

This ecoregion covers parts of the Canadian provinces of Saskatchewan and Alberta. It is defined by the watersheds of the North Saskatchewan and South Saskatchewan Rivers, to the point where they converge to form the Saskatchewan River. Major tributaries to these two rivers include the Vermilion, Battle, and Deer Rivers. Several lakes in this ecoregion occupy much of southern Alberta and southwestern Saskatchewan. Because of its situation on the front range, this ecoregion experiences a rainshadow effect and is consequently subject to periodic droughts.

Biological Distinctiveness

The Upper Saskatchewan [50] ecoregion, like most of the rest of Canada, was subjected to heavy glaciation as recently as ten to fifteen thousand years ago. The area contains no endemic fish, mussels, crayfish, or aquatic herpetofauna. Forty-one native fish species, four unionid mussels, one crayfish, and ten aquatic herpetofauna species are found here. Fish species include the lake sturgeon (*Acipenser fulvescens*), goldeye (*Hiodon alosoides*), lake whitefish (*Coregonus clupeaformis*), mountain whitefish (*Prosopium williamsoni*), arctic grayling (*Thymallus arcticus*), and five salmonids. None of the ecoregion's fish is considered imperiled at the species level by TNC (Natural Heritage Center Databases 1997).

The ecoregion is distinguished by its position on the front range of the Canadian Rockies. High gradient streams coming off the mountains change to slower, wider streams as they flow out over the Great Plains in the eastern part of the ecoregion. Some streams of this ecoregion have become anastomosed (connected), a result of flowing over formerly glaciated land that is in a state of gradual uplift. These streams have braided channels that tend to contain uncommon freshwater habitats.

The well-watered parts of this ecoregion also contain important wetlands. Canada holds one-fourth of the world's wetlands, and the province of Saskatchewan has about 17 percent of Canada's total. These wetlands provide important habitat for rare and imperiled birds, such as piping plovers (*Charadrius melodus*), whooping cranes (*Grus americana*), and trumpeter swans (*Cygnus buccinator*) (Saskatchewan Wetland Conservation Corporation 1997).

Conservation Status

The dangers facing the Upper Saskatchewan [50] ecoregion are primarily due to hydroelectricity, overgrazing, groundwater withdrawal for crop irrigation in dry areas, wheat agriculture in moist regions, pulp mills and logging in the western portion, and pollution from oil refining. Two large dams are located in the North Saskatchewan basin—Big Horn Dam on the North Saskatchewan, which creates Lake Abraham, and Brazeau Dam on the Brazeau River. Multiple canals and reservoirs are also on the South Saskatchewan River's tributary basins (Alberta Environmental Protection 1997).

Suite of Priority Activities to Enhance Biodiversity Conservation

- Promote ecoagriculture.
- Establish strict ecological standards for the protection of forests.
- Extend national parks from the slopes of the Rockies out onto the Great Plains.

Conservation Partners

For contact information, please see appendix H.

- Alberta Wilderness Association
- Canadian Nature Federation
- Ducks Unlimited Canada
- Federation of Alberta Naturalists
- The Nature Conservancy of Canada (Alberta and Saskatchewan Offices)
- Nature Saskatchewan
- Prairie Conservation Forum
- Saskatchewan Wetland Conservation Corporation
- Saskatchewan Wildlife Federation
- Society of Grassland Naturalists
- World Wildlife Fund Canada

Ecoregion Number:	**51**
Ecoregion Name:	**Lower Saskatchewan**
Major Habitat Type:	**Temperate Headwaters and Lakes**
Ecoregion Size:	**523,809 km²**
Biological Distinctiveness:	**Continentally Outstanding**
Conservation Status:	**Snapshot—Relatively Stable**
	Final—Relatively Stable

Introduction

This ecoregion spans eastern Alberta, central Saskatchewan, central Manitoba, and a small portion of northwestern Ontario. The western portion is defined by the watershed of the Lower Saskatchewan, downstream from where the North and South Saskatchewan Rivers come together, and includes all of Cedar Lake in Manitoba. The northern and eastern portions constitute the watersheds of the Churchill River and its tributaries. In the east, the Hayes River and its tributaries define the extent of the ecoregion. Between the Churchill and Hayes Rivers lies the Nelson River. There are numerous lakes throughout.

Biological Distinctiveness

The Lower Saskatchewan [51] ecoregion is located on the southwest edge of Hudson Bay. As a consequence of glaciation and its location on the Canadian Shield, the ecoregion is particularly rich in oligotrophic lakes and clear streams. These habitats support characteristic cold-water species. The thirty-eight native freshwater fish species include the lake sturgeon (*Acipenser fulvescens*), mooneye (*Hiodon tergisus*), lake herring (*Coregonus artedi*), lake whitefish (*C. clupeaformis*), shortjaw cisco (*C. zenithicus*), round whitefish (*Prosopium cylindraceum*), arctic char (*Salvelinus alpinus*), lake trout (*S. namaycush*), arctic grayling (*Thymallus arcticus*), and northern pike (*Esox lucius*). Four species of unionid mussels, one crayfish species, and five species of aquatic herpetofauna also occur here. The Lower Saskatchewan [51] is also a summering region for polar bears.

Conservation Status

The freshwater habitats of the Lower Saskatchewan [51] are threatened by pollution, dam construction, and commercial fishing. Agriculture is the primary cause of pollution, contributing pesticides, fertilizers, and eroded soil by way of runoff.

Historically, a number of dams were built on the Saskatchewan and other rivers. Recent years have shown that new dams can be constructed in this region only at the risk of public protest and litigation (Pearse and Quinn 1996).

Finally, there is a growing commercial fishery in this region for whitefishes, perch, and lake trout. The shortjaw cisco, listed as threatened in Canada, provides an example of the vulnerability of these deepwater lake fishes to overexploitation.

Suite of Priority Activities to Enhance Biodiversity Conservation

- Establish protected areas based on freshwater habitats.
- Educate indigenous and nonindigenous peoples about the fragility of the Arctic ecosystem.
- Prevent further impoundment of free-flowing waters.
- Enforce strict pollution and sedimentation controls.

Conservation Partners

For contact information, please see appendix H.

- Canadian Arctic Resources Committee
- Ecology North
- The Nature Conservancy of Canada (Manitoba and Saskatchewan Offices)
- World Wildlife Fund Canada

Ecoregion Number: **52**
Ecoregion Name: **English-Winnipeg Lakes**
Major Habitat Type: **Temperate Headwaters and Lakes**
Ecoregion Size: **639,319 km²**
Biological Distinctiveness: **Continentally Outstanding**
Conservation Status: **Snapshot—Vulnerable Final—Vulnerable**

Introduction

Stretching from the extreme eastern part of southern Alberta to southwestern Ontario, this ecoregion includes much of southeastern Saskatchewan, southern Manitoba, northern and eastern North Dakota, and northern Minnesota. It is defined by the numerous rivers that drain the area surrounding Lake Winnipeg and by abundant other shallow glacial lakes, including Lakes Winnepegosis and Manitoba and the lakes of the Boundary Waters Canoe Area Wilderness. Among the larger rivers are the Assiniboine and Red Rivers.

Biological Distinctiveness

Like most northern ecoregions in North America, the English-Winnipeg Lakes [52] ecoregion was heavily glaciated as recently as ten to fifteen thousand years ago, limiting its current biodiversity. The area has no endemic fish, mussels, crayfish, or aquatic herpetofauna, but it is biologically distinctive because of its unusual geography. The combination of glaciation, its landscape, and its geology provide a unique set of habitats.

Seventy-nine native freshwater fish species are found in the ecoregion. Among these are three species of lamprey (*Ichthyomyzon castaneus, I. fossor,* and *I. unicuspis*), one sturgeon (*Acipenser fulvescens*), one gar (*Lepisosteus osseus*), four whitefish and ciscos (*Coregonus artedi, C. clupeaformis, C. zenithicus,* and *Leucichthys nipigon*), two salmonids (*Oncorhynchus clarki* and *Salvelinus namaycush*), and two pike (*Esox lucius* and *E. masquinongy*).

The English-Winnepeg Lakes [52] ecoregion also serves as a transition area between the eastern and western halves of the North American continent. For example, *Bufo hemiophrys* is found in the eastern part of the area, and *Bufo boreas* is restricted to the west.

Conservation Status

The English-Winnipeg Lakes [52] ecoregion is currently considered vulnerable and faces the prospect of increasing degradation. Agriculture in the area threatens the biodiversity of watercourses and lakes through the addition of pesticides, fertilizers, and sediment. The three major lakes are productive fisheries but are being commercially overfished. Two species, the goldeye (*Hiodon alosoides*) and the lake sturgeon (*Acipenser fulvescens*), were formerly found in this ecoregion but have been extirpated by fishing. The rivers are threatened by water diversion and damming, and introduced species imperil natives in all freshwater habitats. A potential threat is posed by the underground storage of uranium at the Whiteshell nuclear facility near Pinawa, Manitoba.

Suite of Priority Activities to Enhance Biodiversity Conservation

- Encourage ecoagriculture.
- Encourage ecofisheries.
- Educate inhabitants, fishermen, farmers, and tourists about the ecoregion's aquatic biodiversity.
- Reconsider all proposed dams and water diversions in the area.

Conservation Partners

For contact information, please see appendix H.

- Friends of the Boundary Water Wilderness
- Manitoba Naturalists Society
- The Nature Conservancy (Minnesota and North Dakota Field Offices)
- The Nature Conservancy of Canada (Manitoba, Saskatchewan, and Ontario Offices)
- Nature Saskatchewan
- World Wildlife Fund Canada

Ecoregion Number:	**53**
Ecoregion Name:	**South Hudson**
Major Habitat Type:	**Arctic Rivers and Lakes**
Ecoregion Size:	**601,216 km²**
Biological Distinctiveness:	**Nationally Important**
Conservation Status:	**Snapshot—Relatively Intact**
	Final—Relatively Intact

Introduction

This ecoregion, defined by a number of rivers that drain into Hudson Bay and James Bay, covers most of northern Ontario but includes two disjunct areas in Manitoba and a small portion of western Quebec. The northwestern extent is defined by the Machichi River watershed in Manitoba and the western boundary is largely defined by the Severn River watershed. The eastern extent includes the lower portion of the Harricana River, below the point where it is joined by the Wawagosic. Other large rivers include the Attawapiskat, the Winisk, the Albany, and the Moose.

Biological Distinctiveness

The South Hudson [53] ecoregion is located on the ancient rock of the Canadian Shield, where deglaciation at the end of the last ice age left hundreds of lakes and watercourses. The central part of the South Hudson [53] ecoregion, particularly the watersheds of the Winisk, Severn, Attawapiskat, and Albany Rivers, is dominated by wetlands.

The presence of the continental ice sheet so recently in its development has severely limited the region's aquatic biodiversity. It has no endemic fish, mussels, crayfish, or aquatic herpetofauna. However, it is characterized by greater biodiversity than other ecoregions along Hudson Bay. With forty-three species of native fish and three species of unionid mussels, it is more species rich than adjacent bayside ecoregions. It is also a summering region for polar bears. It does not, however, support the anadromous fish that are characteristic of the East Hudson [54] region.

Conservation Status

The South Hudson [53] ecoregion is relatively intact. Sparsely settled by humans, it does not face direct threats from population centers. However, several threats could imperil this ecoregion in the near future if not addressed. Because of its many watercourses, it is well suited for the development of hydropower. Some dams have already been constructed, and more are planned.

Logging is a particularly onerous problem in this ecoregion because of the delicacy of the northern boreal forests. The forests in this region are mostly intact, which makes them an attractive untapped resource. Some forestry operations are already in place, such as at Iroquois Falls, Cochrane, and Smooth Rock Falls. These northern boreal forests are extremely fragile, with potential regeneration times of hundreds or thousands of years.

Another activity that has had serious consequences for this ecoregion, and may have even more serious effects in the future, is mining. The ancient Canadian Shield rocks of this ecoregion are rich in minerals, particularly uranium. In addition to uranium mining, this ecoregion includes the world's largest commercial uranium refinery and Canada's only uranium conversion facility (Cameco 1999).

Suite of Priority Activities to Enhance Biodiversity Conservation

- Prohibit destructive logging practices in the boreal forest.
- Establish protected areas based on aquatic habitats and wetlands.
- Prohibit the development of the region's streams for hydropower.
- Discourage overfishing and settlement of the South Hudson.

Conservation Partners

For contact information, please see appendix H.

- Federation of Ontario Naturalists
- Grand Council of the Crees
- The Nature Conservancy (Ontario and Québec Offices)
- Northwatch
- The Wildlands League
- World Wildlife Fund Canada

Ecoregion Number:	**54**
Ecoregion Name:	**East Hudson**
Major Habitat Type:	**Arctic Rivers and Lakes**
Ecoregion Size:	**552,051 km²**
Biological Distinctiveness:	**Nationally Important**
Conservation Status:	**Snapshot—Relatively Stable**
	Final—Relatively Stable

Introduction

This ecoregion, found entirely within northwestern Quebec, is defined predominantly by the rivers of the province that drain into the Hudson Bay and James Bay. The western extent of the ecoregion is defined by the watersheds of Rivière Bell and the Upper Harricana. Among the major rivers are the Chukotat and Rivière de Povungnituk in the north, the Grande Rivière de la Baleine in the central portion, and the Rivière de Rupert in the south.

Biological Distinctiveness

Partly because of recent glaciation, the ecoregion contains no endemic fish, mussels, crayfish, or aquatic herpetofauna. This ecoregion is similar to the North Arctic [58] and East Arctic [59] ecoregions in the number of species it supports. Unlike these ecoregions, however, the East Hudson also supports fallfish (*Semotilus corporalis*) and runs of anadromous Atlantic salmon (*Salmo salar*).

Conservation Status

Despite the sparse human population within this ecoregion, the aquatic biota faces a wide range of threats from human activity. The two most notorious projects are the proposed Great Recycling and Northern Development (GRAND) Project and the James Bay Project, the latter of which has been implemented.

GRAND was proposed in 1964 to convert James Bay from a saltwater to a freshwater body. First, a dike would be built at the mouth of the bay, allowing the water to freshen over time as the rivers dilute it. Then the fresh water would be piped into the Great Lakes basin to supply water to the dry American Southwest and Midwest. This project has been hotly debated and has not been implemented. However, increased demands for water in the United States make it a very real peril in the future.

The other project, which has been partially implemented, is HydroQuebec's James Bay Project. This is a massive hydropower project which, if fully implemented, would flood more than 15,000 km² of land and displace up to 16,000 people of indigenous Cree and Inuit descent. When the first phase was implemented in 1973, 10,000 km² of land were flooded for hydropower. Implementation of the second phase, on the Great Whale River, is expected to flood another 5,000 km².

In a move related to these two massive projects, the land of the East Hudson [54] ecoregion is being opened up to provide the infrastructure needed for execution and maintenance of the projects. These changes are primarily in the form of roads and transmission lines.

This ecoregion also supports sustenance fishing and a small but growing commercial fishery.

Suite of Priority Activities to Enhance Biodiversity Conservation

- Stop further dam construction.
- Obtain environmental impact data for James Bay Project from HydroQuebec, and take measures to ameliorate the impacts or prevent future disturbances.
- Obtain material on the environmental impact of the Great Whale phase of the James Bay Project, including future threats and a complete survey of aquatic fauna.
- Promote conservation of electric energy, and use existing energy locally to avoid the huge losses that occur with long-distance transmission.
- Prevent implementation of GRAND, noting the crippling effects that such a project would have on the region's ecosystem.

Conservation Partners

For contact information, please see appendix H.

- Grand Council of the Crees
- Environmental Defense Fund
- The Nature Conservancy of Canada (Québec Office)
- Sierra Club
- Union Québecoise pour la Conservation de la Nature (UQCN)
- World Wildlife Fund Canada

Ecoregion Number:	**55**
Ecoregion Name:	**Yukon**
Major Habitat Type:	**Arctic Rivers and Lakes**
Ecoregion Size:	**1,567,884 km²**
Biological Distinctiveness:	**Continentally Outstanding**
Conservation Status:	**Snapshot—Relatively Intact**
	Final—Relatively Intact

Introduction

Part of the Arctic complex, this ecoregion is the largest in North America. It covers the vast majority of Alaska, most of the Yukon Territory, a small portion of northern British Columbia, and the extreme northwest corner of the Northwest Territories. Included in this ecoregion are the Aleutian Islands, the Pribilof Islands, Nunivak Island, St. Matthew Island, and St. Lawrence Island. The ecoregion is primarily defined by the watershed of the Yukon River but also by several other rivers draining into the Bering and Beaufort Seas. Among these rivers are the Kuzitrin, Kuskokwim, Colville, and Sagavanirktak.

Biological Distinctiveness

While most of this ecoregion was glaciated, a small portion remained unglaciated. Today it harbors the ecoregion's small number of endemic freshwater species. The endemic fish are the Squanga whitefish (*Coregonus* sp.), which is unnamed to date, the Angayukaksurak char (*Salvelinus anaktuvukensis*), the Alaska blackfish (*Dallia pectoralis*), and the Bering cisco (*Coregonus laurettae*). The Yukon [55] ecoregion also contains one endemic unionid mussel, *Anodonta beringiana*. In addition, it is distinguished for its anadromous fish migrations. Salmon have been known to migrate more than 2,000 km up the Yukon system from the Bering Sea.

Conservation Status

Even though this ecoregion is relatively intact, threats exist. Notable among these is mining, especially placer mining, which extracts minerals from alluvial sediments by physically disturbing the sediments and then sifting the material. This encourages increased runoff, which in turn increases turbidity, particularly because water is commonly used to disturb the sediments.

Other threats to the freshwater habitats of this ecoregion include tourism, the impacts associated with growing human settlements, dam construction, and a small level of forestry. Aerial pollution from the industrial regions to the south and in Asia is evidenced by contaminated fish; in Lake Laberge, lake trout have been found to contain elevated concentrations of toxaphene, PCBs, DDT, and mercury (Schindler 1998). Both forests and freshwater habitats in the taiga may have very long regeneration times after having been disturbed.

Suite of Priority Activities to Enhance Biodiversity Conservation

- Establish protected areas that are designed around the needs of freshwater habitats.

Conservation Partners

For contact information, please see appendix H.

- Alaska Boreal Forest Council
- Canadian Arctic Resources Committee
- Ecology North
- The Nature Conservancy (Alaska Field Office)
- The Nature Conservancy of Canada
- World Wildlife Fund Canada

Ecoregion Number:	**56**
Ecoregion Name:	**Lower Mackenzie**
Major Habitat Type:	**Arctic Rivers and Lakes**
Ecoregion Size:	**1,209,314 km²**
Biological Distinctiveness:	**Nationally Important**
Conservation Status:	**Snapshot—Relatively Intact,**
	Final—Relatively Intact

Introduction

This large ecoregion in the Arctic complex includes two disjunct areas in the northern and eastern Yukon Territory, part of northeastern British Columbia, a small portion of northwestern Alberta, and the western part of the Northwest Territories. It is largely defined by the lower portion of the Mackenzie River below the Great Slave Lake. The Great Bear Lake and several smaller lakes occur within this ecoregion.

Biological Distinctiveness

Relatively distinct species assemblages are found in the warm, muddy mainstem of the Mackenzie River and the cleaner, colder waters of its upland tributaries and the lakes. As with most of the Arctic rivers, the Mackenzie supports a range of anadromous fish. These include the arctic cisco (*Coregonus autumnalis*), Alaska whitefish (*C. nelsoni*), broad whitefish (*C. nasus*), inconnu (*Stenodus leucichthys*), and arctic lamprey (*Lampetra japonica*). Several of these fish swim fairly far up the river because the water temperature is 1–3° warmer than in the Beaufort Sea in the winter. In total, the ecoregion supports forty-one species of native freshwater fish, as well as two species of unionid mussels and three aquatic herpetofauna species.

Conservation Status

The ecoregion is relatively intact, with no major threats at present. Impending threats to the lower reaches of the Mackenzie River include aerial pollution from the industrial areas to the south and in Asia. Oil wells at the mouth of the Mackenzie River may threaten brackish-water habitats, and mining imperils water quality (Kavanagh pers. comm.).

Suite of Priority Activities to Enhance Biodiversity Conservation

- Establish protected areas based on freshwater habitats.
- Complete biological surveys and mapping of the ecoregion.

Conservation Partners

For contact information, please see appendix H.

- Canadian Arctic Resources Committee
- Canadian Nature Federation
- Canadian Parks and Wilderness Society
- Ecology North
- The Nature Conservancy of Canada (Headquarters and British Columbia Office)
- World Wildlife Fund Canada

Ecoregion Number:	**57**
Ecoregion Name:	**Upper Mackenzie**
Major Habitat Type:	**Arctic Rivers and Lakes**
Ecoregion Size:	**702,806 km²**
Biological Distinctiveness:	**Nationally Important**
Conservation Status:	**Snapshot—Relatively Stable**
	Final—Vulnerable

Introduction

Part of the Arctic complex, this ecoregion stretches from eastern British Columbia through north-central Alberta and northern Alberta into the southern part of the Northwest Territories and part of northwestern Manitoba. Its delineation is determined predominantly by the watershed of Lake Athabasca. In addition to numerous lakes, the major river is the Slave, including its tributaries the Peace, the Taltson, and the Athabasca Rivers.

Biological Distinctiveness

The Upper Mackenzie [57] is characterized by low diversity and a low degree of endemism among freshwater species. As part of the Arctic complex of ecoregions, it has been subjected to heavy glaciation, which precluded the development of a high level of biological diversity. Forty-one species of native freshwater fish, three unionid mussels, and eight aquatic herpetofauana are found in this ecoregion.

Two important features of the Upper Mackenzie [57] ecoregion are the Peace and Athabasca River deltas. Lake whitefish (*Coregonus clupeaformis*) and other fish species are found in the deltas, which also provide critical nesting sites for whooping cranes (*Grus americana*), ducks, and the tundra swan (*Cygnus buccinator*). The Northern River Basins Study (NRBS)(1996) reports that because the river channels and the water they carry are so changeable, the fish inhabiting them are "confronted with extreme seasonal and environmental fluctuations" and consequently "a key strategy for many species is extensive movement." Additionally, according to the study, different species exhibit "extraordinary variation" in these movements from one habitat to another.

Conservation Status

The Upper Mackenzie is faced with a variety of immediate threats, many of which have been documented by the Northern River Basins Study (1996), an investigation sponsored by the governments of Canada, Alberta, and the Northwest Territories in response to rapidly expanding forestry and associated ecological impacts.

Mining and timber exploitation are among the major sources of disturbance for freshwater habitats. Renewed interest in the rich mineral deposits in the region has led to development of new mining operations, including extraction of uranium, diamonds, and tar sands for oil. For example, McArthur River and Cigar Lake in northern Saskatchewan are the two largest uranium deposits in the world, and uranium production operations are located in the ecoregion as well.

Timber operations in the Upper Mackenzie [57] lead to increased runoff and sedimentation. Associated pulp mills have been implicated in discharging effluents containing high phosphorus loads, which in turn can lead to eutrophication of normally oligotrophic waters. Elevated phosphorus levels in the Athabasca River and the Wapiti/Smoky River system have been attributed primarily to pulp mills and to a lesser extent to inadequate sewage treatment facilities. Both of these basins have lower flows but greater development than the Peace and Slave Rivers (Northern River Basins Study 1996).

The river deltas that are part of Lake Athabasca have historically been highly productive areas for fish and birds, but flow-altering dams have disrupted the system. According to the Northern River Basins Study (1996), the Peace River's Bennett Dam has had "a significant impact on the flow patterns, sediment transport, river morphology, ice formation and habitat along the mainstem Peace River . . . Changes to flow and ice patterns are at least partly responsible for the lack of ice-jam induced floods in the Peace–Athabasca Delta. In the absence of these floods, the delta is slowly drying out—profoundly affecting the natural environment and the traditional lifestyles of local resi-

dents. NRBS research also suggests that flow regulation by the dam is affecting the rate of growth of the Slave River Delta into Great Slave Lake."

Suite of Priority Activities to Enhance Biodiversity Conservation

- Establish protected areas based on freshwater habitats.
- Increase efforts to decrease the effects of logging, oil sands, and other mining development.
- In accordance with the recommendations of the Northern River Basins Study (1996), regulatory agencies should make pollution prevention a primary objective, including the elimination of persistent toxic substances, nutrient loadings, and other wastes. In particular, discharges from municipal sewage treatment facilities and pulp mills must be closely monitored.
- Resume operation of Bennett Dam on the Peace River to reestablish the natural flow regime and consequently rehabilitate the Peace-Athabasca Delta and the riparian and aquatic conditions of the Peace River system (Northern River Basins Study 1996).

Conservation Partners

For contact information, please see appendix H.

- Canadian Arctic Resources Committee
- Canadian Parks and Wilderness Society
- Ecology North
- The Nature Conservancy of Canada (Headquarters, Alberta, British Columbia, Manitoba, and Saskatchewan Offices)
- World Wildlife Fund Canada

Ecoregion Number:	**58**
Ecoregion Name:	**North Arctic**
Major Habitat Type:	**Arctic Rivers and Lakes**
Ecoregion Size:	**598,863 km²**
Biological Distinctiveness:	**Continentally Outstanding**
Conservation Status:	**Snapshot—Relatively Intact**
	Final—Relatively Intact

Introduction

This expansive ecoregion of the Arctic complex falls entirely along the extreme northern mainland coast of Canada. Among the larger rivers in this ecoregion are the Back, the Burnside, the Coppermine, the Hornaday, and the Anderson.

Biological Distinctiveness

The entire North Arctic [58] ecoregion harbors only twenty-two freshwater fish species and contains no native mollusks, crayfish, or herpetofauna. None of the fish species is endemic. This low diversity is largely due to periods of recent glaciation ten to fifteen thousand years ago and the region's current cold climate.

The North Arctic [58] ecoregion is an important transition area. It marks the eastern limit of species such as the arctic cisco (*Coregonus autumnalis*), broad whitefish (*C. nasus*), and least cisco (*C. sardinella*). These species are found from the west of this region to the White Sea and the Bering Strait, but not east of the North Arctic [58] ecoregion.

The migratory patterns of some of the fish also distinguish this region. Arctic char and cisco, as well as other fish normally associated with the Mackenzie River Delta, sometimes migrate via the ocean to streams to feed.

Conservation Status

The North Arctic [58] ecoregion is relatively intact because of its remoteness. Little commercial or industrial development occurs. In fact, there are very few roads. However, this delicate ecoregion is not necessarily safe from all disturbances. Preliminary evidence suggests that the region is experiencing impacts from global warming. Deposition of airborne pollutants from industrialized regions to the south and west (Asia) has been recorded. Increased sportfishing and tourism also threaten native species and habitats. Damage inflicted in this area could not easily be mitigated because of the extreme fragility of the Arctic.

One serious threat that could imperil water quality and habitat integrity in the North Arctic [58] ecoregion is the mining development taking place within the region. To support new mining interests, roads are being built and hydropower plants developed. Copper mining is the primary threat here, but lakes in the area are in danger of being drained for diamonds (Kavanagh pers. comm.).

Suite of Priority Activities to Enhance Biodiversity Conservation

- Establish protected areas based on freshwater biodiversity.
- Educate indigenous and nonindigenous peoples about the fragility of Arctic ecosystems.

Conservation Partners

For contact information, please see appendix H.

- Canadian Arctic Resources Committee
- Ecology North
- The Nature Conservancy of Canada
- World Wildlife Fund Canada

Ecoregion Number: **59**
Ecoregion Name: **East Arctic**
Major Habitat Type: **Arctic Rivers and Lakes**
Ecoregion Size: **624,339 km²**
Biological Distinctiveness: **Nationally Important**
Conservation Status: **Snapshot—Relatively Intact**
Final—Relatively Intact

Introduction

Part of the Arctic complex, this ecoregion stretches from northern Manitoba through the eastern mainland portion of the Northwest Territories. The southern extent is defined by the watershed of the Seal River and the western portion by the Thelon River. Among the other rivers are the Lorillard, the Tha-anne, and the Thlewiaza.

Biological Distinctiveness

This far-northern ecoregion is the most north-westerly section of the Hudson Bay drainage basin. The watershed boundary between the Arctic Ocean coastal drainage and Hudson Bay forms the border between it and the North Arctic [58] ecoregion. Faunistically, this ecoregion is distinguished by the absence of the anadromous fish associated with the western Arctic.

Like the rest of the far North, the East Arctic [59] ecoregion was heavily glaciated as recently as ten to fifteen thousand years ago. Only eighteen native freshwater fish, and no unionid mussel, crayfish, or aquatic herpetofauna species occur here. None of the fish species is endemic.

Conservation Status

The East Arctic [59] ecoregion is relatively intact. Threats that imperil the ecoregion include increasing mining development and aerial pollution from industrial sources to the south and west. Because of the fragility of the Arctic ecosystem, any damage to the environment here has a significant and long-lasting effect.

Suite of Priority Activities to Enhance Biodiversity Conservation

- Encourage the new Nunavut territorial government to establish protected areas based on freshwater biodiversity.
- Educate residents about the fragility of Arctic ecosystems.

Conservation Partners

For contact information, please see appendix H.

- Canadian Arctic Resources Committee
- Ecology North
- The Nature Conservancy of Canada (Headquarters and Manitoba Office)
- Nunavut Wildlife Management Board
- World Wildlife Fund Canada

Ecoregion Number:	**60**
Ecoregion Name:	**Arctic Islands**
Major Habitat Type:	**Arctic Rivers and Lakes**
Ecoregion Size:	**1,407,571 km²**
Biological Distinctiveness:	**Continentally Outstanding**
Conservation Status:	**Snapshot—Relatively Intact**
	Final—Relatively Intact

Introduction

This ecoregion includes the majority of the islands in the Northwest Territories north of the Arctic Circle. These include Baffin Island, the Queen Elizabeth Islands, and the Parry Islands.

Biological Distinctiveness

These islands represent the most northerly extent of fresh water on this continent. They do not support a wide variety of freshwater species. Only eight freshwater fish species are known from this area, five of which are salmonids.

There are no endemic fish species. Mollusks, however, show some degree of endemism. *Stagnicola kennicotti*, a gastropod, is endemic to Victoria Island adjacent to the North American mainland (Crossman and McAllister 1986).

Most of this ecoregion has been heavily glaciated. Spots such as Banks Island and the east coast of Baffin Island escaped glaciation altogether and may have served as refugia for freshwater species including fish and mollusks. Alpine glaciers are still very much part of the landscape. The Arctic Islands [60] ecoregion displays very long ice resident times, with up to six months of polar night.

Normally glaciation has a stark effect on the number of freshwater species in an area. In short, when there is a glacier, liquid water is scarce. What little there is generally supports relatively few species. However, glaciers may support the formation of at least one type of specialized habitat. When coastal areas are glaciated, the land is depressed. Then, as the glacier retreats, saltwater pools up on formerly dry land. When the weight of the glacier has sufficiently decreased, the land begins to rebound, raising some of this saltwater in pools above sea level. Over time, these saltwater pools freshen from rainfall and glacial runoff. The species in these pools, formerly marine species, adapt over generations to fresh water. They are called lacustrine marine relict species. One species of fish in the Arctic Islands [60] ecoregion, the fourhorn sculpin (*Myoxocephalus quadricornis*), and at least one crustacean, *Mysis* sp., exhibit this adaptation to fresh water.

Conservation Status

The freshwater habitats of the Arctic Islands [60] ecoregion do not face any major immediate threats, though they may be affected by air pollution from distant urban sources. A few mining operations exist within the archipelago as well. These are limited in scope but may severely affect local biota through pollution and sedimentation. Increased tourism and sportfishing may put a future burden on the Arctic ecosystem.

Regardless of the absence of severe threats, the Arctic ecosystem is extremely fragile. Damaged environments regenerate over extremely long periods. Plants grow more slowly, and animal populations may crash with a minimum of impact. Extreme care must be taken to ensure that development, tourism, and resource extraction do not damage the biodiversity of this region.

Suite of Priority Activities to Enhance Biodiversity Conservation

- Educate indigenous and nonindigenous peoples about the fragility of Arctic ecosystems.
- Establish protected areas for lacustrine marine relicts.

Conservation Partners

For contact information, please see appendix H.

- Canadian Arctic Resources Committee
- Ecology North
- The Nature Conservancy of Canada
- Nunavut Wildlife Management Board
- World Wildlife Fund Canada

Ecoregion Number:	**61**
Ecoregion Name:	**Sonoran**
Major Habitat Type:	**Xeric-Region Rivers, Lakes, and Springs**
Ecoregion Size:	**203,936 km²**
Biological Distinctiveness:	**Continentally Outstanding**
Conservation Status:	**Snapshot—Vulnerable**
	Final—Endangered

Introduction

This coastal ecoregion consists of several disjunct regions in southern Arizona, most of Sonora, and a small portion of eastern Chihuahua. The northern extent of the ecoregion is defined by a series of mountain ranges, including the Gila, Cabeza Prieta, Sierra Pinta, Comobabi, and Baboquivari in Arizona. The Continental Divide forms the eastern limit of the ecoregion. Sparse rain in these areas drains to the Gulf of California by way of the Río Concepción, Río Sonora, Río Yaqui, and Río Montezuma. The watershed of the Río Mayo defines the southern extent.

Biological Distinctiveness

The Sonoran Desert is one of four deserts in North America. Unlike the Chihuahuan ecoregions further to the east, the Sonoran [61]

ecoregion does not exhibit the same remarkable aquatic biodiversity and endemism. In fact, the freshwater biota of this region is quite depauperate. It includes only eight endemic fish species, including a stoneroller (*Campostoma pricei*), the Yaqui sucker (*Catostomus bernardini*), the Yaqui beautiful shiner (*Notropis formosus mearnsi*), and the Yaqui catfish (*Ictalurus pricei*). At least three endemic snails occur here: the San Bernardino springsnail (*Fontelicella* sp.), Yepomera springsnail (*Fontelicella* sp.), and Yepomera tryonia (*Tryonia* sp.). There are no known endemic crayfish or aquatic herpetofauna.

For the most part, the aquatic fauna of the Sonoran [61] ecoregion closely resembles that of the Rio Grande complex. The most likely reason is that at some previous time, the former headwaters of the Río Conchos were captured by the Yaqui system (Hendrickson et al. 1980).

Aquatic life in the Sonoran [61] ecoregion has developed some remarkable adaptations to the extreme lack of rainfall. Tadpole shrimp (*Triops* sp.) eggs lie dormant in the parched soil for years, waiting for one of the region's infrequent rains. Upon wetting, the eggs hatch, and the shrimp mate in the temporary rain pools and lay their eggs to wait for another rainfall (Sonoran Arthropod Studies Institute 1998).

Conservation Status

The Sonoran [61] ecoregion faces several immediate, severe threats to its aquatic biodiversity. Among the most serious is degradation of water quality. The area has historically been a center for mining, leading to the contamination of stream sediments with heavy metals. Also affecting the water quality of the region is the use of land for grazing and pasturage. This overuse has led to sedimentation and animal-waste contamination. The productive riparian zones along the waterways have been largely modified or destroyed (McIntyre 1997).

As would be expected in an area as arid as the Sonoran [61] ecoregion, freshwater species are in competition with humans for water. Water overuse and aquifer drawdown endanger obligate subterranean species. Surface water has also decreased; this is the chief reason for the extirpation of the Yaqui beautiful shiner from the waters of the United States (Williams et al. 1985). Several large reservoirs, such as the Presa Plutarco Elias Calles on the Yaqui and the Presa de la Angostura on the Bavispe, provide water for municipalities, industry, and agriculture. These structures alter habitat and flow regimes and degrade water quality for native species.

Suite of Priority Activities to Enhance Biodiversity Conservation

- Maintain adequate riparian-zone cover along waterways and restrict livestock, to protect banks from erosion; provide a pollution buffer; and preserve vital terrestrial and aquatic habitats.
- Enforce strict water-use limits within the ecoregion, minimizing aquifer drawdown and elimination of surface water habitats.
- Remediate the damage to sediments caused by heavy metal contamination.
- Conduct a complete biological survey of all exotic and native aquatic species in the Sonoran [61] ecoregion.

Conservation Partners

For contact information, please see appendix H.

- La Comisión Nacional para el Conocimiento y uso de la Biodiversidad (CONABIO)
- Instituto Tecnologico y de Estudios Superiores de Monterrey (ITESM)
- The Nature Conservancy
- Universidad Autónoma de Chihuahua (UACH)
- Universidad Autónoma de Nuevo León (UANL)
- Universidad Autónoma de Sinaloa (UAS)
- Universidad de Occidente
- Universidad de Sonora (UNISON)
- Universidad Nacional Autónoma de México (UNAM)
- University of Arizona
- World Wildlife Fund México

Ecoregion Number: **62**
Ecoregion Name: **Sinaloan Coastal**
Major Habitat Type: **Subtropical Coastal Rivers, Lakes, and Springs**
Ecoregion Size: **126,886 km²**
Biological Distinctiveness: **Continentally Outstanding**
Conservation Status: **Snapshot—Vulnerable Final—Endangered**

Introduction

This ecoregion covers southern Sonora, southwestern Chihuahua, all of Sinaloa, much of western Durango, and part of northern Nayarit. The eastern boundary is formed by the Continental Divide. Moving north to south, the following rivers drain into the Gulf of California and the Pacific Ocean: the Río Fuerte, Río Sinaloa, Río Humaya, Río Tama, Río San Lorenzo, and Río Presidio.

Biological Distinctiveness

The Sinaloan Coastal [62] ecoregion does not exhibit extraordinarily high levels of aquatic endemism. It has only two endemic fish (the Mexican golden trout, *Oncorhynchus chrysogaster*, and the clearfin livebearer, *Poeciliopsis lucida*) and one endemic species of crayfish. The ecoregion contains only an estimated nineteen species of native fish and two crayfishes, but eighty-one aquatic herpetofauna species. Threatened fish of this ecoregion include the roundtail chub (*Gila robusta*) and *Gila purpurea*. In general, the freshwater biota of this ecoregion is poorly known.

Conservation Status

The aquatic biota of this ecoregion is threatened primarily by water-quality degradation and channel modification. This region, with its high-gradient streams flowing from the high mountains of the Sierra Madre Occidental, has been used extensively for the generation of hydropower. Hydropower dams are notorious for changing water chemistry, flow regime, and temperature both upstream and downstream, and for blocking the linear movement of aquatic species. Dams have also been built in this ecoregion for irrigation and flood control.

Water quality is affected on the coastal plain by saltwater intrusion, caused by the overpumping of groundwater. Agriculture is at fault for the degradation of water quality as well, because of the application of agrochemicals and the sedimentation of adjacent streams. In the upper portions of the watersheds that form this ecoregion, mining activities play a large role in pollution, as do domestic and municipal wastes.

Suite of Priority Activities to Enhance Biodiversity Conservation

- Enforce efficient municipal use of water, such as flow-limiting devices and proper waste treatment.
- Enforce limits to agrochemical application.
- Enforce responsible, ecologically friendly agricultural practices.
- Complete a biological inventory of the aquatic habitats of the entire ecoregion.
- Control pollution by mining interests.
- Place a moratorium on dam construction.

Conservation Partners

For contact information, please see appendix H.

- La Comisión Nacional para el Conocimiento y uso de la Biodiversidad (CONABIO)
- Instituto Tecnologico y de Estudios Superiores de Monterrey (ITESM)
- The Nature Conservancy
- Universidad Autónoma de Sinaloa (UAS)
- Universidad de Occidente
- Universidad de Sonora (UNISON)
- Universidad Nacional Autónoma de México (UNAM)
- World Wildlife Fund México

Ecoregion Number:	**63**
Ecoregion Name:	**Santiago**
Major Habitat Type:	**Subtropical Coastal Rivers, Lakes, and Springs**
Ecoregion Size:	**106,988 km²**
Biological Distinctiveness:	**Continentally Outstanding**
Conservation Status:	**Snapshot—Vulnerable Final—Endangered**

Introduction

The extent of this ecoregion is largely defined by the watersheds of the Río San Pedro and Río Santiago and their tributaries, including the Río Mezquital, Río Grande de Santiago, Río Bolaños, and Río Juchipila. The ecoregion extends from the coast through most of central Nayarit, across northeastern Jalisco, and into southern Zacatecas and western Guanajuato, and includes the majority of Aguascalientes.

Biological Distinctiveness

The Santiago [63] ecoregion is located on the western slope of the Sierra Madre Occidental and harbors relatively few known endemic aquatic species. It has five endemic species of fish, one endemic crayfish, and one endemic species of aquatic herpetofauna. The endemic fish of the region include a species of *Cichlasoma* and one species of *Ictalurus*. A total of forty-five known species of native freshwater fish, three crayfish, and eighty-four aquatic herpetofauna occur here.

This region is distinguished in part by its subterranean ecology. Aguascalientes is a warm aquifer located in the Mexican state that bears its name. It harbors unique populations of aquatic invertebrates such as rotifers.

Conservation Status

The integrity of the aquatic ecology of the Santiago [63] ecoregion is in danger. Channel modification, such as damming, and overuse of the region's water supplies have severely affected the freshwater biodiversity. Water quality has also been compromised through the unrestricted use of agrochemicals and by heavy-metal pollution from mining operations in the mountains. Deforestation increases sedimentation of watercourses, and domestic and municipal waste contribute to nutrient loading and fecal coliform contamination. Exotic species introductions have affected the fish biodiversity of the ecoregion. Species such as blue tilapia (*Oreochromis aureus*) and common carp (*Cyprinus carpio*) have been introduced into the waterways of the Santiago [63] ecoregion and threaten native fauna. Finally, as in the Sinaloan Coastal [62] ecoregion to the immediate northwest, the aquatic fauna of this ecoregion is poorly known. A complete biological inventory has never been accomplished.

Suite of Priority Activities to Enhance Biodiversity Conservation

- Complete a biological inventory of the Santiago [63] ecoregion.
- Enforce responsible agricultural use of water resources.
- Enforce pollution regulations on mining operations in the upper part of the watershed.

- Maintain water levels at Aguascalientes to preserve unique fauna.
- Use responsible forestry practices to minimize sedimentation, and prohibit clear-cutting.
- Ban further channel modification, and consider the decommisioning of some extant dams.

Conservation Partners

For contact information, please see appendix H.

- La Comisión Nacional para el Conocimiento y uso de la Biodiversidad (CONABIO)
- The Nature Conservancy
- Universidad Autónoma de Aguascalientes
- Universidad Autónoma de Nayarit (UAN)
- Universidad Nacional Autónoma de México (UNAM)
- World Wildlife Fund México

Ecoregion Number:	**64**
Ecoregion Name:	**Manantlan-Ameca**
Major Habitat Type:	**Subtropical Coastal Rivers, Lakes, and Springs**
Ecoregion Size:	**49,563 km²**
Biological Distinctiveness:	**Globally Outstanding**
Conservation Status:	**Snapshot—Relatively Stable**
	Final—Relatively Stable

Introduction

This ecoregion encompasses a small part of southern Nayarit, most of southern Jalisco, two disjunct sections of western Michoacán, and the entire state of Colima. Its extent is determined by the watersheds of the Río Ameca in the north and the Río Ahuijallo in the south.

Biological Distinctiveness

The Manantlan-Ameca [64] ecoregion is distinctive because of its endemism, especially among fish. Fourteen of its twenty-five fish are endemic, and it has one endemic species of aquatic herpetofauna. Endemic species of fish in this region include the golden livebearer (*Poeciliopsis baenschi*), the dwarf molly (*Poecilia chica*), the blackspotted livebearer (*Poeciliopsis turneri*), and one splitfin (*Ilyodon* sp.). This remarkable

fish endemism contributes to the Manantlan-Ameca's [64] rating of globally outstanding.

Livebearers are well represented in this ecoregion. Species of *Allodontichthys*, *Ilyodon*, *Poecilia*, *Poeciliopsis*, and *Xenotoca* are common in the waters of Río Coahuajana and Río Armeria, and *Xenotaenia resolanae* is located in Río Ayuquila, Río Armeria, and Sierra de Manantlan. An endemic minnow, *Algansea aphanea*, is also found in Río Coahuayana and Río Armeria. Goby genera *Awaous*, *Sicydium*, *Dormitator*, *Eleotris*, and *Gobiomorus* are also present (Azas 1991b).

Conservation Status

Among the human activities that threaten the aquatic biodiversity of this region are population growth, tourism, agriculture (application of agrochemicals), forestry, overuse of the region's aquifers (largely to support municipal and tourist use), and overfishing.

Introduction of exotic species such as the blue tilapia (*Oreochromis aureus*) has affected the freshwater biodiversity of this ecoregion. These exotic species outcompete or prey on native fish species. The aquatic fauna of this region is also very poorly known.

Suite of Priority Activities to Enhance Biodiversity Conservation

- Complete a biological inventory of this ecoregion, including native and exotic species.
- Control forestry in the upper and middle parts of this ecoregion, and preserve riparian zones along streams to limit sedimentation.
- Minimize infrastructure improvements for tourism.
- Encourage responsible agricultural and municipal water use, especially of aquifers, to prevent depletion and saltwater intrusion.

Conservation Partners

For contact information, please see appendix H.

- La Comisión Nacional para el Conocimiento y uso de la Biodiversidad (CONABIO)
- Fundación Ecológica de Cuixmala
- The Nature Conservancy
- Universidad Nacional Autónoma de México (UNAM)
- World Wildlife Fund México

Ecoregion Number: **65**
Ecoregion Name: **Chapala**
Major Habitat Type: **Xeric-Region Rivers, Lakes,**
and Springs
Ecoregion Size: **7,486 km²**
Biological Distinctiveness: **Globally Outstanding**
Conservation Status: **Snapshot—Vulnerable**
Final—Endangered

Introduction

Lake Chapala, the largest natural lake in
Mesoamerica, is located in the Mexican states of
Jalisco and Michoacán on the country's west
coast. Situated in the Mexican altiplano, the lake
is formed in an east-west depression bound by
two faults (Ferrusquía-Villafranca 1993). The
lake sits at the bottom of the Lerma River Basin
and discharges into the Santiago River, which
flows to the Pacific Ocean.

Biological Distinctiveness

Lake Chapala is distinguished by a fish fauna that
is nearly entirely endemic. Endemic or near-
endemic fish species are *Lampetra spadicea*,
Algansea popoche, *Skiffia bilineata*, *Ictalurus dugesi*,
Chapalichthys encaustus, *Chirostoma aculeatum*, *C.
arge*, *C. chapalae*, *C. consocium*, *C. estor*, *C. labarcae*, *C. lucius*, *C. promelas*, and *C. sphyraena*. At
the genus level, both *Skiffia* and *Chapalichthys* are
endemics shared with the Lerma and Santiago
Basins. Also endemic are the frogs *Rana megapo-
da* and *R. montezumae*.

Conservation Status

Lake Chapala suffers from being situated downstream from a highly industrialized and agricultural basin, as well as from being in a region where
water is in short supply for a dense human population. As water flows through the Lerma Basin, it
picks up pollutants from chemical, petroleum,
and food-processing industries; pig farms; and
other agricultural sources. Much of this water has
been diverted for industry and agriculture before
reaching Lake Chapala, and the remaining flow
has often failed to meet minimum water-quality
standards. By 1989, 90 percent of the lake was
classified as having unacceptable water quality for
both human and wildlife uses (Gutierrez 1997;
Mestre 1997; Arriaga et al. 1998).

Additional withdrawals from the lake supply 70
percent of the city of Guadalajara's drinking water.
As a result of these withdrawals and the overexploitation of the Lerma Basin's aquifers, the area
of the lake has decreased in size by 14 percent and
in volume by 54 percent over the past century. The
lake is also besieged by the exotic water hyacinth
(*Eichhornia crassipes*) and threatened by introduced fish species (*Ictalurus dugesi*, *Oreochromis*
spp., *Micropterus salmoides*), and native species are
subject to overfishing (Gutierrez 1997; Arriaga et
al. 1998).

In response to these threats to Lake Chapala
and the Lerma catchment, concerned stakeholders formed the Lerma-Chapala River Basin
Council in 1993. This council adopted a River
Basin Hydraulic Program for distributing surface
water, treating wastewater, improving water-use
efficiency, and correcting land-use practices. The
water quality of Lake Chapala has subsequently
improved, but serious concerns about the future
of the lake's freshwater habitats and its biota
remain as development and population growth in
the area intensify (Mestre 1997).

Suite of Priority Activities to Enhance
Biodiversity Conservation

Based on recommendations from Gutierrez
(1997), Mestre (1997), and Arriaga et al. (1998):

- Work with the Lerma-Chapala River Basin
  Council to achieve its goals of improving
  water allocation policies, water-use efficiency
  (especially for irrigation), and wastewater
  treatment programs upstream of the lake.
- Control cattle grazing in open areas.
- Reforest upstream areas to decrease sedimentation and erosion.
- Prevent introductions of new species and
  control populations of established species,
  without resorting to massive poisoning
  schemes.
- Work with the indigenous Huichol Indians,
  who want to declare the region a protected
  cultural area.

Conservation Partners

For contact information, please see appendix H.

- Comisión Nacional del Agua

- La Comisión Nacional para el Conocimiento y uso de la Biodiversidad (CONABIO)
- Estación Ecológica Chapala (Universidad Autónoma de Guadalajara)
- Lerma-Chapala River Basin Council
- The Nature Conservancy
- World Wildlife Fund México

Ecoregion Number:	**66**
Ecoregion Name:	**Llanos el Salado**
Major Habitat Type:	**Endorheic Rivers, Lakes, and Springs**
Ecoregion Size:	**55,330 km²**
Biological Distinctiveness:	**Continentally Outstanding**
Conservation Status:	**Snapshot—Critical**
	Final—Critical

Introduction

This interior ecoregion of Mexico extends from southeastern Zacatecas, northeastern Aguascalientes, extreme northeastern Jalisco, and extreme northern Guanajuato through most of central San Luis Potosí into southern Nuevo León and southwestern Tamaulipas.

Biological Distinctiveness

The Llanos el Salado [66] ecoregion is an endorheic basin—streams flowing into the basin have no outlet to the ocean. Instead, the water is lost primarily through evaporation. This type of drainage is fairly common throughout very arid, mountainous regions of the world. The harshness of the habitats, the tectonism of the rock on which these basins sit, and the relative genetic isolation of the aquatic species of the basins often lead to development of a highly specialized and unique fauna.

The Llanos el Salado's [66] species richness is fairly low, but its endemism is high. Eight of the nine fish species found in the area are endemic. The southern portion of the region has two endemic species from the genus *Xenoophorus*. The northern portion has six more endemic fish species, all of them pupfish: the Potosi pupfish (*Cyprinodon alvarezi*), the memorial pupfish (*C. inmemoriam*), the long-finned pupfish (*C. longidorsalis*), the charco azul pupfish (*C. veronicae*), the violet pupfish (*C. ceciliae*), and the catarina pupfish (*Megupsilon aporus*). There is also one endemic crayfish, *Procambarus* sp.

Conservation Status

The unique freshwater fauna of the Llanos el Salado [66] ecoregion is particularly vulnerable to water overuse in this xeric basin. The deep aquifers, which have been developing for millennia from the meager rainfall of the region, are being depleted at a rapid rate for agricultural and municipal use. Population growth within this fragile basin has put pressures on its water resources that cannot be borne by the region much longer. Also threatening the biota of the Llanos el Salado [66] ecoregion are introductions of exotic species, which may be more tolerant of disturbed habitats than the species endemic to the region.

Suite of Priority Activities to Enhance Biodiversity Conservation

- Regulate pumping in the region.
- Establish limits to land clearing for agricultural use and settlement.

Conservation Partners

For contact information, please see appendix H.

- La Comisión Nacional para el Conocimiento y uso de la Biodiversidad (CONABIO)
- The Nature Conservancy
- World Wildlife Fund México

Ecoregion Number:	**67**
Ecoregion Name:	**Río Verde Headwaters**
Major Habitat Type:	**Xeric-Region Rivers, Lakes, and Springs**
Ecoregion Size:	**4,859 km²**
Biological Distinctiveness:	**Globally Outstanding**
Conservation Status:	**Snapshot—Relatively Stable**
	Final—Relatively Stable

Introduction

This ecoregion is distinguished primarily by La Medialuna, an extensive series of constant-temperature aquifers, marshes, and outflows that lie in the Río Verde intermontane basin some 10 km south-southwest of the town of Río Verde in the state of San Luis Potosí, at an altitude of about 1,000 m above sea level. In its unmodified state, La Medialuna enjoyed only intermittent contact with the remainder of the Río Verde-Río Panuco drainage during periods of exceptionally heavy rainfall. Its name derives from a large crescent-shaped laguna fed by six spring holes that provides drinking and irrigation water for three adjacent towns. The climate of the Río Verde intermontane basin is subtropical, with moderate, markedly seasonal precipitation.

Biological Distinctiveness

The ichthyofauna of the Río Verde Headwaters [67] ecoregion comprises eleven species, nine of which are endemic. Among these are the rare flatjaw minnow (*Dionda mandibularis*), known from only two spring-fed locations in the ecoregion, and the bicolor minnow (*D. dichroma*), also restricted to spring-fed headwater habitats. Within La Medialuna are the Medialuna killie (*Cualac tessellatus*) and the striped goodeid (*Ataeniobius toweri*), both of which represent monotypic genera. La Medialuna is also home to the Mojarra aracolera (*Cichlasoma bartoni*) and the yellow mojarra (*Cichlasoma* sp.), the La Medialuna shrimp (*Palaemonetes lindsayi*), the La Medialuna crayfish (*Procambarus roberti*), and the crayfish's obligatory parasite, the La Medialuna ostracod (*Ankylocythere barbouri*). According to Williams et al. (1985), the crayfish is "a highly disjunct member of its subgenus and may represent a relict crayfish stock that migrated southward into Mexico during the Pliocene."

Conservation Status

The greatest threat to the integrity of La Medialuna is large-scale pumping of groundwater for commercial agricultural use. Attempts to pump water directly from the main laguna have to date proven unsuccessful for technical reasons. However, the economic pressures driving the commercialization of Mexican agriculture continue unabated and further efforts to directly tap the aquifer that feeds La Medialuna in the near future are highly likely. Such efforts, if met with success, would lead to an immediate loss of the peripheral marsh habitat occupied by the cyprinodont, *Cualac tesselatus*, and the goodeid, *Ataeniobius toweri*.

Since 1972, five exotic species have become established in the La Medialuna system. The blue tilapia (*Oreochromis aureus*) was deliberately introduced. Three poeciliids and a cichlid, *Herichthys*

carpintis, all native to the Río Panuco Basin, appear to have entered the basin via a series of canals dug to divert a portion of the spring flow for irrigation. While both the lagunas and their effluents now support large populations of the exotic poeciliids, numbers of the endemic cyprinodont and goodeid have declined precipitously. Numbers of both endemic cichlids declined dramatically following the establishment of the blue tilapia and *H. carpintis*. Their populations appear to have rebounded somewhat since 1985 and now appear to be stable. However, hybridization between the endemic *Nandopsis labridens* and *H. carpintis* has been documented. In the absence of precise data on the extent of such miscegenation, it is impossible to estimate how much of a threat *H. carpintis* poses to the genetic integrity of *N. labridens*.

Additional threats to the ecoregion include tourism, use of water for recreation, and intensive livestock grazing (Arriaga et al. 1998).

Suite of Priority Activities to Enhance Biodiversity Conservation

- Limit groundwater pumping in the region, and prevent pumping of water from the main laguna of La Medialuna.
- Prevent introductions of new exotic species and develop a program to control or eradicate populations of established exotics, which seriously threaten the ecoregion's few native species.
- Continue to maintain populations of natives in captivity for possible reintroduction.
- Limit tourism and use of the laguna at La Medialuna as a bathing area.
- Limit pasturing of livestock.

Conservation Partners

For contact information, please see appendix H.

- The Nature Conservancy
- Universidad del Noreste
- Universidad de Tampico
- World Wildlife Fund México

NOTE: The above description was quoted directly from Loiselle (1993), except where otherwise noted. Consult Loiselle for original sources of information.

Ecoregion Number:	**68**
Ecoregion Name:	**Tamaulipas-Veracruz**
Major Habitat Type:	**Subtropical Coastal Rivers, Lakes, and Springs**
Ecoregion Size:	**132,524 km²**
Biological Distinctiveness:	**Globally Outstanding**
Conservation Status:	**Snapshot—Vulnerable**
	Final—Endangered

Introduction

This ecoregion encompasses southern Tamaulipas, a small portion of southeastern Nuevo León, eastern San Luis Potosí, northeastern Guanajuato, northeastern Queretaro, most of Hidalgo, northern Mexico, a very small portion of Tlaxcala, northern Puebla, and the northern half of Veracruz. The rivers of this ecoregion drain into the Gulf of Mexico. Among them are the Río Soto la Marina, Río Chihue, and Río Santa Maria.

Biological Distinctiveness

This ecoregion forms the beginning of the transition between the xeric northern portion of Mexico and the moist, subtropical southern extent. The precipitation in this ecoregion is not as scarce as in the Chihuahuan region to the west and northwest, though net evaporation still exceeds precipitation. It contains a particularly rich aquatic fauna, and the ecological state of the region is still relatively unknown. Twenty-nine of its ninety-four known fish species are endemic. Most of these endemics are poeciliids and cichlids, including the black lyre (*Poecilia latipinna*), the Amazon molly (*P. formosa*), and the shortfin molly (*P. mexicana*). Also endemic are species of *Xiphophorus*, *Cyclosoma*, and *Gambusia*. Additionally, this ecoregion has seventeen endemic species of crayfish and sixteen endemic species of aquatic herpetofauna.

Conservation Status

The freshwater habitats of this ecoregion are under pressure from human activities. CONABIO has identified several sites within the Tamaulipas-Veracruz [68] ecoregion as priority sites for conservation. Among them is the Río Tamesí, which flows across the coastal plain of eastern Mexico after descending from the Sierra Madre Oriental. It contains numerous endemic fish and is threatened by pollution from agrochemicals, municipal discharge, thermal discharge, and salinization. Channel modification and overuse of water resources have threatened many riparian habitats, and exotic species (e.g., tilapia [*Oreochormis* spp.], hydrilla [*Eichhornia crassipes*]) exclude native fauna.

Suite of Priority Activities to Enhance Biodiversity Conservation

- Establish continuous study and cataloging of the region's aquatic fauna, including exotic species.
- Address overexploitation of water resources, especially marshes.
- Enforce responsible agricultural practices, such as terracing, to minimize pollution.
- Enact stricter regulations for effluent discharge from municipal and industrial sources.

Conservation Partners

For contact information, please see appendix H.

- Comisión Nacional del Agua
- The Nature Conservancy
- Universidad del Noreste
- World Wildlife Fund México

Ecoregion Number:	**69**
Ecoregion Name:	**Lerma**
Major Habitat Type:	**Endorheic Rivers, Lakes, and Springs**
Ecoregion Size:	**69,780 km²**
Biological Distinctiveness:	**Globally Outstanding**
Conservation Status:	**Snapshot—Endangered**
	Final—Critical

Introduction

This interior Mexican ecoregion encompasses eastern Jalisco, northern Michoacán, most of Guanajuato, southwestern Queretaro, northern Mexico, all of the Federal District, northern Morelos, southeastern Hidalgo, northern Tlaxcala, central Puebla, and western Veracruz. The ecoregion corresponds with the Lerma River Basin, excluding Lake Chapala, into which the waters of the basin drain (see description for the

Chapala [65] ecoregion). The main branch of the Lerma River is more than 700 km long, with numerous tributaries and aquifers along its length. Precipitation in the region is average for Mexico, but during the dry season there can be extreme low-flow events.

Biological Distinctiveness

The Lerma [69] ecoregion is considered globally outstanding for its highly endemic fish fauna. Endemic fish include *Lampetra geminis, Algansea barbata, A. lacustris, Algansea popoche, Chirostoma aculeatum, C. arge, C. attenuatum, C. bartoni, C. chapalae, C. consocium, C. estor, C. labarcae, C. lucius, C. ocatlane, C. patzcuaro, C. promelas, C. riojai, C. sphyraena, Poblana alchichica, P. ferdebueni, P. letholepis, Allotoca meeki, Chapalichthys encaustus, Goodea luitpoldi, Hubbsina turneri, Ilyodon furcidens, Neoophorus diazi, Skiffia bilineata, S. lermae,* and *S. multipunctata.* Four endemic and near-endemic fish genera are also known from this ecoregion—*Hubbsina, Skiffia, Chapalichthys,* and the extinct *Evarra.* The ecoregion also supports one near-endemic salamander genus, *Ryacosiredon.* This degree of taxonomic endemism is virtually unsurpassed in North American ecoregions and is rare worldwide.

Conservation Status

The combined Lerma and Santiago Basins (including Lake Chapala) have been the subject of recent, concerted conservation efforts because of the severity of threats facing the area's water supply and consequently its biota. Historically, the water supply has been subject to intense pressure from agriculture, industry, and human population growth. Irrigated agriculture accounts for approximately 78 percent of the Lerma-Santiago Basin's water use, which includes groundwater extraction. In comparison, industrial and domestic water use is small at the regional scale but can have localized impacts, particularly on groundwater. Naturally low water levels during the dry season are exacerbated by these water withdrawals, such that the volume of water reaching Lake Chapala has declined substantially over time (Gutierrez 1997; Mestre 1997; Arriaga et al. 1998).

Lower water levels have also intensified declines in water quality. Wastewater from chemical, petroleum, and food-processing indus-tries in Mexico and Queritaro, and from pig farms in Guanajuato and Michoacán, has largely been untreated and flushed into the region's rivers and streams. Rapid development in the region has led to deforestation and erosion on plains and slopes, resulting in serious sedimentation problems. By 1989, three major reaches of the Río Lerma and several major tributaries were classified as highly polluted (Gutierrez 1997; Mestre 1997; Arriaga et al. 1998).

To begin to address these problems, multiple stakeholders formed the Lerma-Chapala River Basin Council in 1991. Through its River Basin Hydraulic Program, the council has made some progress in improving water-use efficiency and water quality, but there are still major hurdles to overcome before the basin's health is restored and protected (Mestre 1997).

Suite of Priority Activities to Enhance Biodiversity Conservation

Based on recommendations from Gutierrez (1997), Mestre (1997), and Arriaga et al. (1998):

- Work with the Lerma-Chapala River Basin Council to achieve its goals of improving water allocation policies, water-use efficiency (especially for irrigation), and wastewater treatment programs upstream of the lake.
- Control cattle grazing in open areas.
- Reforest upstream areas to decrease sedimentation and erosion.
- Prevent introductions of new species and control populations of established species, without resorting to massive poisoning schemes.
- Work with the indigenous Huichol Indians, who want to declare the region a protected cultural area.

Conservation Partners

For contact information, please see appendix H.

- Comisión Nacional del Agua
- Estación Ecológica Chapala (Universidad Autónoma de Guadalajara)
- Lerma-Chapala River Basin Council
- The Nature Conservancy
- World Wildlife Fund México

Ecoregion Number:	**70**
Ecoregion Name:	**Balsas**
Major Habitat Type:	**Subtropical Coastal Rivers, Lakes, and Springs**
Ecoregion Size:	**114, 547 km²**
Biological Distinctiveness:	**Continentally Outstanding**
Conservation Status:	**Snapshot—Vulnerable Final—Endangered**

Introduction

The extent of this coastal ecoregion is largely defined by the Río Balsas and its tributaries, including the Tepalcatepec River. Other rivers that drain into the Pacific Ocean include the Coalcoman and Nexpa. The ecoregion extends from the coast northward through central Michoacán, northward into eastern Jalisco, then eastward through northern Guerrero, southern México, most of Morelos, southern Tlaxcala, and southern Puebla, reaching its eastern limit in western Oaxaca.

Biological Distinctiveness

The Balsas [70] ecoregion harbors a relatively small number of fish species, but a high percentage of these are endemic. This pattern is consistent with the distribution of fish in the other ecoregions of the mountainous, subtropical Pacific coast of Mexico (Sinaloan Coastal [62], Santiago [63], Manantlan-Ameca [64], and Tehuantepec [74]). This ecoregion is also remarkable for its populations of endemic frogs and other aquatic herpetofauna, such as *Rana madrensis*, *Rana sweifeli*, and *Rana forrers*. In total, there are about 100 aquatic reptile and amphibian species.

Seven endemic fish are found in the Balsas [70] ecoregion. These are the Balsas splitfin (*Ilyodon whitei*), Balsas allotoca (*Allotoca regalis*), Balsas livebearer (*Poeciliopsis balsas*), Balsas molly (*Poecilia maylandi*), Balsas silverside (*Atherinella balsana*), the Balsas catfish (*Ictalurus balsanus*), and the Balsas shiner (*Hybopsis boucardi*). This ecoregion does not have the richness of cichlid species that most of southern Mexico exhibits. Only one of these fish, the redside cichlid (*Cichlasoma istlanum*), appears in the Balsas [70] ecoregion (Azas 1991b).

Conservation Status

CONABIO has identified several sites within the Río Balsas watershed that are of particular concern for aquatic conservation. Among these is the Cuenca Baya del Río Balsas, which is imperiled by pollution and by extensive modification of riparian habitat. River modifications have included at least two dams, forming one very large reservoir near the mouth of the Balsas, and one smaller impoundment on the Río Sultepec, a tributary of the Balsas. These dams are used to supply water and hydroelectric power.

Pollution in this area comes primarily from intensive mining operations in the upper Balsas watershed. Heavy metals and acid runoff create high levels of contamination. Also contributing to pollution is runoff from agricultural activities, especially cattle waste, sedimentation, and agrochemicals. Deforestation creates additional pollution.

Suite of Priority Activities to Enhance Biodiversity Conservation

- Place stricter controls on the pollution of the Río Balsas from acid runoff and heavy metals from mining operations.
- Control deforestation by banning clear-cutting and preserving riparian zones along watercourses.
- Enforce responsible agricultural practices to minimize water use and runoff of agrochemicals and sediment.
- Prohibit further channel modification and diversion.

Conservation Partners

For contact information, please see appendix H.

- Instituto Nacional de Ecología
- The Nature Conservancy
- Universidad Michoacana
- Universidad Nacional Autónoma de México (UNAM), Instituto de Biología
- World Wildlife Fund México

Ecoregion Number: **71**
Ecoregion Name: **Papaloapan**
Major Habitat Type: **Subtropical Coastal Rivers, Lakes, and Springs**
Ecoregion Size: **55,870 km²**
Biological Distinctiveness: **Continentally Outstanding**
Conservation Status: **Snapshot—Vulnerable Final—Endangered**

Introduction

This coastal ecoregion encompasses most of central Veracruz, two disjunct portions of eastern Puebla, and part of northern Oaxaca. The extent of the ecoregion is defined largely by the watersheds of the two major rivers, the Río Santo Domingo and Río Trinidad, that drain into the Gulf of Mexico.

Biological Distinctiveness

The Papaloapan [71] ecoregion is characterized by relatively high endemism within its ichthyofauna. Twelve (17 percent) of the seventy native fish species are endemic to this region. Endemics include the graceful priapella (*Priapella bonita*), species of *Gambusia* and *Heterandria*, and two endemic species of crayfish. A remarkable twenty-seven (26 percent) endemic aquatic herpetofauna species, such as *Anotheca spinosa*, occur here as well.

Also within this ecoregion is Laguna de Catemaco [72], a remarkable lake with flocks of endemic fish. Because the lake's biota is so distinctive, it is treated as a separate ecoregion.

Conservation Status

The greatest threats to the integrity of the aquatic ecosystems of the Papaloapan [71] ecoregion are channel modification for agricultural purposes and agricultural pollution. There are several large reservoirs on the Río Papaloapan, and water is continually diverted from natural channels for irrigation of sugarcane crops. The production of sugarcane, a water-intensive crop, encourages excessive use and pumping of water and leads to agrochemical contamination. Industrial and urban discharge also contribute to poor water quality in this relatively low-elevation region of southern Mexico.

Suite of Priority Activities to Enhance Biodiversity Conservation

- Reduce the amount of sugarcane that is grown in this ecoregion. This alone would measurably improve the water quality of the Papaloapan [71] through decreased use of water and nutrient loading.
- Enforce responsible agricultural practices, such as terracing, to minimize water loss and contamination.
- Ensure effective treatment of municipal and industrial effluents.

Conservation Partners

For contact information, please see appendix H.

- La Comisión Nacional para el Conocimiento y uso de la Biodiversidad (CONABIO)
- Instituto de Ecología A.C.
- The Nature Conservancy
- Universidad Veracruzana
- World Wildlife Fund México

Ecoregion Number: **72**
Ecoregion Name: **Catemaco**
Major Habitat Type: **Subtropical Coastal Rivers, Lakes, and Springs**
Ecoregion Size: **192 km²**
Biological Distinctiveness: **Globally Outstanding**
Conservation Status: **Snapshot—Relatively Stable Final—Relatively Stable**

Introduction

Lago Catemaco is a volcanic crater lake located on the Atlantic Coastal Plain of Mexico, 176 km south of the city of Veracruz. The lake lies at an altitude of 335 m above sea level in a basin 11 km long by 8 km wide. A number of small streams flow down the interior face of the crater into the lake, whose outlet to the northeast is into a river known locally as the Río Grande. Twelve kilometers downstream of the outflow, a 45-m high waterfall, the Salto de Eyiplanta, effectively isolates the lake from the lower reaches of the Río Papaloapan system. The climate along the coastal plain in the state of Veracruz is tropical, with heavy but seasonal precipitation.

The lake is extremely turbid because it contains suspended particles of basaltic dust.

Biological Distinctiveness

The ichthyofauna of Lago Catemaco comprises ten species, eight of which are endemic (Contreras-Balderas pers. comm.). The most important families are the Cichlidae, the Poeciliidae, and the Characidae. The lake's endemic cichlid, an undescribed sister species of the widespread fluviatile, *Theraps fenestratus*, is characterized by color polymorphism.

Conservation Status

Five exotic species, among them the blue tilapia (*Oreochromis aureus*), are established in the lake basin. However, the principal threat to this ecosystem is anthropogenic eutrophication. The interior of the crater is heavily settled, the city of Catemaco on the northern shore of the lake boasting a population of 250,000. The lake is a popular vacation spot and the focus of ongoing real-estate development.

Other threats include deforestation and associated erosion, road construction, heavy-metal contamination, and overexploitation of native species (Arriaga et al. 1998).

Suite of Priority Activities to Enhance Biodiversity Conservation

- Prevent introductions of new exotic species and develop a program for control of populations of established exotics.
- Educate residents about their impacts on the lake.
- Strengthen wastewater treatment to reduce eutrophication.
- Limit forestry surrounding the lake.
- Develop and enforce fishing regulations.

Conservation Partners

For contact information, please see appendix H.

- Instituto de Ecología A.C.
- The Nature Conservancy
- Universidad Nacional Autónoma de México (UNAM), Instituto de Biología
- Universidad Veracruzana
- World Wildlife Fund México

NOTE: The above description was quoted directly from Loiselle (1993), except where otherwise noted. Consult Loiselle for original sources of information.

Ecoregion Number:	**73**
Ecoregion Name:	**Coatzacoalcos**
Major Habitat Type:	**Subtropical Coastal Rivers, Lakes, and Springs**
Ecoregion Size:	**29,294 km²**
Biological Distinctiveness:	**Continentally Outstanding**
Conservation Status:	**Snapshot—Vulnerable Final—Endangered**

Introduction

This coastal ecoregion encompasses northeastern Oaxaca, eastern Veracruz, and western Tabasco. The extent of the ecoregion is defined by the watershed of the Río Coatzacoalcos, its lower lacustrine and swamp system, and some minor coastal drainages. Major tributaries of the Río Coatzacoalcos are the Jaltepec and Uspanapa Rivers. Together these rivers drain an area north of the Continental Divide to the Gulf of Mexico.

Biological Distinctiveness

The Coatzacoalcos [73] ecoregion exhibits very high endemism among its fish. Eighteen of its fifty-eight (31 percent) native fish species are endemic to the region. These fish are mostly cichlids and livebearers (poeciliids). The area does not support the same numbers of aquatic herpetofauna as does the montane Pacific ecoregion of Tehuantepec [74] to the south. In fact, only two of its seventy-one native herpetofauna are endemic to the region. This ecoregion has no endemic crayfish.

The montane headwaters of this ecoregion are particularly rich in species. Eighteen fish species are distributed among eight families in the headwaters, with a concentration in the Río Grande. These families are Characidae (*Axtyanax* sp.), Pimelodidae (*Rhamdia* sp.), Belontiidae (*Strogylura* sp.), Poeciliidae (two *Mollienesia* species, two species of *Poeciliopsis*, and one *Pseudoxiphophorus*), Atherinidae (*Archomenidia* sp. and *Xenatherina* sp.), Mugilidae (*Agonostomus* sp.), Eleotridae (*Gobiomorus* sp.), and Cichlidae (six species) (Azas 1991a).

Conservation Status

This ecoregion is threatened by deforestation and other changes to the landscape. Additionally, pollution imperils its freshwater ecosystems. Agrochemicals, solid waste, and fecal coliform from agricultural activity contaminate the streams.

An additional problem in this ecoregion is the introduction of exotic species for aquaculture or other purposes. The most common exotic found in the waters of Coatzacoalcos is the blue tilapia (*Oreochromis aureus*).

Suite of Priority Activities to Enhance Biodiversity Conservation

- Monitor and control pollution caused by livestock and agrochemicals.
- Control deforestation, and maintain riparian zones around streams to minimize sedimentation. Keep riparian zones off-limits for forestry and livestock.
- Complete a biological inventory of the freshwater fauna of the area, including exotic species.

Conservation Partners

For contact information, please see appendix H.

- Centro Interdisciplinario de Investigación para el Desarrollo Integral Regional (CIIDIR)
- La Comisión Nacional para el Conocimiento y uso de la Biodiversidad (CONABIO)
- The Nature Conservancy
- Proyecto UNAM-PEMEX
- Universidad Juarez Autónoma de Tabasco (UJAT)
- World Wildlife Fund México

Ecoregion Number:	**74**
Ecoregion Name:	**Tehuantepec**
Major Habitat Type:	**Subtropical Coastal Rivers, Lakes, and Springs**
Ecoregion Size:	**963,703 km²**
Biological Distinctiveness:	**Continentally Outstanding**
Conservation Status:	**Snapshot—Vulnerable Final—Vulnerable**

Introduction

This coastal ecoregion stretches all the way from Mexico to the northern half of Costa Rica and encompasses numerous rivers on the southern slopes of the Continental Divide that drain into the Pacific Ocean. In this book, only the portion in Mexico covering southern Guerrero, southern Oaxaca, and southern Chiapas is described. Among the major rivers in this ecoregion are the Petatlán, Papagayo, Verde, and Tehuantepec.

Biological Distinctiveness

This ecoregion, encompassing the southern extent of the Pacific coast of Mexico, has a fairly typical fish distribution for the mountainous coastal stretch. It has only twenty-nine known native fish species, but six of these (21 percent) are endemic, including cichlids. More remarkable are this ecoregion's populations of endemic aquatic herpetofauna. The area harbors 125 native species, far more than any other ecoregion in North America. Of these species, forty-two are endemic. This is the highest level of endemism of aquatic herpetofauna in North America. Among these endemics are the frogs *Rana sierramadrensis*, *Rana zluefeli*, and *Rana omiltemana*.

Conservation Status

The freshwater ecosystems of this ecoregion have been extensively degraded by human factors such as deforestation, channel modification, tourism, water waste, overpumping of groundwater, and agricultural contamination. The popularity of this area as a tourist destination suggests that these pressures will continue in the future.

One of the major problems has been sedimentation and contamination of streams by the conversion of land for cattle grazing. This may contribute to topsoil loss, soil impaction, fecal coliform contamination, sedimentation of watercourses, and ultimately a change in the flow and morphology of streams. Tourism has been particularly detrimental for the ecosystem because of the garbage and organic waste discharge that originates from hotels and cities in general.

Suite of Priority Activities to Enhance Biodiversity Conservation

- Establish protected areas within the Tehuantepec [74] ecoregion that are based on freshwater habitats.
- Place a cap on further infrastructure development for the support of more tourism.
- Maintain riparian zones along waterways that are off-limits to livestock and forestry operations.
- Encourage efficient municipal use of water resources to minimize water withdrawal.

Conservation Partners

For contact information, please see appendix H.

- Centro Interdisciplinario de Investigación para el Desarrollo Integral Regional (CIIDIR)
- La Comisión Nacional para el Conocimiento y uso de la Biodiversidad (CONABIO)
- The Nature Conservancy
- Universidad Autónoma de Chihuahua (UACH)
- Universidad Autónoma de Guerrero (UAG)
- Universidad del Mar
- Universidad Nacional Autónoma de México (UNAM)
- World Wildlife Fund México

Ecoregion Number:	**75**
Ecoregion Name:	**Grijalva-Usumacinta**
Major Habitat Type:	**Subtropical Coastal Rivers, Lakes, and Springs**
Ecoregion Size:	**136,290 km²**
Biological Distinctiveness:	**Continentally Outstanding**
Conservation Status:	**Snapshot—Vulnerable**
	Final—Vulnerable

Introduction

The extent of the Grijalva-Usumacinta [75] ecoregion encompasses extreme southeastern Veracruz, most of Chiapas, eastern Tabasco, most of southern Campeche, and southeastern Quintana Roo. Portions of the ecoregion also extend into Guatemala, but these areas are not discussed. The boundaries of this ecoregion are largely defined by the watersheds of the Grijalva and Usumacinta Rivers and their tributaries.

These rivers lie north of the Continental Divide and drain into the Gulf of Mexico. The endorheic basins of Río Comitan and Lagunas de Montebello are within this ecoregion.

Biological Distinctiveness

The Grijalva-Usumacinta [75] ecoregion is characterized by a wide variety of freshwater habitats. It has fairly extensive riparian habitats, huge expanses of wetlands, and even endorheic basins. Given this habitat diversity, and especially the isolation provided by the endorheic habitats, one would expect this region to have a high level of endemism among fish. Indeed, twenty-nine of this ecoregion's seventy fish species (41 percent) are endemic. Some of the endemic fish of this ecoregion are species of *Xenodexia*, *Cichlasoma callolepus*, *C. ellioti*, and *C. pozolera*. Twelve of the eighty-two (15 percent) species of native aquatic herpetofauna are endemic as well.

Conservation Status

Although this is a large, relatively sparsely settled portion of Mexico, water and other natural resources of the region are being exploited at an ever-increasing rate. One major problem is deforestation. Large expanses of timber are being taken for firewood and construction. Deforestation leads to increased sedimentation and nutrient loading of receiving waters, endangering the natural biota of the streams. Also problematic is agriculture. Intensive use of agrochemicals such as fertilizers and pesticides guarantees that these chemicals will end up in waterways. Industrial waste from petroleum operations in the area contributes to pollution as well.

Dams on the Río Grijalva, including Presa Nezahualcoyotl and Presa de la Angostura, have significantly altered its natural flow regime and created several large reservoirs. This river's dams provide a substantial proportion of Mexico's hydroelectric power.

Suite of Priority Activities to Enhance Biodiversity Conservation

- Ban clear-cutting. Ensure that responsible forestry practices are used, including strict regulations protecting riparian zones.
- Establish protected areas based on freshwater habitats, including endorheic basins and large wetlands.
- Enforce responsible agricultural practices to minimize water wastage and pollution.

Conservation Partners

For contact information, please see appendix H.

- Comisión Nacional del Agua
- La Comisión Nacional para el Conocimiento y uso de la Biodiversidad (CONABIO)
- The Nature Conservancy
- Secretaria de Medio Ambiente Recursos Naturales y Pesca (SEMARNAP)
- Universidad Juarez Autónoma de Tabasco (UJAT)
- World Wildlife Fund México

Ecoregion Number: **76**
Ecoregion Name: **Yucatán**
Major Habitat Type: **Subtropical Coastal Rivers, Lakes, and Springs**
Ecoregion Size: **79,602 km²**
Biological Distinctiveness: **Continentally Outstanding**
Conservation Status: **Snapshot—Relatively Stable**
Final—Relatively Stable

Introduction

Restricted to the Yucatán Peninsula, this ecoregion encompasses northern Campeche, all of Yucatán, and northern Quintana Roo. The Yucatán [76] ecoregion includes many coastal streams that drain into the Gulf of Mexico and Caribbean, grottos, *cenotes*, and *aquadas* in the interior of the peninsula.

Biological Distinctiveness

From a freshwater perspective, the Yucatán [76] ecoregion is relatively unexplored biologically; thus it is estimated that its true biodiversity is richer than the described species would suggest.

Nonetheless, information exists regarding a number of endemic freshwater species. Among the endemic freshwater fish are five pupfish, all of which are endemic to Laguna Chichancanab in the Mexican state of Quintana Roo (Espinosa Perez et al. 1993). These species are *Cyprinodon beltrani, C. simus, C. maya, C. labiosus,* and *C. verecundus.* Two frogs, *Triprion petasatus* and *Bolitoglossa yucatana,* as well as the Creaser's mud turtle, *Kinosternon creaseri,* are also endemic to the Yucatán [76] ecoregion. The ecoregion's karst-derived subterranean habitats support a number of unusual species, such as the fish *Ogilbia pearsi* and *Ophisternon infernale,* and several subterranean amphipod crustaceans (Holsinger 1990, 1993; Proudlove 1997). This ecoregion is also distinguished by the presence of *rías,* streams that flow inland from the sea.

Conservation Status

Several freshwater areas within the Yucatán [76] ecoregion have been identified by CONABIO as important and requiring conservation intervention. These are Laguna Chichancanab, Cono Sur-Peto, Zona Citricol, Anillo de Cenotes, Corredor Cancun-Tulum, Cenotes Tulum-Coba, and Sian Ka'an (Arriaga et al. 1998). The endemic fauna of Laguna Chichancanab are threatened by introduced tilapia (*Oreochromis mossambicus*), and the habitat is degraded as a result of flow modification and contamination by organic matter. Other sites are threatened by agricultural pollution, aquifer contamination, illegal fishing, extraction of water and wood, urban and industrial pollution, saltwater intrusion, loss of surrounding vegetation, and the effects of unregulated tourism (Arriaga et al. 1998).

Suite of Priority Activities to Enhance Biodiversity Conservation

- Further investigate freshwater biodiversity of the ecoregion, including that of subterranean systems.
- Prevent introductions of new exotic species, and develop management plans for established exotics.
- Promote more efficient water use, and discourage further modification of natural water flow.
- Reduce pollution from human settlements that could lead to eutrophication.

Conservation Partners

For contact information, please see appendix H.

- Amigos de Sian Ka'an
- Centro de Investigación y de Estudios Avanzados (CINVESTAV)
- La Comisión Nacional para el Conocimiento y uso de la Biodiversidad (CONABIO)
- El Colegio de la Frontera Sur (ECOSUR)
- Instituto Tecnológico Regional de Chetumal
- The Nature Conservancy
- Pronatura Peninsula de Yucatan
- Universidad de Quintana Roo
- World Wildlife Fund México

APPENDIX H

Conservation Partner Contact Information

Note: The following list of conservation partners, compiled from the preceding ecoregion descriptions, is not intended to be a comprehensive list of organizations or individuals working for freshwater conservation. The number of groups working at local, regional, and national scales, including those focused on particular watersheds, is too large for inclusion in this book. For more extensive lists of such groups, we recommend consulting the following information clearinghouses:

International Rivers Network
1847 Berkeley Way
Berkeley, CA 94703 USA
Tel: 510-848-1155, Fax: 510-848-1008
Email: irn@irn.org
Website: http://www.irn.org

River Network
P.O. Box 8787
Portland, OR 97207, USA
Tel: 503-241-3506
Email: info@rivernetwork.org
Website: http://www.rivernetwork.org

Additionally, a multitude of governmental agencies is involved in conservation initiatives throughout the United States, Canada, and Mexico. Contact information for these agencies is readily available through local directories and on the internet.

Alabama Natural Heritage Program
Huntington College, Massey Hall
1500 East Fairview Avenue
Montgomery, AL 36106-2148, USA
Tel: 334-834-4519, Fax: 334-834-5439
Email: alnhp@wsnet.com
Website: http://www.heritage.tnc.org/nhp/us/al/

Alabama Rivers Alliance
700 28th Street South, Suite 202G
Birmingham, AL 35233, USA
Tel: 205-322-6395, Fax: 205-322-6397
Email: alabamariv@aol.com

Alaska Boreal Forest Council
Hidden Drive, Number 3
PO Box 84530
Fairbanks, AK 99708, USA
Tel: 907-457-8453, Fax: 907-457-5185
Website: http://www.ptialaska.net/~abfc/

Alberta Wilderness Association
P.O. Box 6398, Station D
Calgary, AB T2P 2E1, Canada
Tel: 403-283-2025, Fax: 403-270-2743
Website: http://www.web.net/~awa/index.htm

American Rivers
1025 Vermont Avenue, NW
Suite 720
Washington, DC 20005, USA
Tel: 202-347-9224, Fax: 202-347-9240
Email: amrivers@amrivers.org
Website: http://www.amrivers.org/

Amigos de Sian Ka'an
Website:
    http://www.coa.edu/HEJourney/yucatan/SianKaan/

Anacostia Watershed Society
The George Washington House
4302 Baltimore Avenue
Bladensburg, MD 20710, USA
Tel: 301-699-6204, Fax: 301-699-3317
Website: http://www.anacostiaws.org/

The Arctic Institute of North America
201 Rasmuson Library
University of Alaska
Fairbanks, AK 99775-1005, USA

Arkansas Natural Heritage Commission
1500 Tower Building
323 Center Street
Little Rock, AR 72201, USA
Tel: 501-324-9150, Fax: 501-324-9618
Website:
    http://www.heritage.state.ar.us:80/nhc/heritage.html

Atlantic Salmon Federation
PO Box 429
St. Andrews, NB E0G 2X0, Canada
Tel: 506-529-4581, Fax: 506-529-4985
Email: asfpub@nbnet.nb.ca
Website: http://www.asf.ca/

Banff/Bow Valley Naturalists
Box 1693
Banff, AB T0L 0C0, Canada
Tel: 403-762-4160, Fax: 403-762-4160

Bioconservación, A.C.
Apartado Postal 504
San Nicolas, N.L. 66450, México
Tel/Fax: 8-376-2231

British Columbia Ministry of Environment, Lands, and
    Parks
Fifth Floor, 2975 Jutland Road
Victoria, BC V8T 5J9, Canada
Tel: 250-387-1161

Cahaba River Society
2717 7th Avenue South
Suite 205
Birmingham, AL 35233-3421, USA
Tel: 205-322-5326, Fax: 205-324-8346
Email: cahaba@igc.apc.org
Website: http://home.judson.edu/ers.html

California Department of Fish and Game
1416 9th Street
Sacramento, CA 95814, USA
Tel: 916-653-7664, Fax: 916-653-1856

California Native Grass Association
P.O. Box 566
Dixon, CA 95620, USA
Tel: 916-78-6282

The California Native Plant Society
1772 J Street, Suite 17
Sacramento, CA 95814, USA
Tel: 916-447-2677, Fax: 916-447-2727
Website:
    http://www.calpoly.edu/~dchippin/society.html

Canadian Arctic Resources Committee
7 Hinton Avenue North, Suite 200
Ottawa, ON K1N 4P1, Canada
Tel: 613-759-4284, Fax: 613-722-3318
Website: http://www.carc.org/

Canadian Nature Federation
1 Nicholas Street, Suite 606
Ottawa, ON K1N 7B7, Canada
Tel: 613-562-3447, Fax: 613-562-3371
Website: http://www.cnf.ca/

Canadian Parks and Wilderness Society
880 Wellington Street, Suite 506
Ottawa, ON K1R 6K7, Canada
Tel: 613-569-7226, Fax: 613-569-7098
Website: http://www.cpaws.org/

Canyon Preservation Trust
1401 Canyon Road
Santa Fe, NM 87501, USA
Tel: 505-989-8606

Central Cascades Alliance
P.O. Box 1104
Hood River, OR 97031, USA
Tel/Fax: 541-387-2274
Email: cascades@linkport.com

Le Centre de Donnees sur le Patrimoine Naturel du
    Québec
Ministère de l'Environnement et de la Faune
Direction de la conservation et du patrimoine
    écologique
675, boulevard René-Lévesque Est, 10e étage
Québec (Québec) G1R 4Y1, Canada
Tel: 418-644-3361, Fax: 418-646-6169
Email: cdpnq@mef.gouv.qc.ca
Website: http://www.mef.gouv.qc.ca/fr/environn/
    devdur/centre.htm

Centro Interdiscipcinario de Investigacion para el
    Desarrollo Integral Regional (CIIDIR)
Tel/Fax: 52/91-353-30218
Email: cidirm@vmredipn.ipn.mx

Centro de Investigaciones Biológicas del Noreste, S.C.
    (CIBNOR)
Aptdo 128, LaPaz, B.C.S. 23000, México
Tel: 682-5-3633, Fax: 682-5-3625
Website: http://www.main.conacyt.mx/conacyt/
    sepconacyt/cibnor.html

Centro de Investigación y de Estudios Avanzados (CIN-
    VESTAV)
A.P. Cordemex
C.P. 97310
Mérida, Yucatán, México
Tel: 52/81-29-60 or 52/81-29-31, Fax: 52/81-29-23 or
    52/81-29-19
Telex: 753654 CIEMME
Website:
    http://kin.cieamer.conacyt.mx/LaUnidad/Home.html

Chesapeake Bay Foundation
Headquarters
162 Prince George Street
Annapolis, MD 21401, USA
Tel: 1-888-SAVEBAY (728-3229)
Website: http://www.cbf.org/

Chattahoochee Riverkeeper
P.O. Box 1492
Columbus, GA 31902, USA
Tel: 706-663-2774, Fax: 706-323-9809

Coastal Plains Institute and Land Conservancy
1313 North Duval Street
Tallahassee, FL 32301, USA
Tel: 850-681-6208

El Colegio de la Frontera Sur (ECOSUR)
Website: http://www.ecosur.mx

Colorado Rivers Alliance
P.O. Box 40065
Denver, CO 80204, USA
Tel: 303-212-2405, Fax: 303-758-8976
Email: info@coloradorivers.org
Website: http://www.coloradoriversalliance.org

Columbia Environmental Research Center
4200 New Haven Road
Columbia, MO 65201, USA
Tel: 573-875-5399, Fax: 573-876-1896
Website: http://www.ecrc.cr.usgs.gov/

Columbia River Inter-Tribal Fish Commission
729 NE Oregon, Suite 200
Portland, OR 97232, USA
Tel: 503-238-0667, Fax: 503.235.4228
Email: croj@critfc.org
Website: http://www.critfc.org/

Comisión Nacional del Agua
Gerencia de Saneamiento y Calidad del Agua
Comisión Nacional del Agua
Avenida San Bernabé 549
San Jerónimo Lídice
10200 México, D.F., México
Website: http://www.can.gob.mx/

La Comisión Nacional para el Conocimiento y uso de
    la Biodiversidad (CONABIO)
Fernández Leal No. 43 Barrio de la Concepción
Coyoacán, D.F. C.P. 04020, México
Tel: 52/422-35-00, Fax: 52/422-35-31
Email: conabio@xolo.conabio.gob.mx
Website: http://www.conabio.gob.mx/

Confederated Tribes of the Umatilla Indian Reservation
PO Box 638
Pendleton, OR 97801, USA
Tel: 541-276-3165, Fax: 541-276-3095

Connecticut River Watershed Council
1 Ferry Street
Easthampton, MA 01027, USA
Tel: 413-529-9500, Fax: 413-529-9501
Website: http://www.ctriver.org/crwc.html

Conservation Council of New Brunswick
180 St. John Street
Fredericton, NB E3B 4A9, Canada
Tel: 506-458-8747, Fax: 506-458-1047
Website: http://www.webnet/~ccnb/

Coosa River Society
816 Chestnut Street
Gadsden, AL 35999, USA
Tel: 205-546-4429, Fax: 205-546-8173

Defenders of Wildlife
1101 14th Street, NW #1400
Washington, D.C. 20005, USA
Tel: 202-682-9400
Website: http://www.defenders.org/

Desert Fishes Council
P.O. Box 337
Bishop, CA 93515, USA
Tel/Fax: 760-872-8751
Website:http://www.utexas.edu/depts/tnhc/.
    www/fish/dfc

Ducks Unlimited Canada
Box 1160, Oak Hammock Marsh
Stonewall, MB R0C 2Z0, Canada
Tel: 204-467-3000, Fax: 204-467-9436
Website: http://www.ducks.ca/

Ecology Action Centre
Suite 31, 1568 Argyle Street
Halifax, NS B3J 2B3, Canada
Tel: 902-429-2202
Website: http://www.chebucto.ns.ca/Environment/EAC/
    EAC-Home.html

Ecology North
4807 49th Street, Suite 8
Yellowknife, NT X1A 3T5, Canada
Tel: 403-873-6019, Fax: 403-873-3654
Website:
    http://www.ssimicro.com/%Eecononorth/about.htm

Environment Canada
Inquiry Centre
351 Saint Joseph Boulevard
Hull, Québec K1A Oh3, Canada
Tel: 819-997-2800
Email: enviroinfo@ec.gc.ca
Website: http://www.ec.gc.ca/

Environment North
704 Holly Crescent
Thunder Bay, ON P7E 2T2, Canada
Tel: 807-475-5267, Fax: 807-577-6433

Environmental Defense Fund
257 Park Avenue South
New York, NY 10010, USA
Website: http://www.edf.org/

Estación Ecológica Chapala (Universidad Autónoma de
    Guadalajara)
Contact through Chapala Ecology Station
Department of Biology
Baylor University
PO Box 97388
Waco, TX 76798-7388, USA
Website: http://www.baylor.edu/~ces/

Everglades Coalition (an alliance of nearly 30 Florida
    and national environmental organizations)
Co-chair: Mary Munson (Defenders of Wildlife)
Tel: 202-682-9400
Co-chair: David Guggenheim (Conservancy of South-
    west Florida)
Tel: 941-262-0304

Federation of Alberta Naturalists
Box 1472
Edmonton, AB T5J 3V9, Canada
Tel: 780-427-8124, Fax: 780-422-2663
Email: fan@connect.ab.ca
Website: http://www.connect.ab.ca/~fan/

Federation of Nova Scotia Naturalists
c/o Nova Scotia Museum
1747 Summer Street
Halifax, NS B3H 3A6, Canada
Website:
    http://www.chebucto.ns.ca/Environment/FNSN/

Federation of Ontario Naturalists
355 Lesmill Road
Don Mills, ON M3B 2WB, Canada
Tel: 416-444-8419, Fax: 416-444-9866
Website: http://www.ontarinature.org/

Florida Audubon Society
1331 Palmetto Avenue, Suite 110
Winter Park, FL 32789, USA
Tel: 407-539-5700, Fax: 407-539-5701
Website: http://www.audubon.usf.edu/

Florida Defenders of the Environment
4424 N.W. 13th Street, Suite C-8
Gainesville, FL 32609-1885, USA
Tel: 352-378-8465, Fax: 352-377-0869
Website: http://www.afn.org/~fde/

Florida Natural Areas Inventory
1018 Thomasville Road
Suite 200-C
Tallahassee, FL 32303, USA
Tel: 850-224-8207, Fax: 850-681-9364
Website: http://www.fnai.org/

Fondo Mexicano para la Conservación de la
    Naturaleza (FMCN)
Damas 49
San Jose Insurgentes
03900 México, D.F., México
Tel: 5-661-9779

Forest Guardians
1411 Second Street, Suite One
Santa Fe, NM 87505, USA
Tel: 505-988-9126, Fax: 505-989-8623
Website: http://www.fguardians.org/

Friends of the Animas River
P.O. Box 3685
Durango, CO 81302, USA
Tel/Fax: 970-259-1120

Friends of the Boundary Waters Wilderness
1313 Fifth Street, SE, Suite 329
Minneapolis, MN 55414, USA
Tel: 612-379-3835
Website: http://www.friends-bwca.org/

Friends of the Earth
1025 Vermont Avenue NW, Third Floor
Washington, DC 20005, USA
Tel: 202-783-7400, Fax: 202-783-0444
Email: foe@foe.org
Website: http://www.foe.org

Friends of the Los Angeles River
P.O. Box 292134
Los Angeles, CA 90029, USA
Tel: 213-223-0585, Fax: 818-980-0700
Website: http://www.folar.org

Friends of the White Salmon
367 Oakridge Road
White Salmon, WA 98672, USA
Tel: 360-493-3891

Fundación Ecológica de Cuixmala
Cuixmala, Jalisco, México
Tel: 335-10-040

Georgia Natural Heritage Program
Wildlife Resources Division
Georgia Department of Natural Resources
2117 U.S. Highway 278 SE
Social Circle, GA 30279, USA
Tel: 706-557-3032, Fax: 706-557-3040
Email: natural_heritage@mail.dnr.state.ga.us
Website: http://www.heritage.tnc.org/nhp/us/ga/

Grand Canyon Trust
2601 N. Fort Valley Road
Flagstaff, AZ 86001, USA
Tel: 520-774-7488, Fax: 520-774-7570
Website: http://www.kaibab.org/gct/

Grand Council of the Crees
2 Lakeshore Road
Nemaska, James Bay (Québec) J0Y 2B0, Canada
Tel: 819-673-2600, Fax: 819-673-2606
Website: http://www.gcc.ca/

Great Lakes Commission
400 Fourth Street
Ann Arbor, MI 48103, USA
Tel: 734-665-9135, Fax: 734-665-4370
Email: glc@great-lakes.net
Website: http://www.glc.org

Greater Yellowstone Coalition
13 South Willson, Suite 2
P.O. Box 1874
Bozeman, MT 59771, USA
Tel: 406-586-1593, Fax: 406-586-0851
Email: gyc@greateryellowstone.org
Website: http://www.greateryellowstone.org/

Headwaters Environmental Center
84 Fourth Street
PO Box 729
Ashland, OR 97520, USA
Tel: 541-482-4459, Fax: 541-482-7282
Email: headwtrs@mind.net
Website: http://www.headwaters.org

Hill Country Foundation
1800 Guadalupe, Suite C
Austin, TX 78701, USA
Tel: 512-478-5743

Humboldt State University
Arcata, CA 95521, USA
Tel: 707-826-3256
Website: http://www.humboldt.edu

Idaho Rivers United
731 N. 15th Street
Boise, ID 83702, USA
Tel: 208-343-7481, Fax: 208-343-8184

Illinois Department of Natural Resources
524 South 2nd Street
Room 400 LTP
Springfield, IL 62764, USA
Tel: 217-782-5597
Website: http://dnr.state.il.us/

Illinois Natural Heritage Division
Department of Natural Resources
524 South Second Street
Springfield, IL 62701, USA
Tel: 217-785-8774, Fax: 217-785-8277
Website: http://dnr.state.il.us/

Indiana Department of Natural Resources
402 West Washington Street
Room W255B
Indianapolis, IN 46204-2748, USA
Tel: 317-232-4200, Fax: 317-232-8036
Website: http://www.state.in.us/dnr/

Indiana Natural Heritage Data Center
Division of Nature Preserves
Indiana Department of Natural Resources
402 W. Washington Street, Room W267
Indianapolis, IN 46204, USA
Tel: 317-232-4052, Fax: 317-233-0133
Website:
    http://www.state.in.us/dnr/naturepr/center.htm

Instituto de Ecología A.C.
Km 2.5 Antigua carretera a Coatepec
Xalapa 91000, Veracruz, México
Tel: 28-18-6000
Email: ieco@sun.ieco.conacyt.mx/

Instituto Tecnologico Regional de Chetumal
Ave. Andres Quintana Roo y Ave. Insurgentes
Chetumal, Quintana Roo, México
Tel: 2-10-19

Instituto Nacional de Ecología (affiliated with
    SEMARNAP)
Avenue Revolución 1425
Col. Campestre Tlacopac, C.P.
01040, México, D.F., México
Website: http://www.ine.gob.mx/

Instituto Tecnológico y de Estudios Superiores de
    Monterrey (ITESM)
Av. Eugenio Garza Sada #2501
Sur. Sucursal de Correos "J" C.P. 64849
Monterrey, N.L., México
Tel: 8-358-2000
Website: http://www.mty.itesm.mx/

Kansas Natural Heritage Inventory
Kansas Biological Survey
2041 Constant Avenue
Lawrence, KS 66047-2906, USA
Tel: 785-864-7698, Fax: 785-864-5093
Website: http://www.heritage.tnc.org/nhp/us/ks/

Kentucky Natural Heritage Program
Kentucky State Nature Preserves Commission
801 Schenkel Lane
Frankfort, KY 40601, USA
Tel: 502-573-2886, Fax: 502-573-2355
Email: ksnpc@mail.state.ky.us
Website: http://www.state.ky.us/agencies/nrepc/ksnpc/
    index.htm

Klamath Forest Alliance
PO Box 820
Etna, CA 96027, USA
Tel: 530-467-5405, Fax: 530-467-3130
Email: klamath@sisqtel.net
Website: http://www.sisqtel.net/users/klamath

Labrador Inuit Association
302-240 Water Street
St. John's, NF A1C 1B7, Canada
Tel: 709-722-6160, Fax: 709-722-6185
Website: http://www.cancom.net/~franklia/main.html

Lerma-Chapala River Basin Council (Consejo de la Cuenca Lerma-Chapala)
Email: erick@ciateq.mx
Website: http://www.ciateq.mx/~lermaham/lerma.htm

Louisiana Natural Heritage Program
Department of Wildlife & Fisheries
PO Box 98000
Baton Rouge, LA 70898-9000, USA
Tel: 225-765-2821, Fax: 225-765-2607
Website: http://www.heritage.tnc.org/nhp/us/la/

Lower Mississippi River Conservation Committee
2524 South Frontage Road, Suite C
Vicksburg, MS 39180, USA
Tel: 601-629-6602, Fax: 601-636-9541
Website: http://www.lmrcc.org/index.htm

Maine Natural Areas Program
159 Hospital Street
No. 93 State House Station
Augusta, ME 04333-0093, USA
Tel: 207-287-8044, Fax: 207-287-8040
Website: http://www.state.me.us/doc/nrimc/mnap/
    home.htm#welcome

Manitoba Naturalists Society
63 Albert Street, Suite 401
Winnipeg, MB R3B 1G4, Canada
Tel: 204-943-9029
Website: http://www.mbnet.mb.ca/mns/

Michigan Environmental Council
119 Pere Marquette Drive, Suite 2A
Lansing, MI 48912, USA
Tel: 517-487-9539, Fax: 517-487-9541
Website: http://www.mienv.org/

Michigan Land Use Institute
845 Michigan Avenue
P.O. Box 228
Benzonia, MI 49616, USA
Tel: 616-882-4723
Website: http://www.mlui.org/

Michigan Natural Areas Council
University of Michigan
c/o Matthaei Botanical Gardens
1800 N. Dixboro Road
Ann Arbor, MI 48109-9741, USA
Tel: 313-435-2070
Website: http://www.cyberspace.org/~mnac/

Minnesota Department of Natural Resources
500 Lafayette Road
St. Paul, MN 55155-4001, USA
Tel: 612-296-6157, Fax: 612-297-3618
Website: http://www.dnr.state.mn.us/

Mississippi Natural Heritage Program
Museum of Natural Science
111 North Jefferson Street
Jackson, MS 39201-2897, USA
Tel: 601-354-7303, Fax: 601-354-7227
Email: heritage@mmns.state.ms.us
Website: http://www.heritage.tnc.org/nhp/us/ms/

Mississippi Wildlife Federation
P.O. Box 1814
Jackson, MS 39215, USA
Tel: 601-353-6922, Fax: 601-353-3437

Missouri Department of Conservation
PO Box 180
Jefferson City, MO 65102-0180, USA
Tel: 573-751-4115, Fax: 573-751-4467
Website: http://www.conservation.state.mo.us/

Missouri Natural Heritage Database
(same contact information as above for Missouri
    Department of Conservation)

Mono Lake Committee
PO Box 29
Lee Vining, CA 93541, USA
Tel: 760-647-6595, Fax: 760-647-6377
Website: http://www.monolake.org/

MoRAP
Midwest Science Center
4200 New Haven Road
Columbia, MO 65201, USA
Tel: 314-875-5399, Fax: 314-876-1896
Website: http://www.msc.nbs.gov/morap/

National Audubon Society—Main Office
700 Broadway
New York, NY 10003, USA
Tel: 212-979-3000
Website: http://www.audubon.org/nas/

National Cattlemen's Beef Association
Environmental Stewardship Award
PO Box 3469
Englewood, CO 80155, USA
Tel: 303-694-0305
Website: http://www.beef.org/

National Wildlife Federation
8925 Leesburg Pike
Vienna, VA 22184, USA
Tel: 703-790-4000
Website: http://www.nwf.org/nwf/

Natural Resources Council of Maine
3 Wade Street
Augusta, ME 04330, USA
Tel: 207-622-3101, Fax: 207-622-4343

Natural Resources Conservation Service
USDA
14th and Independence Avenue, SW
PO Box 2890
Washington, DC 20013, USA
Tel: 202-720-3210 (call for regional information)
Website: http://www.nrcs.usda.gov/

Natural Resources Defense Council
40 West 20th Street
New York, NY 10011, USA

The Nature Conservancy
1815 North Lynn Street
Arlington, VA 22209, USA
Tel: 703-841-5300, Fax: 703-841-1283
Website: http://www.tnc.org

The Nature Conservancy, Alabama Field Office
2821-C 2nd Avenue S.
Birmingham, AL 35233, USA
Tel: 205-251-1155

The Nature Conservancy, Alaska Field Office
421 West First Avenue, Suite 200
Anchorage, AK 99501, USA
Tel: 907-276-3133

The Nature Conservancy, Arizona Field Office
300 E. University Boulevard
Suite 230
Tucson, AZ 85705, USA
Tel: 520-622-3861

The Nature Conservancy, Arkansas Field Office
601 N. University Avenue
Little Rock, AR 72205, USA
Tel: 501-663-6699

The Nature Conservancy, California Regional Office
201 Mission Street, 4th Floor
San Francisco, CA 94105, USA
Tel: 415-777-0487

The Nature Conservancy, Colorado Field Office
1244 Pine Street
Boulder, CO 80302, USA
Tel: 303-444-2950

The Nature Conservancy, Connecticut Field Office
55 High Street
Middletown, CT 06457, USA
Tel: 860-344-0716

The Nature Conservancy, Delaware Field Office
260 Chapman Road, Suite 201D
Newark, DE 19702, USA
Tel: 302-369-4144

The Nature Conservancy, Eastern Regional Office
201 Devonshire Street, 5th Floor
Boston, MA 02110, USA
Tel: 617-542-1908

The Nature Conservancy, Florida Regional Office
222 S. Westmonte Drive, Suite 300
Altamonte Springs, FL 32714, USA
Tel: 407-682-3664

The Nature Conservancy, Georgia Field Office
1401 Peachtree Street, N.E., Suite 236
Atlanta, GA 30309, USA
Tel: 404-873-6946

The Nature Conservancy, Great Lakes Program
8 South Michigan Avenue, Suite 2301
Chicago, IL 60603, USA
Tel: 312-759-8017, Fax: 312-759-8409

The Nature Conservancy, Idaho Field Office
P.O. Box 165
Sun Valley, ID 83353, USA
Tel: 208-726-3007

The Nature Conservancy, Illinois Field Office
8 S. Michigan Avenue, Suite 900
Chicago, IL 60603, USA
Tel: 312-346-8166

The Nature Conservancy, Indiana Field Office
1330 West 38th Street
Indianapolis, IN 46208, USA
Tel: 317-923-7547

The Nature Conservancy, Iowa Field Office
108 Third Street, Suite 300
Des Moines, IA 50309-4758, USA
Tel: 515-244-5044, 515-244-8890

The Nature Conservancy, Kansas Field Office
820 SE Quincy, Suite 301
Topeka, KS 66612-1158, USA
Tel: 913-233-4400, Fax: 913-233-2022

The Nature Conservancy, Kentucky Field Office
642 W. Main Street
Lexington, KY 40508, USA
Tel: 606-259-9655

The Nature Conservancy, Louisiana Field Office
P.O. Box 4125
Baton Rouge, LA 70821, USA
Tel: 504-338-1040

The Nature Conservancy, Maine Field Office
14 Maine Street, Suite 401
Brunswick, ME 04011, USA
Tel: 207-729-5181

The Nature Conservancy, Maryland/DC Field Office
Chevy Chase Metro Building
2 Wisconsin Circle, Suite 300
Chevy Chase, MD 20815, USA
Tel: 301-656-8673

The Nature Conservancy, Massachusetts Field Office
79 Milk Street, Suite 300
Boston, MA 02109, USA
Tel: 617-423-2545

The Nature Conservancy, Michigan Field Office
2840 E. Grand River, Suite 5
East Lansing, MI 48823, USA
Tel: 517-332-1741

The Nature Conservancy, Midwest Regional Office
1313 Fifth Street, S.E., No. 314
Minneapolis, MN 55414, USA
Tel: 612-331-0700

The Nature Conservancy, Minnesota Field Office
1313 Fifth Street, S.E.
Suite 320
Minneapolis, MN 55414, USA
Tel: 612-331-0750

The Nature Conservancy, Mississippi Field Office
P.O. Box 1028
Jackson, MS 39215-1028, USA
Tel: 601-355-5357

The Nature Conservancy, Missouri Field Office
2800 S. Brentwood Boulevard
Street Louis, MO 63144, USA
Tel: 314-968-1105

The Nature Conservancy, Montana Field Office
32 South Ewing
Helena, MT 59601, USA
Tel: 406-443-0303

The Nature Conservancy, Nebraska Field Office
1722 Street Mary's Avenue
Suite 403
Omaha, NE 68102, USA
Tel: 402-342-0282

The Nature Conservancy, Nevada Field Office
1771 East Flamingo, Suite 111B
Las Vegas, NV 89119, USA
Tel: 702-737-8744

The Nature Conservancy, New Hampshire Field Office
2 1/2 Beacon Street, Suite 6
Concord, NH 03301, USA
Tel: 603-224-5853

The Nature Conservancy, New Jersey Field Office
200 Pottersville Road
Chester, NJ 07930, USA
Tel: 908-879-7262

The Nature Conservancy, New Mexico Field Office
212 East Marcy, Suite 200
Santa Fe, NM 87501, USA
Tel: 505-988-3867

The Nature Conservancy, New York Office, Adirondack
   Chapter
P.O. Box 65, Route 73
Keene Valley, NY 12943, USA
Tel: 518-576-2082

The Nature Conservancy, New York Office, Central and
   Western NY Chapter
315 Alexander Street, 2nd Floor
Rochester, NY 14604, USA
Tel: 716-546-8030

The Nature Conservancy, New York Office, Eastern NY
   Chapter
200 Broadway, 3rd Floor
Troy, NY 12180, USA
Tel: 518-272-0195

The Nature Conservancy, New York Office, Long Island
   Chapter
250 Lawrence Hill Road
Cold Spring Harbor, NY 11724, USA
Tel: 516-367-3225

The Nature Conservancy, New York Office, Lower Hudson Chapter
41 South Moger Ave
Mt. Kisco, NY 10549, USA
Tel: 914-244-3271

The Nature Conservancy, New York Office, South
   Fork/Shelter Island Chapter
P.O. Box 5125
E. Hampton, NY 11937, USA
Tel: 516-329-7689

The Nature Conservancy, New York City Office
570 Seventh Avenue No. 601
New York, NY 10018, USA
Tel: 212-997-1880

The Nature Conservancy, New York Regional Office
415 River Street, 4th floor
Troy, NY 12180, USA
Tel: 518- 273-9408, Fax: 518-273-5022

The Nature Conservancy, North Carolina Field Office
4011 University Drive, Suite 201
Durham, NC 27707, USA
Tel: 919-403-8558

The Nature Conservancy, North Dakota Field Office
2000 Schafer Street, Suite B
Bismarck, ND 58501, USA
Tel: 701-222-8464

The Nature Conservancy, Ohio Field Office
6375 Riverside Drive, Suite 50
Dublin, OH 43017, USA
Tel: 614-717-2770

The Nature Conservancy, Oklahoma Field Office
23 West Fourth, Suite 200
Tulsa, OK 74103, USA
Tel: 918-585-1117

The Nature Conservancy, Oregon Field Office
821 SE 14th Avenue
Portland, OR 97214, USA
Tel: 503-230-1221

The Nature Conservancy, Pennsylvania Field Office
Lee Park, Suite 470
1100 East Hector Street
Conshohocken, PA 19428, USA
Tel: 610-834-1323, Fax: 610-834-6533

The Nature Conservancy, Rhode Island Field Office
45 South Angell Street
Providence, RI 02906, USA
Tel: 401-331-7110

The Nature Conservancy, South Carolina Field Office
P.O. Box 5475
Columbia, SC 29250, USA
Tel: 803-254-9049

The Nature Conservancy, South Dakota Field Office
1000 West Avenue North, Suite 100
Sioux Falls, SD 57104, USA
Tel: 605-331-0619

The Nature Conservancy, Southeast Regional Office
P.O. Box 2267
Chapel Hill, NC 27515-2267, USA
Tel: 919-967-5493

The Nature Conservancy, Tennessee Field Office
50 Vantage Way, Suite 250
Nashville, TN 37228, USA
Tel: 615-255-0303

The Nature Conservancy, Texas Field Office
P.O. Box 1440
San Antonio, TX 78295-1440, USA
Tel: 210-224-8774

The Nature Conservancy, Utah Field Office
559 E. South Temple
Salt Lake City, UT 84102, USA
Tel: 801-531-0999

The Nature Conservancy, Vermont Field Office
27 State Street
Montpelier, VT 05602-2934, USA
Tel: 802-229-4425

The Nature Conservancy, Virginia Field Office
1233-A Cedars Court
Charlottesville, VA 22903-4800, USA
Tel: 804-295-6106

The Nature Conservancy, Washington Field Office
217 Pine Street, Suite 1100
Seattle, WA 98101, USA
Tel: 206-343-4344

The Nature Conservancy, West Virginia Field Office
723 Kanawha Boulevard East Suite 500
Charleston, WV 25301, USA
Tel: 304-345-4350

The Nature Conservancy, Western Regional Office
2060 Broadway, Suite 230
Boulder, CO 80302, USA
Tel: 303-444-1060

The Nature Conservancy, Wisconsin Field Office
633 West Main Street
Madison, WI 53703, USA
Tel: 608-251-8140

The Nature Conservancy, Wyoming Field Office
258 Main Street, Suite 200
Lander, WY 82520, USA
Tel: 307-332-2971

The Nature Conservancy of Canada, National Office
110 Eglinton Avenue West, Suite 400
Toronto, Ontario M4R 1A3, Canada
Tel.: (416) 932-3202, Fax: (416) 932-3208
Toll-free: 1-800-465-0029
Website: http://www.natureconservancy.ca/

The Nature Conservancy of Canada, Alberta Office
602 - 11th Avenue S.W., Suite 320
Calgary, Alberta T2R 1J8, Canada
Tel: 403-262-1253, Fax: 403-515-6987

The Nature Conservancy of Canada, Atlantic Region
  Office
108 Prospect Street Suite, Suite 2
Fredericton, New Brunswick E3B 2T9, Canada
Tel: 506-450-6010, Fax: 506-450-6013

The Nature Conservancy of Canada, Manitoba Office
298 Garry Street
Winnipeg, Manitoba R3C 1H3, Canada
Tel: 204-942-6156, Fax: 204-947-2591

The Nature Conservancy of Canada, Ontario Office
121 Wyndam Street N., Suite 202-204,
Guelph, ON N1H 4E9, Canada
Tel: (519)826-0068, Fax:(519)826-9206

The Nature Conservancy of Canada, Québec Office
800 René-Lévesque Blvd. West, Suite 2450
Montreal, Québec H3B 4V7, Canada
Tel: 514-876-1606, Fax: 514-871-8772

The Nature Conservancy of Canada, Saskatchewan
　　Office
1845 Hamilton Street, 7th Floor
Regina, Saskatchewan S4P 2C7, Canada
Tel: 306-777-9885, Fax: 306-569-9444

The Nature Conservancy of Canada, British Columbia,
　　Vancouver Office
827 West Pender Street, 2nd Floor
Vancouver, British Columbia V6C 3G8, Canada
Tel: 604-684-1654, Fax: 250-479-0546

The Nature Conservancy of Canada, British Columbia,
　　Victoria Office:
3960 Quadra Street, Suite 404
Victoria, BC V8X 4A3, Canada
Tel: 250-479-3191, Fax: 250-479-0546

Nature Saskatchewan
1860 Lorne Street, Suite 206
Regina, SK S4P 2L7, Canada
Website: http://www.unibase.com/~naturesk/

Nebraska Natural Heritage Program
Game and Parks Commission
2200 North 33rd Street
PO Box 30370
Lincoln, NE 68503, USA
Tel: 402-471-5469, Fax: 402-471-5528
Website: http://www.heritage.tnc.org/nhp/us/ne/

New Brunswick Federation of Naturalists
277 Douglas Avenue
Saint John, NB E2K 1E5, Canada
Tel: 506-532-3482

New Hampshire Natural Heritage Program
Department of Resources and Economic Development
172 Pembroke Road
PO Box 1856
Concord, NH 03302-1856, USA
Tel: 603-271-3623, Fax: 603-271-2629
Website: http://www.heritage.tnc.org/nhp/us/nh/

New Mexico Department of Game and Fish
Villagra Building
PO Box 25112
Santa Fe, NM 87504, USA
Tel: 505-827-7911, Fax: 505-827-7915
Website: http://www.gmfsh.state.nm.us/

New York Natural Heritage Program
Department of Environmental Conservation
700 Troy-Schenectady Road
Latham, NY 12110-2400, USA
Tel: 518-783-3932
Website: http://www.heritage.tnc.org/nhp/us/ny/

North Carolina Coastal Federation
3609 Highway 24 (Ocean)
Newport, NC 28570, USA
Tel: 800-232-6210, Fax: 252-393-7508
Email: nccf@nccoast.org
Website: http://www.nccoast.org/

North Carolina Natural Heritage Program
NC Deparment of Environment, Health & Natural
　　Resources
Division of Parks and Recreation
PO Box 27687
Raleigh, NC 27611, USA
Tel: 919-733-4181, Fax: 919-715-3085
Website: http://www.ncsparks.net/nhp

Northcoast Environmental Center
879 9th street
Arcata, CA 95521, USA
Tel: 707-822-6918, Fax: 707-822-0827
Email: nec@igc.apc.org

Northern Appalachian Restoration Project
PO Box 6
Lancaster, NH 03584, USA
Tel: 603-636-2952

Northwatch
Box 264
North Bay, ON P1B 8H2, Canada
Tel: 705-497-0373, Fax: 705-476-7060

Northwest Power Planning Council
851 SW 6th Avenue, Suite 1100
Portland, OR 97204, USA
Tel: 800-222-3355, Fax: 503-795-3370
Website: http://www.nwppc.org/welcome.htm

Nunavut Wildlife Management Board
P.O. Box 1379
Iqaluit, NWT, Canada
Tel: 819-979-6962
Website: http://pooka.nunanet.com/~nwmb/index.html

Ohio Natural Heritage Database
Division of Natural Areas & Preserves
Department of Natural Resources
1889 Fountain Square, Building F-1
Columbus, OH 43224, USA
Tel: 614-265-6543, Fax: 614-267-3096
Website:
　　http://www.dnr.state.oh.us/odnr/dnap/dnap.html

Ohio Valley Environmental Coalition
Post Office Box 6753
Huntington, WV 25773-6753, USA
Tel: 304-522-0246
Website: http://www.ohvec.org/

Oklahoma Natural Heritage Inventory
111 East Chesapeake Street
University of Oklahoma
Norman, OK 73019-0575, USA
Tel: 405-325-1985, Fax: 405-325-7702
Website: http://www.biosurvey.ou.edu/onhi.html

Oregon Lakes Association
P.O. Box 345
Portland, OR 97207, USA
Email: breiling@worldnet.att.net
Website: http://www.ola.pdx.edu

Oregon Natural Desert Association
16 NW Kansas
Bend, OR 97701, USA
Tel: 541-330-2638, Fax: 541-385-3370
Website: http://www.onda.org

Oregon Natural Resources Council
5825 North Greeley
Portland, OR 97217-4145, USA.
Tel: 503-283-6343, Fax: 503-283-0756
Website: www.onrc.org

Oregon Trout
117 S.W. Naito Parkway
Portland, OR 97204, USA
Tel: 503-222-9091, Fax: 503-222-9187
Email: info@ortrout.org
Website: http://www.ortrout.org

The Ozark Society
PO Box 2914
Little Rock, AR 72203, USA

Pacific Rivers Council
PO Box 10798
Eugene, OR 97440, USA
Tel: 541-345-0119, Fax: 541-345-0710
Email: info@pacrivers.org
Website: http://www.pacrivers.org

Pennsylvania Natural Diversity Inventory-West
Western Pennsylvania Conservancy
409 Fourth Avenue
Pittsburgh, PA 15222, USA
Tel: 412-288-2777, Fax: 412-281-1792
Website: http://www.paconserve.org/

Prairie Conservation Forum
Bag 3014, Third Floor, YPM Place
530-8th Street South
Lethbridge, AB T1J 4C7, Canada
Tel: 403-381-5430, Fax: 403-381-5723

PROFAUNA-Protección de la Fauna Mexicana A.C.
Morelos Sur 371
Saltillo, Coah. 25000, México
Tel: 528-412-5404, Fax: 528-410-5714

Pronatura Peninsula de Yucatan
Calle 1-D #254 entre 36 y 38
Col. Campestre, CP 97120
Merida, Yucatan, México
Tel/Fax: 52-99-44-2290, 44-3580
Email ppy@pibil.finred.com.mx
Website:
    http://www.coa.edu/ACADEMICPROGRAM/IntStud-
    ies/pronaturaquestions.html

Protected Areas Association of Newfoundland and
    Labrador
P.O. Box 1027, Station C
St. John's, NF A1C 5M5, Canada
Tel: 709-726-2603
Website: http://www.web.net/~paa/paa.html

Proyecto UNAM-PEMEX
See Universidad Nacional Autónoma de México

Pyramid Lake Paiute Tribe
P.O. Box 256
Nixon, NV 89424, USA
Tel: 702-574-1000, Fax: 702-574-1008
Website:
    http://thecity.sfsu.edu/NativeWeb/home/plpt.html

Restore the Northwoods
23 Bradford Street, Floor 3
Concord, MA 01742, USA
Tel: 978-287-0320

Rio Grande Alliance
Office of Border Affairs, MC 121
P.O. Box 13087
Austin, TX 78711, USA
Tel: 512-239-3600, Fax: 512-239-3515
Website: http://www.riogrande.org/

Rivers Unlimited-Mill Creek Restoration Project
2 Centennial Plaza, Suite 610
805 Central Avenue
Cincinnati, OH 45202, USA
Tel: 513-352-1588, Fax: 513-352-4970

Sacramento River Preservation Trust
PO Box 5366
Chico, CA 95927, USA
Tel: 916-345-1865

Saskatchewan Wetland Conservation Corporation
2050 Cornwall Street, Suite 202
Regina, Saskatchewan S4P 2K5, Canada
Tel: 306-787-0726, Fax: 306-787-0780
Website: http://www.wetland.sk.ca/

Saskatchewan Wildlife Federation
Box 788
Moose Jaw, SK S6H 4P5, Canada
Tel: 306-692-8812, Fax: 306-692-4370
Website: http://www.wwwdi.com/swf/

Save Barton Creek Association
P O Box 5923
Austin, TX 78763, USA
Fax: 512-328-3001

Save Our Streams
258 Scotts Manor Drive
Glen Burnie, MD 21061, USA
Tel: 410-969-0084, Fax: 410-969-0135
Website: http://www.saveourstreams.org/

Save Our Wild Salmon
975 John Street, Suite 204
Seattle, WA 98109, USA
Tel: 206-622-2904
Website: http://www.wildsalmon.org

Secretaria de Medio Ambiente Recursos Naturales y
   Pesca (SEMARNAP)
Dirección General de Comunicación Social
Periférico Sur 4209, Fraccionamiento Jardines en la
   Montaña
Delegación Tlalpan, C.P. 14210
México, D.F., México
Tel: 5/628-0600, ext. 2104
Website: http://www.semarnap.gob.mx

Sierra Club
85 Second Street, Second Floor
San Francisco, CA 94105, USA
Tel: 415-977-5500, Fax: 415-977-5799
Website: http://www.sierraclub.org

Sierra Club, Cascade Chapter
8511 15th Avenue NE, Suite 201
Seattle, WA 98115, USA
Tel: 206-523-2147, Fax: 206-523-2079
Email: cascade.chapter@sierraclub.org
Website: http://www.cascadechapter.org/

Sierra Club, Rocky Mountain Chapter
1410 Grant Street, Suite B205
Denver, CO 80203, USA
Website: http://www.rmc.sierraclub.org/

Siskiyou Regional Education Project
PO Box 220
Cave Junction, OR 97523, USA

Society of Grassland Naturalists
Box 2491
Medicine Hat, AB T1A 8G8, Canada
Tel: 403-526-6443

Society for the Protection of New Hampshire Forests
54 Portsmouth Street
Concord, NH 03301-5400, USA
Tel: 603-224-9945, Fax: 603-228-0423
Website: http://www.spnhf.org/

Society for Range Management
1839 York Street
Denver, CO 80206, USA
Tel: 303-355-7070, Fax: 303-355-5059
Email: srmden@ix.netcom.com
Website: http://www.srm.org/

Sonoran Institute
7650 E. Broadway, Suite 203
Tucson, AZ 85710
Tel: 520- 290-0828, Fax: 520- 290-0969
Email: si_info@sonoran.org
Website: http://www.sonoran.org/si/

South Carolina Heritage Trust
South Carolina Department of Natural Resources
PO Box 167
Columbia, SC 29202, USA
Tel: 803-734-3893, Fax: 803-734-6310
Website: http://water.dnr.state.sc.us/wild/heritage/pre-
   serve.html

Southeast Alaska Conservation Council
419 6th Street, No. 328
Juneau, AK 99801, USA
Tel: 907-586-6942, Fax: 907-463-3312
Email: info@seacc.org
Website: http://www.seacc.org

Southern Utah Wilderness Alliance
1471 S. 1100 E.
Salt Lake City, UT 84105-2423, USA
Tel: 801-486-3161, Fax: 801-486-4233
Website: http://www.suwa.org/

Southwest Center for Biological Diversity
PO Box 710
Tucson, AZ 85702, USA
Tel: 520-624-7893, Fax: 520-623-9797
Website: http://www.sw-center.org/swcbd/

Strategies Saint-Laurent, Inc.
690 Grande-Allée, 4th Floor
Québec, QC G1R 2K5, Canada
Tel: 418-648-8079, Fax: 418-648-0991

Tennessee Aquarium
One Broad Street
Chattanooga, TN 37401-2048, USA
Tel: 800-262-0695 or 423-265-0698
Website: http://www.tennis.org/

Tennessee Division of Natural Heritage
Tennessee Department of Environment &
   Conservation
410 Church Street, Life and Casualty Tower, 8th Floor
Nashville, TN 37243-0447, USA
Tel: 615-532-0431, Fax: 615-532-0046
Website: http://www.state.tn.us/environment/nh/

Texas Center for Policy Studies
PO Box 2618
Austin, TX 78768, USA
Tel: 512-474-0811

Texas Conservation Data Center
The Nature Conservancy of Texas
PO Box 1440
San Antonio, TX 78295-1440, USA
Tel: 210-224-8774, Fax: 210-228-9805
Website:
    http://www.tnc.org/infield/State/Texas/texas.htm

Texas Parks and Wildlife
4200 Smith School Road
Austin, TX 78744, USA
Tel: 800-792-1112 or 512-389-4800
Website: http://www.tpwd.state.tx.us/

Trout Unlimited
1500 Wilson Boulevard, Suite 310
Arlington, VA 22209, USA
Tel: 703-522-0200, Fax: 703-284-9400

Union Québecoise pour la Conservation de la Nature
    (UQCN)
690 Grande Allée, 4th Floor
Québec, QC G1R 2K5, Canada
Tel: 418-648-2104, Fax: 418-648-0991
Website: http://uqen.qc.ca/

University of Wisconsin Sea Grant Program
Sea Grant Institute
University of Wisconsin-Madison
1975 Willow Drive, Second Floor
Madison, WI 53706-1103, USA
Tel: 608-262-0905, Fax: 608-262-0591
Website: http://www.seagrant.wisc.edu/

U.S. Bureau of Land Management
Office of Public Affairs
1849 C Street, Room 406-LS
Washington, DC 20240, USA
Tel: 202-452-5125, Fax: 202-452-5124
Website: http://www.blm.gov/

U.S. Fish and Wildlife Service
1849 C Street, NW
Washington, DC 20240, USA
Website: http://www.fws.gov/

U.S. Forest Service
P.O. Box 96090
Washington, DC 20090-6090, USA
Tel: 202-205-1760, Fax: 202-205-0885
Email: oc/wo@fs.fed.us
Website: http://www.fs.fed.us/

Universidad Autónoma Agraria Antonio Narro (UAAAN)
Buenavista, Coahuila, México
Tel: 52/84-17-30-22, ext. 401, 402, or 407 (Saltillo)
Tel. 52/17-33-10-90, ext. 110, 111 (Torreón)
Website: http://www.uaaan.mx/index_az.html

Universidad Autónoma de Aguascalientes
Avenida Universidad
Esq. Avenida Aguascalientes, México
Tel: 52/91-49-12-33-45 or 52/91-49-12-32-84
Email:wwwadm@correo.uaa.mx
Website: http://www.uaa.mx:8001/

Universidad Autónoma de Baja California (UABC)
S/N Edificio de Rectoría, C.P. 21100
Mexicali, B.C., México
Tel: 01-65-518200 de México (within México)
Tel: 011-526-551-8200 (from outside México)

Universidad Autónoma de Baja California Sur (UABCS)
Carretera al Sur km. 4-5
23080 La Paz, B. C. S. México
Tel: 52/112-107-55 or 52/112-111-40, ext. 124

Universidad Autónoma de Chiapas
Tel: 52/5-10-21, Fax: 52/5-06-64 or 52/5-04-05
Tuxtla Gutierrez, Chiapas, México
Website: http://www.unach.mx/

Universidad Autónoma de Chihuahua
Avenida Escorza #900
Colonia Centro
C.P. 31000
Chihuahua, Chihuahua, México
Tel: 011-439-1550
Website: http://www.uach.mx/

Universidad Autónoma de Guerrero
Website: http://uag.uagfm.mx/

Universidad Autónoma de Nayarit
Ciudad de la Cultura Amado Nervo
C.P. 63190, Tepic, Nayarit, México
Tel: 32-14-85-12
Website: http://www.uan.mx/

Universidad Autónoma de Nuevo León
Website: http://www.uanl.mx/

Universidad Autónoma de Sinaloa (UAS)
Tel: 01 (67) 15-65-20, 13-93-91 or 12-54-41

Universidad Juarez Autónoma de Tabasco (UJAT)
Website: http://México.ujat.mx/

Universidad de Occidente
Website: http://200.33.16.97/

Universidad de Quintana Roo
Website: http://www.uqroo.mx/uqroo/index1.html

Universidad de Sonora (UNISON)
Ext. 340 or 341
Tel. 52/59-21-36 or 52/59-21-37, Fax: 52/59-21-35

Universidad de Tampico
Instituto Tecnologico y de Estudios Superiores de
  Monterrey
Campus Tampico
Blvd. Petrocel Km. 1.3 Puerto Industrial Altmira,
  Tamaulipas, México
Tel/Fax: 52-12-29-16-00
Website: http://www.tam.itesm.mx/

Universidad del Mar
Km. 1.5 Carr. a Zipolite
Puerto Angel, Oaxaca, México
Tel: 52/958-4-30-78 or 52/958-4-30-57,
Fax: 52/958-4-30-49
Website: http://www.umar.mx/

Universidad del Noreste
Pro. Av. Hidalgo 6315
Col. Nuevo Aeropuerto
Tampico, Tamualipas C.P. 89337, México
Tel/Fax: 12-28-11-82, 12-28-11-56, 12-28-11-77

Universidad Michoacana
Edificio de Rectoría, segundo piso
Tel: 52/43-16-7020
Website: http://www.ccu.umich.mx/

Universidad Nacional Autónoma de México (UNAM)
Costado Norte del Edificio "D" de la Facultad de
  Química
Circuito de la Investigación Científica
Ciudad Universitaria, D.F. 045 México
Tel: 52-550-9192, 52-622-5204, or 52-622-4999,
Fax: 52-622-5223

Universidad Veracruzana
Dirección General de Informática
Dirección de Cómputo Académico
Lomas del Estadio S/N, Edificio "E"
Zona Centro, Xalapa Veracruz, México
Tel: 52/42-17-99 or 52/12-21-87
Website: http://www.coacode.uv.mx/

The University of Arizona
Tucson, AZ 85721, USA
Tel: 520-621-2211 (main switchboard)
Website: http://www.arizona.edu/

Upper Chattahoochee Riverkeeper
1900 Emery Street, Suite 450
Atlanta, GA 30318, USA
Tel: 404-352-9828, Fax: 404-352-8676
Email: rivrkeep@mindspring.com
Website: http://www.chattahoochee.org

Utah Rivers Council
1471 S. 1000 E.
Salt Lake City, UT 84105, USA
Tel: 801-486-4776
Website: http://www.wasatch.com/~urc

Virginia Natural Heritage Program
217 Governor Street
Richmond, VA 23219, USA
Tel: 804-786-7951, Fax: 804-371-2674
Website: http://www.state.va.us/~der/vaher.html

Washington Environmental Council
1100 Second Avenue, Suite 102
Seattle, WA 98101, USA
Tel: 206-622-8103, Fax: 206-622-8113

West Central Research and Extension Station,
  University of Nebraska
Route 4 Box 46A
North Platte, NE 69101-9495, USA
Tel: 308-532-3611
Website: http://www.ianr.unl.edu/ianr/wcrec/index.htm

West Virginia Highlands Conservancy
Tel: 304-924-6263
Website: http://www.wvhighlands.org/

West Virginia Natural Heritage Program
Division of Natural Resources
Ward Road, PO Box 67
Elkins, WV 26241, USA
Tel: 304-637-0245, Fax: 304-637-0250
Website: http://www.heritage.tnc.org/nhp/us/wv/

Western Pennsylvania Conservancy
PNDI-West, 209 Fourth Avenue
Pittsburgh, PA 15222, USA
Tel: 412-288-2777, Fax: 412-281-1792
Website: http://www.paconserve.org/

Wild Earth
PO Box 455
Richmond, VT 05477, USA
Tel: 802-434-4077

The Wilderness Society
900 Seventeenth Street NW
Washington, DC 20006, USA
Tel: 1-800-THE-WILD
Website: http://www.wilderness.org

The Wilderness Society-Alaska Region
430 West 7th Avenue, No. 210
Anchorage, AK 99501, USA
Tel: 907-272-9453

The Wilderness Society-Northern Rockies Region
105 W. Main Street, Suite E
Bozeman, MT 59715, USA
Tel: 406-587-7331

The Wilderness Society-Pacific Northwest Region
1424 Fourth Avenue, Suite 816
Seattle WA 98101-2217, USA
Tel: 202-624-6430

The Wilderness Society-Southeast Region
1447 Peachtree Street, N.E., Suite 812
Atlanta, GA 30309, USA
Tel: 404-872-9453

The Wildlands League
401 Richmond Street West, Suite 380
Toronto, ON M5V 3A8, Canada
Tel: 416-971-9453, Fax: 416-979-3155
Website: http://www.wildlandsleague.org/

The Wildlands Project
1955 W. Grant Road, Suite 148
Tucson, AZ 85745, USA
Tel: 520-884-0875
Website: http://www.wildlandsproject.org/

Wisconsin Department of Natural Resources
Box 7921
Madison, WI 53707, USA
Tel: 608-266-2621
Website: http://www.dnr.state.wi.us/

World Wildlife Fund Canada
245 Eglinton Avenue East, Suite 410
Toronto, ON M4P 3J1, Canada
Tel: 416-489-8800, Fax: 416-489-3611
Website: http://www.wwfcanada.org/

World Wildlife Fund México
Ave. Mexico No. 51
Col. Hipodromo Condesa
06170 Mexico, D.F., México
Tel: 525-286-5631/5634, Fax: 525-286-5637

GLOSSARY

allochthonous   Originating outside and transported into a given system or area (Lincoln et al. 1982).

allotopic   Referring to populations or species that occupy different macrohabitats (Lincoln et al. 1982).

alpha diversity   Species diversity within a habitat.

amphibian   A member of the vertebrate class Amphibia (frogs, toads, and salamanders).

amphipod   Any of a large group of small crustaceans with laterally compressed bodies, belonging to the order Amphipoda.

anadromous   Diadromous species that spawn in fresh water and migrate to marine habitats to mature (e.g., salmon).

analysis of variance   Tests the hypothesis that means from several samples are equal. Generally, analysis of variance (ANOVA) is a statistical procedure used to determine whether means from two or more samples are drawn from populations with the same mean.

anthropogenic   Caused or produced through the agency of humans (Lincoln et al. 1982).

aquatic   Living in water.

aquifer   A formation, group of formations, or part of a formation that contains sufficient saturated permeable material to yield significant quantities of water to wells and springs (Maxwell et al. 1995).

arctic   Referring to all nonforested areas north of the coniferous forests in the Northern Hemisphere (Brown and Gibson 1983).

artesian   Referring to underground water that moves under pressure and flows to the surface naturally.

basin   See *catchment*.

benthic   Living at, in, or associated with structures on the bottom of a body of water (Brown and Gibson 1983).

beta-diversity   Species diversity between habitats (thus reflecting changes in species assemblages along environmental gradients).

biodiversity   (Also called *biotic* or *biological diversity*.) The variety of organisms considered at all levels, from genetic variants belonging to the same species through arrays of species to arrays of genera, families, and still higher taxonomic levels; includes the variety of ecosystems, which comprise both communities of organisms within particular habitats and the physical conditions under which they live (Wilson 1992).

biodiversity conservation   The goal of conservation biology, which is to retain indefinitely as much of the earth's biodiversity as possible, with emphasis on biotic elements most vulnerable to human impacts (Angermeier and Schlosser 1995).

biogeography   The study of the geographic distribution of organisms, both past and present (Brown and Gibson 1983).

biological distinctiveness   Scale-dependent assessment of the biological importance of an ecoregion based on species richness, endemism, relative scarcity of major habitat type, and rarity of ecological phenomena. The biological distinctiveness classes are *globally* *outstanding, continentally outstanding, bioregionally outstanding,* and *nationally important.*

bioregion   A geographically related assemblage of ecoregion complexes that share a similar biogeographic history and thus have strong affinities at higher taxonomic levels (e.g., genera, families). North America has three freshwater biogeographic divisions, consisting of contiguous complexes: the Pacific, Arctic-Atlantic, and Mexican Transition bioregions.

bioregionally outstanding   A *biological distinctiveness* class.

biota   The combined flora, fauna, and microorganisms of a given region (Wilson 1992).

biotic   Biological, especially referring to the characteristics of faunas, floras, and ecosystems (Wilson 1992).

catadromous   *Diadromous* species that spawn in marine habitats and migrate to fresh water to mature (e.g., eels).

catchment   All lands enclosed by a continuous hydrologic-surface drainage divide and lying upslope from a specified point on a stream (Maxwell et al. 1995); or, in the case of closed-basin systems, all lands draining to a lake.

cavernicoles   Animals that inhabit caves for all or part of their lives.

cenotes   (Spanish) Natural wells.

chemoautotrophic   Obtaining metabolic energy by the oxidation of inorganic substrates, such as sulphur, nitrogen, or iron (Lincoln et al. 1982).

clear-cut   A logged area where all or virtually all the forest canopy trees have been eliminated.

community   A collection of organisms of different species that co-occur in the same habitat or region and interact through trophic and spatial relationships (Fiedler and Jain 1992).

complex   A geographically related assemblage of ecoregions that share a similar biogeographic history and thus have strong affinities at higher taxonomic levels (e.g., genera, families). There are nine freshwater complexes in North America, not including

much of Mexico (derived from Maxwell et al. 1995).

conservation status   Assessment of the status of ecological processes and of the viability of species or populations in an ecoregion. The different status categories used are *critical, endangered, vulnerable, relatively stable,* and *relatively intact.* The snapshot conservation status is based on an index derived from values of four landscape-level variables. The final conservation status is the snapshot assessment modified by an analysis of threats to the ecoregion over the next twenty years.

continentally outstanding   A *biological distinctiveness* class.

convergence   The independent evolution of structural or functional similarity in two or more unrelated or distantly related lineages or forms that is not based on genotypic similarity (Lincoln et al. 1982).

conversion   Habitat that is no longer "intact" is considered to be converted.

creek   The smallest size class of a lotic system, typically associated with headwaters.

critical   The conservation status category one level below "extinct" and characterized by low probability that remaining intact habitat will persist.

degradation   The loss of native species and processes due to human activity such that only certain components of the original biodiversity still persist, often including significantly altered natural communities.

diadromous   Species that migrate between freshwater and marine habitats, spawning in one habitat and maturing in another (Nyman 1991).

disturbance   Any relatively discrete event in time that disrupts ecosystem, community, or population structure and changes resources, substrate availability, or the physical environment (Fiedler and Jain 1992).

diversion   The removal of water from a water body.

drainage basin   See *catchment.*

ecological processes   A complex mix of interactions between animals, plants, and their environment that ensures maintenance of an ecosystem's full range of biodiversity. Examples include population and predator-prey dynamics, pollination and seed dispersal, nutrient cycling, migration, and dispersal.

ecoregion A geographically distinct assemblage of natural communities that (a) share a large majority of their species and ecological dynamics; (b) share similar environmental conditions; and (c) interact ecologically in ways that are critical for their long-term persistence.

ecoregion-based conservation (ERBC) Conservation strategies and activities whose efficacy is enhanced through close attention to larger (landscape- or aquascape-level) spatial and temporal-scale patterns of biodiversity, ecological dynamics, threats, and strong linkages of these issues to fundamental goals and targets of biodiversity conservation.

ecosystem A system resulting from the integration of all living and nonliving factors of the environment (Tansley 1935).

ecosystem service A benefit or service provided free by an ecosystem or by the environment, such as clean water, flood mitigation, or groundwater recharge.

endangered The conservation status category between *critical* and *vulnerable* and characterized by medium to low probability of persistence of remaining intact habitat.

endemic A species or race native to a particular place and found only there (Wilson 1992).

endemism Degree to which a geographically circumscribed area, such as an ecoregion or a country, contains species not naturally occurring elsewhere.

endorheic A closed basin (no exterior drainage).

epigean Above ground.

erosion The wearing away of land surface by wind, water, ice, or other geologic agents. Erosion occurs naturally from weather or runoff but is often intensified by human land-use practices (Eckhardt 1998).

estuary A deepwater tidal habitat and its adjacent tidal wetlands, which are usually semi-enclosed by land but have open, partially obstructed, or sporadic access to the open ocean, and in which ocean water is at least occasionally diluted by freshwater runoff from the land (Maxwell et al. 1995).

eutrophication Over-enrichment of a water body with nutrients, resulting in excessive growth of organisms and depletion of oxygen concentration (Lincoln et al. 1982).

evolutionary phenomenon Within the context of WWF regional conservation assessments, an evolutionary phenomenon refers to a pattern of community structure and taxonomic composition resulting from extraordinary evolutionary processes, such as pronounced adaptive radiations.

exotic species A species that is not native to an area and has been introduced intentionally or unintentionally by humans; not all exotics become successfully established.

extinct Describes a species or population (or any lineage) with no surviving individuals.

extinction The termination of any lineage of organisms, from subspecies to species and higher taxonomic categories from genera to phyla. Extinction can be local, in which one or more populations of a species or other unit vanish but others survive elsewhere, or total (global), in which all the populations vanish (Wilson 1992).

extirpated Status of a species or population that has completely vanished from a given area but that continues to exist in some other location.

extirpation Process by which an individual, population, or species is totally destroyed (Fiedler and Jain 1992).

family In the hierarchical classification of organisms, a group of species of common descent higher than the genus and lower than the order; a group of genera (Wilson 1992).

fauna All the animals found in a particular place.

floodplain Land next to a river that becomes covered by water when the river overflows its banks (Eckhardt 1998).

flora All the plants found in a particular place.

fragmentation The process by which habitats are increasingly subdivided into smaller units (Fiedler and Jain 1992).

freshet A great rise or overflowing of a stream caused by heavy rains or melted snow.

fresh water In the strictest sense, water that has a salt concentration of less than 0.5 percent (Brown and Gibson 1983); in this study, refers to rivers, streams, creeks, springs, and lakes.

genera The plural of *genus*.

genetic drift The occurrence of random changes in the gene frequencies of small, isolated populations, not due to selection, mutation, or immigration (Lincoln et al. 1982).

genus A group of similar species with common descent, ranked below the *family* (Wilson 1992).

Global 200 A set of approximately 200 terrestrial, freshwater, and marine ecoregions around the world that support globally outstanding or representative biodiversity as identified through analyses by World Wildlife Fund–United States.

globally outstanding A *biological distinctiveness* class.

glochidia (Singular: glochidium.) The intermediate larval stage of freshwater mussels. In the spring or summer, glochidia are expelled from a female mussel's gills into the water. The glochidia then attach to an appropriate host, usually a fish, and form numerous cysts.

groundwater Water in the ground that is in the zone of saturation, from which wells and springs and groundwater runoff are supplied (Maxwell et al. 1995).

guild A group of organisms, not necessarily taxonomically related, that are ecologically similar in characteristics such as diet, behavior, or microhabitat preference, or with respect to their ecological role in general.

habitat An environment of a particular kind, often used to describe the environmental requirements of a certain species or community (Wilson 1992).

habitat loss A landscape-level variable that refers to the percentage of the original land area of the ecoregion that has been lost (converted). It underscores the rapid loss of species and disruption of ecological processes

predicted to occur in ecosystems when the total area of remaining habitat declines.

headspring A spring that is the source of a stream.

headwaters The source of a stream or river.

herpetofauna All the species of amphibians and reptiles inhabiting a specified region.

hydrography The study, description, and mapping of oceans, lakes, and rivers, especially with reference to their navigational and commercial uses (Nevada Division of Water Planning 1998).

hydrology The science of waters of the earth: their occurrences, distributions, and circulations; their physical and chemical properties; and their reactions with the environment, including living beings (Nevada Division of Water Planning 1998).

hydrophyte Any plant adapted to live in water or very wet habitats (Lincoln et al. 1982).

hydropower Electrical energy produced by falling water.

ichthyofauna All the species of fishes inhabiting a specified region (Brown and Gibson 1983).

impoundment A body of water such as a pond, confined by a dam, dike, floodgate, or other barrier, which is used to collect and store water for future use (Eckhardt 1998).

indigenous Native to an area.

intact habitat Relatively undisturbed areas characterized by the maintenance of most original ecological processes and by communities with most of their original native species still present.

introduced species See *exotic species*.

invasive species Exotic species (i.e., alien or introduced) that rapidly establish themselves and spread through the natural communities into which they are introduced.

invertebrate Any animal lacking a backbone or bony segment that encloses the central nerve cord (Wilson 1992).

isopod A member of the crustacean order Isopoda, a diverse group of flattened and segmented invertebrates. Pillbugs are an example.

karst   Applies to areas underlain by gypsum, anhydrite, rock salt, dolomite, quarzite (in tropical moist areas), and limestone (Hobbs 1992).

keystone species   Species that are critically important for maintaining ecological processes or the diversity of their ecosystems.

lake   An inland body of fresh or salt water of considerable size occupying a basin or hollow on the earth's surface, and which may or may not have a current or single direction of flow (U.S. Dept. of Agriculture 1995).

landform   The physical shape of the land reflecting geologic structure and processes of geomorphology that have sculpted the structure (Hunt 1974).

landscape   An aggregate of landforms, together with its biological communities (Lotspeich and Platts 1982).

landscape ecology   A branch of ecology concerned with the relationship between landscape-level features, patterns, and processes and the conservation and maintenance of ecological processes and biodiversity in entire ecosystems.

lentic   Referring to standing freshwater habitats, such as ponds and lakes (Brown and Gibson 1983).

levee   A natural or artificial earthen obstruction along the edge of a stream, lake, or river. Usually used to restrain the flow of water out of a river bank (Eckhardt 1998).

life cycle   The entire life span of an organism from the moment it is conceived to the time it reproduces (Wilson 1992).

lotic   Referring to running freshwater habitats, such as springs and streams (Brown and Gibson 1983).

macroinvertebrates   Invertebrates large enough to be seen with the naked eye (e.g., most aquatic insects, snails, and amphipods; Maxwell et al. 1995).

major ecosystem type   A set of major habitat types whose ecoregions (a) share comparable ecosystem dynamics; (b) have similar characteristic responses to disturbance; (c) exhibit similar degrees of beta diversity; and (d) require an ecosystem-specific conservation approach.

major habitat type   A set of ecoregions that (a) experience comparable climatic regimes; (b) have similar vegetation structure; (c) display similar spatial patterns of biodiversity; and (d) contain flora and fauna with similar guild structures and life histories. Eight major habitat types (MHTs) are defined in this study.

marine   Living in salt water (Brown and Gibson 1983).

minimal viable population   The smallest population size necessary for a species' long-term survival, given foreseeable impacts (Peck 1998; Quammen 1996).

mollusk or mollusc   An animal belonging to the phylum Mollusca, such as a snail or clam (Wilson 1992).

morphology   The form and structure of an organism, with special emphasis on external features (Lincoln et al. 1982).

nationally important   A *biological distinctiveness* category.

Nearctic   Referring to the zoogeographical region comprising North America, Greenland, and northern Mexico.

Neotropical   Referring to the zoogeographical region comprising South America, the West Indies, and Central America south of the Mexican plateau.

nonindigenous species   See *exotic species*.

nonpoint source   A diffuse form of water-quality degradation in which wastes are not released at one specific, identifiable point but from a number of points that are spread out and difficult to identify and control (Eckhardt 1998).

non-native species   See *exotic species*.

obligate species   A species that must have access to a particular habitat type to persist.

oligotrophic   Having a low supply of plant nutrients (Eckhardt 1998).

overexploitation   Refers to levels of collection, hunting, or fishing that are not ecologically sustainable.

Pearson correlation   A measure of linear association between two variables. Values of the correlation coefficient range from -1 to 1. The absolute value of the correlation coefficient indicates the strength of the linear relationship between the variables, with larger absolute values indicating stronger relationships. The sign of the coefficient indicates the direction of the relationship.

polymorphism   The co-occurrence of several different forms (Lincoln et al. 1982).

point source   A source of pollution that involves discharge of waste from an identifiable point, such as a smokestack or sewage-treatment plant (Eckhardt 1998).

population   In biology, any group of organisms belonging to the same species at the same time and place (Wilson 1992).

radiation   The diversification of a group of organisms into multiple species, due to intense isolating mechanisms or opportunities to exploit diverse resources.

rarity   Seldom occurring either in absolute number of individuals or in space (Fiedler and Jain 1992).

recharge   Refers to water entering an underground aquifer through faults, fractures, or direct absorption (Eckhardt 1998).

refugia   Habitats that have allowed the persistence of species or communities because of the stability of favorable environmental conditions over time.

relatively intact   The conservation status category indicating the least possible disruption of ecosystem processes. Natural communities are largely intact with species and ecosystem processes occurring within their natural ranges of variation.

relatively stable   The conservation status category between *vulnerable* and *relatively intact* in which extensive areas of intact habitat remain but local species declines and disruptions of ecological processes have occurred.

relictual taxa   A species or group of organisms largely characteristic of a past environment or ancient biota.

representation   The protection of the full range of biodiversity of a given biogeographic unit within a system of protected areas.

reservoir   A pond, lake, tank, or basin (natural or human made) where water is collected and used for storage. Large bodies of groundwater are called reservoirs of water (Eckhardt 1998).

restoration   Management of a disturbed and/or degraded habitat that results in recovery of its original state (Wilson 1992).

riparian   Referring to the interface between freshwater habitats and the terrestrial landscape.

river   A natural stream of water of considerable volume, larger than a brook or creek.

runoff   Surface water entering rivers, freshwater lakes, or reservoirs (Eckhardt 1998).

sediment   Soil particles, sand, and minerals washed from the land into aquatic systems as a result of natural and human activities (Eckhardt 1998).

semiaquatic   Living partially in or adjacent to water (Brown and Gibson 1983).

siltation   The deposition of finely divided soil and rock particles on the bottom of stream- and riverbeds and reservoirs (Eckhardt 1998).

sinkholes   Depressions or cavities created by dissolution of limestone bedrock or collapse of caves. Typically found in karst landscapes.

speciation   The process of species formation, in which one species evolves into two or more species (Quammen 1996).

species   The basic unit of biological classification, consisting of a population or series of populations of closely related and similar organisms (Wilson 1992).

species adaptation   A particular part of the anatomy, physiological process, or behavior pattern of a particular species that increases its chances of survival or ability to reproduce (Wilson 1992).

species assemblage   The combination of particular species that occur together in a specific location and have a reasonable opportunity to interact with one another (Matthews 1998).

species radiation   Refers to the evolution of a single species into several different species within the same geographical range (Wilson 1992). Also referred to as *adaptive radiation*.

species richness   A simple measure of species diversity calculated as the total number of species in a habitat or community (Fiedler and Jain 1992).

spring   A natural discharge of water as leakage or overflow from an aquifer through a natural opening in the soil or rock onto the land surface or into a body of water (Hobbs 1992).

stenothermic   Tolerant of a narrow range of environmental temperatures (Lincoln et al. 1982).

stream   A general term for a body of flowing water (Maxwell et al. 1995); often used to describe a mid-sized tributary (as opposed to a river or creek).

stochastic   Random.

stock   A race, stem, or lineage (Lincoln et al. 1982).

subspecies   Subdivision of a species. Usually defined as a population or series of populations occupying a discrete range and differing genetically from other geographical races of the same species (Wilson 1992).

subtropical   An area in which the mean annual temperature ranges from 13°C to 20°C (Brown and Gibson 1983).

surface water   All waters whose surface is naturally exposed to the atmosphere, for example, rivers, lakes, reservoirs, ponds, streams, impoundments, seas, and estuaries, and all springs, wells, or other collectors directly influenced by surface water (Nevada Division of Water 1998).

systematics   The classification of living organisms into hierarchical series of groups emphasizing their phylogenetic interrelationships; often used as equivalent to taxonomy (Lincoln et al. 1982).

taxon (pl. taxa)   A general term for any taxonomic category, for example, a species, genus, family, or order (Brown and Gibson 1983).

temperate   An area in which the mean annual temperature ranges from 10°C to 13°C.

terrestrial   Living on land.

tributary   A stream or river that flows into a larger stream, river, or lake, feeding it water.

trophic   Pertaining to nutrition (Lincoln et al. 1982).

turbidity   Refers to the extent to which light penetrates a body of water. Turbid waters are those that do not generally support net growth of photosynthetic organisms (Jeffries and Mills 1990).

vagile   Able to be transported or to move actively from one place to another (Brown and Gibson 1983).

vulnerable   The conservation status category between *endangered* and *relatively stable* and characterized by good probability of persistence of remaining intact habitat (assuming adequate protection) but also by loss of some sensitive or exploited species.

xeric   Describes dryland or desert areas.

watershed   See *catchment*.

wetlands   Lands transitional between terrestrial and aquatic systems where the water table is usually at or near the surface or the land is covered by shallow water. These areas are inundated or saturated by surface water or groundwater at a frequency and duration sufficient to support a prevalence of vegetation typically adapted to life in saturated soil conditions (Maxwell et al. 1995).

zoogeography   The study of the distributions of animals (Brown and Gibson 1983).

LITERATURE CITED AND CONSULTED

Abramovitz, J. N. 1996. *Imperiled waters, impoverished future: The decline of freshwater ecosystems.* Worldwatch Paper 128. Washington, DC: Worldwatch Institute.

Adler, K. 1996. The salamanders of Guerrero, Mexico, with descriptions of five new species of pseudoeurycea (Caudata: Plethodontidae). *Occasional Papers of the Natural History Museum, University of Kansas* 177: 1–28.

Alberta Environmental Protection. 1997. "Alberta's River Basins." Available at http://www.gov.ab.ca/env/water/basins.html; INTERNET.

Alberta Wilderness Association. 1998a. "Legal Action Over The Proposed Cheviot Mine." Available at http://www.web.net/~awa/cheviot/cheviot.htm; INTERNET.

———. 1998b. "Public Notice: Cardinal River Coals Ltd. Water Resource Act Notice of Application." Available at http://www.web.net/~awa/cheviot/waterapp.htm; INTERNET.

Alcocer, J., and E. Kato. 1995. Cuerpos Acuaticos de Cuatro Ciénegas, Coahuila. In J. G. de la Lanza-E and J. L. Garcia-C (editors). *Lagos y Présas de México.* México, D. F.: De. Centro do Ecologia y Desarrollo.

Alexander, C. E., M. A. Boutman, and D. W. Field. 1986. *An inventory of coastal wetlands of the USA.* Washington, DC: U.S. Department of Commerce. USA. 25 pp.

Allan, J. D. and A. S. Flecker. 1993. Biodiversity conservation in running waters: Identifying the major factors that threaten destruction of riverine species and ecosystems. *BioScience* 43: 32–43.

American Rivers. 1996. "American Rivers' Most Endangered and Threatened Rivers of 1996." Available at http://www.amrivers.org/10-most96.html; INTERNET.

———. 1997. "American Rivers' Most Endangered and Threatened Rivers of 1997." Available at http://www.amrivers.org/10-most97.html; INTERNET.

———. 1998. "American Rivers' Most Endangered Rivers of 1998." Available at http://www.amrivers.org/98endangered.html; INTERNET.

———. 1999. "American Rivers' Most Endangered Rivers of 1999." Available at http://www.amrivers.org/99endangered.html; INTERNET.

Andrews, W. J. 1994. *Nitrate in ground water and spring water near four dairy farms in north Florida, 1990–1993.* Water-Resources Investigations Report 94-4162. Tallahassee, FL: U.S. Geological Survey.

Angermeier, P. L. 1995. Ecological attributes of extinction-prone species: Loss of freshwater fishes of Virginia. *Conservation Biology* 9: 143–158.

Angermeier, P. L. and I. J. Schlosser. 1995. Conserving aquatic diversity: Beyond species and populations. *American Fisheries Society Symposium* 17: 402–414.

Anonymous. 1997. U.S. and Mexican communities share river, concerns. *Arizona Water Resource* 6(1): 1–2.

Arriaga, L., V. Aguilar, J. Alcocer, R. Jiménez, E. Muñoz, E. Vázquez, and C. Aguilar. 1998. *Programa de Cuencas Hidrológicas Prioritarias y Biodiversidad de México de la Comisión Nacional para el Conocimiento y Uso do la Biodiversidad.* Primer informe técnico. México, D. F.: CONABIO/USAID/FMCN/WWF.

Azas, J. M. A. 1991a. "Rare Jewels of the Tehuantepec Isthmus." Available at http://www.badgerstate.com/JAWS/faqs/tehuantepec.html; INTERNET.

———. 1991b. "On the Balsas living diamond, *Cichlasoma istlanum,* Jordan and Snyder 1899." Available at http://www.badgerstate.com/JAWS/faqs/cich-istlanum.html; INTERNET.

Bailey, R. G. 1976. Ecoregions of the United States. Map scale 1:7,500,000. Ogden, UT: USDA Forest Service, Intermountain Region.

———. 1983. Delineation of ecosystem regions. *Environmental Management* 7: 365–373.

Barr, T. C. Jr. 1961. *Caves of Tennessee.* Bulletin 64. Nashville, TN: Tennessee Department of Conservation and Commerce, Division of Geology.

Barr, T. C., and J. R. Holsinger. 1985. Speciation in cave faunas. *Annual Review of Ecology and Systematics* 16: 313–337.

Bauer, B. H., D. A. Etnier, and N. M. Burkhead. 1995. *Etheostoma* (Ulocentra) *scotti* (Osteichthyes: Percidae), a new darter from the Etowah River system in Georgia. *Bulletin of the Alabama Museum of Natural History* 17: 1–16

Bayley, P. B. 1995. Understanding large river-floodplain ecosystems: Significant economic advantages and increased biodiversity and stability would result from restoration of impaired systems. *BioScience* 45: 153–158.

Behnke, R. J. and D. E. Benson. 1983. *Endangered and threatened fishes of the Upper Colorado River Basin.* Washington, DC: U.S. Fish and Wildlife Service.

Berger, D. 1995. "Precious Resource: Water Issues in the Lower Rio Grande Basin." Available at http://www.txinfinet.com/mader/ecotravel/border/sabal.html; INTERNET.

Berkman, H. E., and C. F. Rabeni. 1987. Effect of siltation on stream fish communities. *Environmental Biology of Fishes* 18: 285–294.

Beschta, R. L. 1997. Restoration of riparian and aquatic systems for improved aquatic habitats in the Upper Columbia River Basin. Pages 475–491 in D. J. Stouder, P. A. Bisson, and R. J. Naiman (editors), *Pacific salmon and their ecosystems: Status and future options.* New York: Chapman and Hall.

Bibby, C. J. 1992. *Putting biodiversity on the map: Priority areas for global conservation.* Washington, DC: ICDP.

Blockstein, D. E. 1992. An aquatic perspective on U.S. biodiversity policy. *Fisheries* 17: 26–30.

Bogan, A. E., and P. W. Parmalee. 1983. *Tennessee's rare wildlife.* Vol. 2, *The mollusks.* Nashville, TN: Tennessee Wildlife Resources Agency.

Boschung, H. T., R. L. Mayden, and J. R. Tomelleri. 1992. *Etheostoma chermocki,* a new species of darter (Teleostei: Percidae), from the Black Warrior River drainage of Alabama. *Bulletin Alabama Museum of Natural History* 13: 11–20.

Botosaneau, L., and J. Notenboom. 1986. *Isopoda: Cirolanidae, Stygofauna Mundi.* Edited by L. Botosaneanu. Leiden, The Netherlands: E. J. Brill.

Bowles, D. E., and T. L. Arsuffi. 1993. Karst aquatic ecosystems of the Edwards Plateau region of central Texas, USA: A consideration of their importance, threats to their existence, and efforts for their conservation. *Aquatic Conservation: Marine and Freshwater Ecosystems* 3: 317–329.

Briggs, J. C. 1986. Introduction to the zoogeography of North American fishes. Pages 1–16 in C. H. Hocutt and E. O. Wiley (editors), *The zoogeography of North American freshwater fishes.* New York: John Wiley.

Brown, J. H., and A. C. Gibson. 1983. *Biogeography.* St. Louis, MO: C. V. Mosby.

Brune, G. 1975. Major and historical springs of Texas. *Texas Water Development Board Report* 189: 1–94.

———. 1981. *Springs of Texas.* Forth Worth, TX: Branch-Smith.

Bryant, D., D. Nielsen, and L. Tangley. 1997. *The last frontier forests: Ecosystems and economies on the edge.* Washington, DC: World Resources Institute.

Bryce, S. A., and S. E. Clarke. 1996. Landscape-level ecological regions: Linking state-level ecoregion frameworks with stream habitat classifications. *Environmental Management* 20: 297–311.

Burch, J. B. 1973. *Freshwater unionacean clams (Mollusca: Pelecypoda) of North America.* Biota of Freshwater Ecosystems Identification Manual No. 11. Washington, DC: U.S. Government Printing Office.

Burkhead, N. M., and R. E. Jenkins. 1991. Fishes. Pages 321–409 in K. Terwilliger (editor), *Virginia's endangered species. Proceedings of a symposium.* Blacksburg, VA: McDonald and Woodward.

Burkhead, N. M., S. J. Walsh, B. J. Freeman, and J. D. Williams. 1997. Status and restoration of the Etowah River, an imperiled southern Appalachian ecosystem. Pages 375–444 in G. A. Benz and D. E. Collins (editors), *Aquatic fauna in peril: The southern perspective.* Special Publication 1, Southeast Aquatic Research Institute. Decatur, GA: Lenz Design and Communications.

Burr, B. M., and R. L. Mayden. 1992. Phylogenetics and North American freshwater fishes. Pages 18–75 in R. L. Mayden (editor), *Systematics, historical ecology, and North American freshwater fishes.* Stanford, CA: Stanford University Press.

Burr, B. M., and L. M. Page. 1986. Zoogeography of fishes of the lower Ohio–upper Mississippi Basin. Pages 287–324 in C. H. Hocutt and E. O. Wiley (editors), *The zoogeography of North American freshwater fishes.* New York: John Wiley.

Butler, M. J., and R. A. Stein. 1985. An analysis of the mechanisms governing species replacements in crayfish. *Oecologia* 66: 168–177.

Camacho, A. I., E. Bello, J. M. Becerra, and N. Vaticon. 1992. A natural history of the subterranean environment and its associated fauna. Pages 171–197 in A. I. Camacho (editor), *The natural history of biospeleology.* Madrid: Monagrafias Museo Nacional de Ciencias Naturales.

Cameco. 1999. "Fuel Services." Available at http://www.cameco.com/fuel/; INTERNET.

Campbell, J. A. and W. W. Lamar. 1989. *The venomous reptiles of Latin America.* Ithaca, NY: Comstock.

Capelli, G. M., and J. F. Capelli. 1980. Hybridization between crayfish of the genus *Orconectes*: morphological evidence (Decapoda, Cambaridae). *Crustaceana* 39: 121–132.

Capelli, G. M. and B. L. Munjal. 1982. Aggressive interactions and resource competition in relation to species displacement among crayfish of the genus *Orconectes*. *Journal of Crustacean Biology* 2: 486–492.

Carillo-Bravo, J. 1971. La plataforma Valles-San Luis Potosí. *Boletin de la Asociación Mexicana de Geólogos Petroleros* 23: 1–113.

Carl, G. C., G. C. Clifford, W. A. Clemens, and C. C. Lindsey. 1967. *The fresh-water fishes of British Columbia.* Vancouver, BC: A. Sutton.

Carpenter, S. R., S. G. Fisher, N. B. Grimm, and J. F. Kitchell. 1992. Global change and freshwater ecosystems. *Annual Review Ecology and Systematics* 23: 119–39.

Carr, M. H. (editor. 1994). *A naturalist in Florida: A celebration of Eden.* New Haven, CT: Yale University Press.

Carranza, J. 1954. Descripción del primer bagre anoftalmo y depigmentado encontrado en aguas Mexicanas. *Ciéncia* (México) 14(7–8): 129–136.

Ceas, P. A., and L. M. Page. 1997. Systematic studies of the *Etheostoma spectabile* complex (Percidae; subgenus *Oligocephalus*), with descriptions of four new species. *Copeia* 1997: 496–522.

Chabreck, R. A. 1988. *Coastal marshes: Ecology and wildlife management.* Minneapolis: University of Minnesota Press.

Clancy, P. 1997. Feeling the pinch: The troubled plight of America's crayfish. *The Nature Conservancy Magazine*, May/June, pp. 12–15.

Clarke, A. H. 1981. *The freshwater molluscs of Canada.* Ottawa, Ontario: National Museum of Natural Science.

Cole, G. A. 1984. Crustacea of the Bolson of Cuatro Ciénegas, Coahuila, Mexico. *Journal Arizona-Nevada Academy Sciences* 19: 3–12.

Cole, G. E., and W. L. Minckley. 1972. Stenasellid Isopod Crustaceans in the Western Hemisphere—a New Genus and Species from Mexico—With a Review of Other North American Freshwater Isopod Genera. *Proceedings Biological Society Washington* 84: 313–326.

Collar, N. J., L. P. Gonzaga, N. Krabbe, A. Madrono Nieto, L. G. Naranjo, T. A. Parker III, and D. C. Wege. 1992. *Threatened birds of the Americas: The ICBP/IUCN Red Data Book*, 3rd ed. Cambridge, England: International Council for Bird Preservation.

Colorado Plateau Forum. 1998. "Colorado River Endangered Fish Recovery Program." Available at http://www.nbs.nau.edu/Forum/Sourcebooks/colriven.html; INTERNET.

Committee on Restoration of Aquatic Ecosystems. 1992. *Restoration of aquatic ecosystems: Science, technology, and public policy.* Washington, DC: National Academy Press.

Committee on the Status of Endangered Wildlife in Canada (COSEWIC). 1998. Subcommittee for reptiles and amphibians. Available at http://www.mcgill.ca/redpath/ cosehome.htm; INTERNET.

CONABIO. 1994. *Norma Oficial Mexicana.* Diario Oficial de la Federación. NOM-059-ECOL-1994. 16 May 1994.

Conant, R. 1977. Semiaquatic reptiles and amphibians of the Chihuahuan Desert and their relationships to drainage patterns of the region. Pages 455–492 in R. H. Wauer and D. H. Riskind (editors), *Transactions of the symposium on the biological resources of the Chihuahuan Desert Region, United States and Mexico*, Sul Ross State University, 1974. National Park Service Transactions and Proceedings Series No. 3. Washington, DC: U.S. Government Printing Office.

Conant, R. and J. T. Collins. 1991. *A field guide to reptiles and amphibians: Eastern and central North America.* Boston, MA: Houghton Mifflin.

Conner, J. V., and R. D. Suttkus. 1986. Zoogeography of freshwater fishes of the western Gulf Slope of North America. Pages 413–456 in C. H. Hocutt and E. O. Wiley (editors), *The zoogeography of North American freshwater fishes.* New York: John Wiley.

Constantz, G. 1994. *Hollows, peepers, and highlanders. An Appalachian Mountain ecology.* Missoula, MT: Mountain Press.

Contreras-Arquieta A., G. Guajardo-Martinez, and S. Contreras-Balderas. 1995. *Thiara* (Melanoides) *tuberculate* (Müller, 1774) (Gastropoda: Thiaridae), su probable impacto ecológico en Mexico. *Publ. Biol. FCB-UANL, Mexico* 8: 17–24.

Contreras-Balderas, S. 1969. Perspectivas de la ictiofauna en las zonas aridas del Norte de Mexico. Mem. Primer Simp. Internacional de Aumento de Producción en Zonas Aridas, ICASALS, *Texas Tech. Publ.* 3: 293–304.

———. 1978a. Biota endemica de Cuatro Ciénegas, México. *Mem. Promer Cong. Nal. Zool.* 1: 106–113.

———. 1978b. Speciation aspects and man-made community composition changes in Chihuahuan desert fishes. *Trans. Biol. Res. Chih. Desert Reg., U.S. Mex. U.S.D.I., N.P.S. Trans. Proc.* 3: 1–658.

———. 1984. Environmental impacts in Cuatro Ciénegas, Coahuila, México: A commentary. *Journal Arizona-Nevada Academy Sciences* 19: 85–88.

———. 1991. Conservation of Mexican freshwater fishes: Some protected sites and species, and recent federal legislation. Pages 191–197 in W. L. Minckley and J. E. Deacon (editors), *Battle against extinction: Native fish management in the American West.* Tucson: University of Arizona Press.

Contreras-Balderas, S., and M. A. Escalante. 1984. Distribution and known impacts of exotic fishes in Mexico. Pages 102–130 in W. R. Courtenay Jr. and J. R. Stauffer Jr. (editors), *Distribution, biology, and management of exotic fishes.* Baltimore, MD: The Johns Hopkins University Press.

Contreras-Balderas, S., and M. D. L. Lozano-Vilano. 1994. Water, endangered fishes, and development perspectives in arid lands of Mexico. *Conservation Biology* 8: 379–387.

———.1996. Extinction of most Sandía and Potosí valleys (Nuevo León, México) endemic pupfishes, crayfishes and snails. *Ichthyol. Explor. Freshwater* 7: 33–40.

Cook, F. R. 1984. *Introduction to Canadian amphibians and reptiles.* Ottawa, Ontario: National Museum of Natural Sciences, National Museums of Canada.

Courtenay, W. R., and G. K. Meffe. 1989. Small fishes in strange places: A review of introduced poeciliids. Pages 319–332 in G. K. Meffe and F. F. Snelson Jr. (editors), *Ecology and evolution of livebearing fishes (Poeciliidae).* Englewood Cliffs, NJ: Prentice-Hall.

Courtenay, W. R., D. A. Hensley, J. N. Taylor, and J. A. McCann. 1984. Distribution of exotic fishes in the continental United States. Pages 41–77 in W. R. Courtenay and J. R. Stauffer Jr. (editors), *Distribution, biology, and management of exotic fishes.* Baltimore, MD: Johns Hopkins University Press,

Courtenay, W. R. Jr., D. A. Hensley, J. N. Taylor, and J. A. McCann. 1986. Distribution of exotic fishes in North America. Pages 675–698 in C. H. Hocutt, and E. O. Wiley, (editors), *The zoogeography of North American freshwater fishes.* New York: John Wiley.

Courtenay, W. R. Jr., J. E. Deacon, D. W. Sada, R. C. Allan, and G. L. Vinyard. 1985. Comparative studies of fishes along the course of the pluvial White River, Nevada. *Southwest Nature* 30: 503–524.

Crandall, K. A. 1995. "Infraorder Astacidea." Available at http://www.utexas.edu/depts/.www/crayfish/astacidea/astacidea.html; INTERNET.

Crawford, A. B. and D. F. Peterson (editors). 1974. *Environmental management in the Colorado River Basin.* Logan, UT: State University Press.

Cross, F. B., R. L. Mayden, and J. D. Stewart. 1986. Fishes of the western Mississippi Basin (Missouri, Arkansas and Red rivers). Pages 363–412 in C. H. Hocutt and E. O. Wiley (editors), *The zoogeography of North American freshwater fishes*. New York: John Wiley.

Crossman, E. J., and D. E. McAllister. 1986. Zoogeography of freshwater fishes of the Hudson Bay drainage, Ungava Bay and the Arctic archipelago. Pages 53–104 in C. H. Hocutt and E. O. Wiley (editors), *The zoogeography of North American freshwater fishes*. New York: John Wiley.

Crowe, J. C., and Sharp, J. M. Jr. 1997. Hydrogeologic delineation of habitats for endangered species: The Comal Springs/River System. *Environmental Geology* 30: 17–28.

Culver, D. C. 1970. Analysis of simple cave communities. 1. Caves as islands. *Evolution* 24: 463–474.

———. 1982. *Cave life evolution and ecology*. Cambridge, MA: Harvard University Press.

———. 1986. Cave faunas. Pages 426–443 in M. E. Soulé (editor), *Conservation biology: The science of scarcity and diversity*. Sunderland, MA: Sinauer Associates.

Cushing, C. E. 1996. The ecology of cold desert spring-streams. *Arch. Hydrobiol.* 135: 499–522.

De La Rosa, C. 1995. Middle American streams and rivers. Pages 189–217 in C. E. Cushing, K. W. Cummins, and G. W. Minshall (editors), *River and stream ecosystems*. Ecosystems of the World 22. Amsterdam: Elsevier.

Deacon, J. E., and C. D. Williams. 1991. Ash Meadows and the legacy of the Devils Hole pupfish. Pages 69–90 in W. L. Minckley and J. E. Deacon (editors), *Battle against extinction: Native fish management in the American West*. Tucson: University of Arizona Press.

Degenhardt, W. G., C. W. Painter, and A.H. Price. 1996. *Amphibians and reptiles of New Mexico*. Albuquerque: University of New Mexico Press.

Deyrup, M., and R. Franz (editors). 1994. *Rare and endangered biota of Florida*. Vol. 4, *Invertebrates*. Gainesville: University Press of Florida.

di Castri, F. 1989. History of biological invasions with special emphasis on the Old World. Pages 1–30 in J. A. Drake, H. A. Mooney, F. di Castri, R. H. Groves, F. J. Kruger, M. Rejmanek, and M. Williamson (editors), *Biological invasions: A global perspective*. New York: John Wiley.

Dinerstein, E., D. M. Olson, D. J. Graham, A. L. Webster, S. A. Primm, M. P. Bookbinder, and G. Ledec. 1995. *A conservation assessment of the terrestrial ecoregions of Latin America and the Caribbean*. Washington, DC: The World Bank.

Dinerstein, E., D. M. Olson, J. Atchley, C. Loucks, S. Contreras-Balderas, R. Abell, E. Inigo, E. Enkerlin, C. E. Wiliams, and G. Castilleja (editors). 1998. *Ecoregion-based conservation in the Chihuahuan Desert: A biological assessment and biodiversity vision, Draft report*. A collaborative effort by World Wildlife Fund, Comision Nacíonal para el Conocimiento y Uso de la Biodiversidad (CONABIO), The Nature Conservancy, PRONATURA Noroeste, and the Instituto Technológico y de Estudios Superiores de Monterrey (ITESM). Washington, DC: World Wildlife Fund–US.

Dobson, A. P, J. P. Rodriguez, W. M. Roberts, and D. S. Wilcove. 1997. Geographic distribution of endangered species in the United States. *Science* 275: 550–553.

Drake, J. A., H. A. Mooney, F. di Castri, R. H. Groves, F. J. Kruger, M. Rejmanek, and M. Williamson (editors). 1989. *Biological invasions: A global perspective*. New York: John Wiley.

Drengson, A. R., and D. M. Taylor. 1997. *Ecoforestry. The art and science of sustainable forest use*. Gabriola Island, BC: New Society Publishers.

Duellman, W. E. 1961. The amphibians and reptiles of Michoacán, México. *University of Kansas Publications, Museum of Natural History* 15: 1–148.

———. 1963. A review of the Middle American tree frogs of the genus *Ptychohyla*. *University of Kansas Publications, Museum of Natural History* 15: 297–349.

———. 1964. A review of the frogs of the *Hyla bistinca* group. *University of Kansas Publications, Museum of Natural History* 15: 469–491.

———. 1965a. A biogeographic account of the herpetofauna of Michoacán, México. *University of Kansas Publications, Museum of Natural History* 15: 627–709.

———. 1965b. Amphibians and reptiles from the Yucatán Peninsula, México. *University of Kansas Publications, Museum of Natural History* 15: 577–614.

———. 1970. *The hylid frogs of Middle America*. Vols. 1 and 2. Monograph of the Museum of Natural History, No. 1. Lawrence: The University of Kansas.

———. 1982. Amphibia. Pages 502–514 in S. H. Hurlbert and A. Villalobos-Figueroa (editors), *Aquatic biota of Mexico, Central America and the West Indies*. San Diego, CA: San Diego State University.

Dunn, E. R. 1926. *The salamanders of the Family Plethodontidae*. Northhampton, MA: Smith College.

Dynesius, M., and C. Nilsson. 1994. Fragmentation and flow regulation of river systems in the northern third of the world. *Science* 266: 753–762.

Eckhardt, G. A. 1998. "Glossary of Water Resource Terms." Available at http://www.edwardsaquifer.net/glossary.html; INTERNET.

Edwards, R. J., and S. Contreras-Balderas. 1991. Historical changes in the ichthyofauna of the lower Rio Grande (Río Bravo del Norte), Texas and Mexico. *Southwestern Naturalist* 36: 201–212.

Edwards, R. J., H. E. Beaty, G. Longley, D. H. Riskind, D. D. Tupa, and B. G. Whiteside. 1984. *The San Marcos recovery plan for San Marcos River endangered and threatened species*. Albuquerque, NM: US Fish and Wildlife Service.

Edwards, R. J., G. Longley, R. Moss, J. Ward, R. Matthews, and B. Stewart. 1989. A classification of Texas aquatic communities with special consideration toward the conservation of endangered and threatened taxa. *Texas Journal of Science* 41: 232–240.

Elton, C. S. 1958. *The ecology of invasions by plants and animals*. New York: John Wiley.

Environment Canada. 1996. *Conserving Canada's natural legacy*. Minister of Public Works and Government Services Canada. Folio Infobase.

Ernst, C. H., and R. W. Barbour. 1989. *Turtles of the world.* Washington, DC: Smithsonian Institution Press.

Ernst, C. H., R. W. Barbour, and J. E. Lovich. 1994. *Turtles of the United States and Canada.* Washington, DC: Smithsonian Institution Press.

Eschmeyer, W. N. (editor). 1998. *Catalog of fishes.* San Francisco: California Academy of Sciences.

Espinosa-Pérez, H., P. Fuentes-Mata, M. T. Gaspar-Dillanes, and V. Arenas. 1993a. Notes on Mexican ichthyofauna. Pages 229–252 in T. P. Ramamoorthy, R. Bye, A. Lot, and J. Fa (editors), *Biological diversity in Mexico: Origins and distribution.* New York: Oxford University Press.

Espinosa-Pérez, H., M. T. Gaspar-Dillanes, and P. Fuentes-Mata. 1993b. *Listados faunisticos de Mexico* 3. *Los peces dulceacuicolas Mexicanos.* Mexico: Universidad Nacional Autonoma de Mexico.

Etnier, D. A., and W. C. Starnes. 1993. *The fishes of Tennessee.* Knoxville: University of Tennessee Press.

Eyton, J. R., and J. I. Parkhurst. 1975. *A re-evaluation of the extraterrestrial origin of the Carolina Bays.* Occasional publications of the Department of Geography No. 9. Urbana-Champaign: Geography Graduate Student Association, University of Illinois.

Federal Emergency Management Agency. 1996. "Water Control Infrastructure: National Inventory of Dams–Updated data: 1995–96." CD-Rom.

Federal Register. 1986. *Endangered and threatened wildlife and plants.* 50 CFR 17.11 and 17.12. Department of Interior, U.S. Fish and Wildlife Service. Washington, DC: U.S. Government Printing Office.

———. 1995. *Endangered and threatened wildlife and plants.* 50 CFR 17.11 and 17.12. Department of Interior, U.S. Fish and Wildlife Service. Washington, DC: U.S. Government Printing Office.

Ferrington, L. C. Jr. 1995. Biodiversity of aquatic insects and other invertebrates in springs: Introduction. *Journal of the Kansas Entomological Society* 68: 1–3.

Ferrusquia-Villafranca, I. 1993. Geology of Mexico: A synopsis. Pages 3–107 in T. P. Ramamoorthy, R. Bye, A. Lot, and J. Fa (editors), *Biological diversity of Mexico: Origins and distribution.* New York: Oxford University Press.

Fiedler, P. L. and S. K. Jain (editors). 1992. *Conservation biology: The theory and practice of nature conservation, preservation, and management.* New York: Chapman and Hall.

Fish, J. E. 1977. "Karst hydrogeology and geomorphology of the Sierra de El Abra and the Valles-San Luis Potosí Region, México." Ph.D. thesis, McMaster University.

Flores Villela, O. A. 1991. "Analysis de la distribucion de la herpetofauna de México." Doctoral thesis, Universidad Nacional Autonoma de México.

———. 1993a. *Herpetofauna Mexicana: Annotated list of the species of amphibians and reptiles of Mexico, recent taxonomic changes, and new species.* Special Publication No. 17. Pittsburgh, PA: Carnegie Museum of Natural History.

———. 1993b. Herpetofauna of Mexico: Distribution and endemism. Pages 253–280 in T. P. Ramamoorthy, R. Bye, A. Lot, and J. Fa (editors), *Biological diversity in Mexico: Origins and distribution.* New York: Oxford University Press.

Florida Geological Survey. 1977. *Springs of Florida,* rev. ed. Bulletin no. 31. Tallahassee, FL: Florida Geological Survey.

Ford, R. G. In prep. *A conservation assessment of the marine ecoregions of the United States.* Washington, DC: World Wildlife Fund.

Forest Guardians. 1998. "Water and the Rio Grande." Available at http://fguardians.org/riogrand.html; INTERNET.

Franz, R., J. Bauer, and T. Morris. 1994. Review of biologically significant caves and their faunas in Florida and south Georgia. *Brimleyana* 20: 1–109.

Fremling, C. R., J. L. Rasmussen, R. E. Sparks, S. P. Cobb, C. F. Bryan, and T. O. Claflin. 1989. Mississippi River fisheries: A case history. Can. Spec. Publ. Fish. Aquat. Sci. 106. Pages 309–351 in D. P. Dodge (editor), *Proceedings of the International Large River Symposium.*

Frest, T. J., and E. J. Johannes. 1995. *Interior Columbia Basin mollusk species of special concern.* Seattle, WA: Deixis.

Frissell, C. A., and D. Bayles. 1996. Ecosystem management and the conservation of aquatic biodiversity and ecological integrity. *Water Resources Bulletin* 32: 229–240.

Frost, S. L., and W. J. Mitsch. 1989. Resource development and conservation history along the Ohio River. *Ohio Journal Science* 89: 143–152.

García, E. (editor). 1988. *Modificaciones al sistema de clasificación climatica de Köppen (para adaptarlo a las condiciones de la Republica Mexicana.* Mexico: Enriqueta García.

García de León, F. J., D. A. Hendrickson, and D. M. Hillis. 1998. Phylogenetic relationships of Mexican blindcats (Prietella: Siluriformes: Ictaluridae). *Journal of Caves and Karst Studies* 59: 166.

Garvey, J. E., and R. A. Stein. 1993. Evaluating how chela size influences the invasion potential of an introduced crayfish (*Orconectes rusticus*). *American Midland Naturalist* 129: 172–181.

Gee, J. H. R. 1991. Speciation and biogeography. Pages 172–185 in R. S. K. Barnes and K. H. Mann (editors), *Fundamentals of aquatic ecology.* Oxford, England: Blackwell Scientific.

Geist, D. R. 1995. The Hanford Reach: What do we stand to lose? *Illahee* 11: 130–141.

Gilbert, C. R. (editor). 1992. *Rare and endangered biota of Florida.* Vol. 2, *Fishes.* Gainesville: University Press of Florida.

Gilpin, M. E., and M. E. Soulé. 1986. Minimum viable populations: Processes of species extinctions. Pages 19–34 in M. E. Soulé (editor), *Conservation biology: The science of scarcity and diversity.* Sunderland, MA: Sinauer Associates.

Gines, A., and J. Gines. 1992. Karst phenomena and biospeleological environments. Pages 27–56 in A. I. Camacho (editor), *The natural history of biospeleology.* Madrid: Monagrafias Museo Nacional de Ciencias Naturales.

Gloyd, H. K., and R. Conant. 1990. *Snakes of the* Agkistrodon *complex: A monographic review.* Contributions to Herpetology, No. 6. Society for the Study of Amphibians and Reptiles.

González, A., J. L. Camarillo, F. Mendoza, and M. Mancilla. 1995. Impact of expanding human populations on the herpetofauna of the Valley of Mexico. *Herpetological Review* 17: 30–31.

Government of Canada and U.S. Environmental Protection Agency. 1995. *The Great Lakes: An environmental atlas and resource book,* 3rd ed. Toronto, Ontario: Government of Canada; and Chicago, IL: U.S. Environmental Protection Agency.

Grand Canyon Trust. 1997. "Protecting the Magnificent Virgin River Watershed." Available at http://www.kaibab.org/gct/html/virgin.htm; INTERNET.

Great Lakes Information Network. 1999. "Lake Michigan Facts and Figures." Available at http://www.greatlakes.net/refdesk/almanac/lakes/michfact.html; INTERNET.

Gregory, S. V., and P. A. Bisson. 1997. Degradation and loss of anadromous salmonid habitat in the Pacific Northwest. Pages 277–314 in D. J. Stouder, P. A. Bisson, and R.J. Naiman (editors), *Pacific salmon and their ecosystems: Status and future options.* New York: Chapman and Hall.

Groombridge, B. (editor). 1992. *Global biodiversity: Status of the earth's living resources.* London: Chapman and Hall.

Gutierrez, R. 1997. "Environmental problems at Lake Chapala." Available at http://www.igc.apc.org/elaw/update_summer_97.html; INTERNET.

Hackney, C. T., S. M. Adams, and W. H. Martin (editors). 1992. *Biodiversity of the southeastern United States: Aquatic communities.* New York: John Wiley.

Ham, L. K., and H. H. Hatzell. 1996. *Analysis of nutrients in the surface waters of the Georgia-Florida Coastal Plain study unit, 1970–1991.* Water-Resources Investigations Report 96–4037. Tallahassee, FL: U.S. Geological Survey.

Hammond, H. 1991. *Seeing the forest among the trees.* Vancouver, BC: Polestar.

Hampson, P. S. 1995. *National Water-Quality Assessment Program: The Upper Tennessee River basin study unit.* FS-150-95. Knoxville, TN: U.S. Geological Survey.

Hamr, P. 1998. *Conservation status of Canadian freshwater crayfishes.* Toronto, Ontario: World Wildlife Fund Canada and the Canadian Nature Federation.

Hardy, L. M., and R. W. McDiarmid. 1969. The amphibians and reptiles of Sinaloa, México. *University of Kansas Publications, Museum of Natural History* 18: 39–252.

Harper, K. T., L. L. St. Clair, K. H. Thorne, and W. M. Hess (editors). 1994. *Natural history of the Colorado Plateau and Great Basin.* Niwot: University Press of Colorado.

Hart, J. 1996. *Storm over Mono: The Mono Lake battle and the California water future.* Berkeley: University of California Press.

Hart, J. L. 1973. *Pacific fishes of Canada.* Ottawa, Ontario: Fisheries Research Board of Canada.

Heard, W. H. 1970. Eastern freshwater mollusks (2): The South Atlantic and Gulf drainages. *Malacologia* 10: 3–31.

Hendricks, A. S. 1998. *The conservation and restoration of the robust redhorse* Moxostoma robustum. Vol. 1. Report prepared for Federal Energy Regulatory Commission. Smyrna, GA: Georgia Power.

Hendrickson, D. A. 1983. Distribution records of native and exotic fishes in Pacific drainages of Northern México. *Journal Arizona-Nevada Academy Sciences* 18: 33–38.

———. 1998. "Desert Fishes Council Index to Fish Images, Maps, and Information." Available at http://www.utexas.edu/ftp/pub/tnhc/.www/fish/dfc/na/index.html; INTERNET.

———. 1998. "TNHC North American Freshwater Fishes Index Images, Maps and Information." Available at http://www.utexas.edu/depts/tnhc/.www/fish/tnhc/na/naindex.html; INTERNET.

Hendrickson, D. A., and J. K. Krejca. 1997. Notes on biogeography, ecology and behavior of Mexican blind catfish, Genus *Prietella* (Ictaluridae). *Journal of Caves and Karst Studies* 59: 166.

Hendrickson, D. A., W. L. Minckley, R. R. Miller, D. J. Siebert, and P. Haddock Minckley. 1980. Fishes of the Rio Yaqui Basin, México and United States. *Journal Arizona-Nevada Academy Sciences* 15: 65–106.

Hershler, R. 1984. The hydrobiid snails (Gastropoda: Rissoacea) of the Cuatro Ciénegas Basin: Systematic relationships and ecology of a unique fauna. Pages 61–76 in P. C. Marsh (editor), *Biota of Cuatro Ciénegas, Coahuila, Mexico.* Transactions Arizona-Nevada Academy Sciences 19.

———. 1985. Systematic revision of the Hydrobiidae (Gastropoda: Rissoacea) of the Cuatro Ciénegas Basin, Coahuila, Mexico. *Malacologia* 26: 31–123.

———. 1998. A systematic review of the hydrobiid snails (Gastropoda: Rissoacea) of the Great Basin, western United States. Part 1, Genus *Pyrgulopsis. Veliger* 41: 1–132.

Hershler, R., and W. L. Minckley. 1986. Microgeographic variation in the banded spring snail (Hydrobiidae: Mexipyrgus) from the Cuatro Ciénegas Basin, Coahuila, Mexico. *Malacologia* 27(2): 357–374.

Hershler, R., and D. W. Sada. 1987. Spring snails (Gastropoda: Hydrobiidae) of Ash Meadows, Amargosa Basin, California-Nevada, USA. *Proceedings of the Biological Society of Washington* 100(4): 776–843.

Hobbs, H. H. Jr. 1988. Crayfish distribution, adaptive radiation, and evolution. Pages 52–82 in D. M. Holdich and R. S. Lowery (editors), *Freshwater crayfish biology, management, and exploitation.* London: Croom Helm.

Hobbs, H. H. III. 1992. Caves and springs. Pages 59–131 in C. T. Hackney, S. M. Adams, and W. H. Martin (editors), *Biodiversity of the southeastern United States: Aquatic communities.* New York: John Wiley.

Hocutt, C. H., and E. O. Wiley (editors). 1986. *The zoogeography of North American freshwater fishes.* New York: John Wiley.

Hocutt, C. H., R. E. Jenkins, and J. R. Stauffer Jr. 1986. Zoogeography of the fishes of the central Appalachians and central Atlantic coastal plain. Pages 161–211 in C. H. Hocutt and E. O. Wiley (editors), *The zoogeography of North American freshwater fishes.* New York: John Wiley.

Holing, D. 1988. *California wild lands: A guide to the Nature Conservancy preserves.* San Francisco, CA: Chronicle Books.

Holland, R. F. 1978. *The geographic and edaphic distribution of vernal pools in the Great Central Valley, California.* Special Publication No. 4. Sacramento: California Native Plant Society.

Holsinger, J. R. 1966. A preliminary study of the effects of organic pollution of Banners Corner Cave, Virginia. *International Journal of Speleology* 2: 75–79.

———. 1972. *The freshwater amphipod crustaceans (Gammaridae) of North America.* Biota of Freshwater Ecosystems Identification Manual No. 5. Washington, DC: U.S. Government Printing Office.

Holsinger, J. R. 1986. Zoogeographic patterns of North American subterranean amphipod crustaceans. Pages 85–106 in R. H. Gore and K. L. Heck (editors), *Crustacean biogeography in crustacean issues.* Rotterdam: Balkema.

———. 1988. Troglobites: The evolution of cave-dwelling organisms. *American Scientist* 76(2): 147–153.

———. 1990. *Tuluweckelia cernua,* a new genus and species of stygobiont amphipod crustacean (Hadziidae) from anchialine caves on the Yucatán Peninsula in Mexico. *Beaufortia* 41: 97–107.

———. 1993. Biodiversity of subterranean amphipod crustaceans: Global patterns of zoogeographic implications. *Journal of Natural History* 27: 821–835.

Holsinger, J. R., and D. C. Culver. 1988. The invertebrate cave fauna of Virginia and a part of eastern Tennessee: Zoogeography and ecology. *Brimleyana* 14: 1–162

Hubbs, C. 1995. Springs and spring runs as unique aquatic systems. *Copeia* 1995: 989–991.

Hubbs, C., and W. F. Hettler. 1964. Observations on the toleration of high temperatures and low dissolved oxygen in natural waters by *Crenichthys baileyi. Southwestern Naturalist* 9: 279–326.

Hubbs, C. L., and R. R. Miller. 1965. Studies of Cyprinodont Fishes. XXII. Variation in *Lucania parva,* its establishment in western United States, and description of a new species from an interior basin in Coahuila, México. *Miscellaneous Publications of the Museum of Zoology, University of Michigan* 659: 1–15.

Hubbs, C., R. C. Baird, and J. W. Gerald. 1967. Effects of dissolved oxygen concentration and light intensity on activity cycles of fishes inhabiting warm springs. *American Midland Naturalist* 77: 104–115.

Hubbs, C. L., R. R. Miller, and L. C. Hubbs. 1974. *Hydrographic history and relict fishes of the North-Central Great Basin.* Memoirs of the California Academy of Sciences, vol. 7. San Francisco: California Academy of Sciences.

Hughes, R. M., and D. P. Larsen. 1988. Ecoregions: An approach to surface water protection. *Journal of the Water Pollution Control Federation* 60: 486–493.

Hughes, R. M., and J. M. Omernik. 1981. A proposed approach to determine regional patterns in aquatic ecosystems. Pages 92–102 in N.B. Armantrout (editor), *Acquisition and utilization of aquatic habitat inventory information.* Proceedings of a symposium held 28–30 October 1981. Bethesda, MD: American Fisheries Society.

Hughes, R. M., E. Rexstad, and C. E. Bond. 1987. The relationship of aquatic ecoregions, river basins and physiographic provinces to the ichthyogeographic regions of Oregon. *Copeia* 1087: 423–432.

Hughes, R. M., T. R. Whittier, C. M. Rohm, and D. P. Larsen. 1990. A regional framework for establishing recovery criteria. *Environmental Management* 14: 673–683.

Hughes, R. M., S. A. Heiskary, W. J. Matthews, and C. O. Yoder. 1994. Use of ecoregions in biological monitoring. Pages 125–151 in S. L. Loeb and A. Spacie (editors), *Biological monitoring of aquatic systems.* Boca Raton, FL: Lewis Publishers.

Hunt, C. B. 1974. *Natural regions of the United States and Canada.* San Francisco, CA: W.H. Freeman.

Ice Age Park and Trail Foundation. 1998. "Wisconsin's Glacial Landscape: How It All Happened." Available at http://www.iceagetrail.org/wgl.htm; INTERNET.

International Rivers Network. 1996. "Mexican Dam Worsens Water Woes Downstream. World Rivers Review. January 1996." Available at http://www.irn.org/pubs/wrr/9601/mexico.html; INTERNET.

Isphording, W. C., and J. F. Fitzpatrick Jr. 1992. Geologic and evolutionary history of drainage systems in the southeastern United States. Pages 19–56 in C. T. Hackney, S. M. Adams, and W. H. Martin (editors), *Biodiversity of the southeastern United States: Aquatic communities.* New York: John Wiley.

IUCN. 1988. *IUCN red list of threatened animals.* Gland, Switzerland: IUCN.

———. 1994. *IUCN red list categories.* Gland, Switzerland: IUCN.

Jagger, T. 1998. "El Camino Real: A Trail of Ideas." Available at http://www.alacranpress.com; INTERNET.

Jeffries, M. and D. Mills. 1990. *Freshwater ecology: Principles and applications.* London: Belhaven Press.

Jenkins, R. E., and N. M. Burkhead. 1994. *The freshwater fishes of Virginia.* Bethesda, MD: American Fisheries Society.

Johnson, J. E. 1986. Inventory of Utah crayfish with notes on current distribution. *Great Basin Naturalist* 46: 625–631.

Johnson, R. I. 1978. Systematics and zoogeography of *Plagiola* (Dysnomia Epioblasma), an almost extinct genus of freshwater mussels (Bivalvia: Unionidae) from middle North America. *Bulletin of the Museum of Comparative Zoology* 148: 239–320.

Jones, R., D. C. Culver, and T. C. Kane. 1992. Are parallel morphologies of cave organisms the result of similar selection pressures? *Evolution* 46: 353–365.

Karr, J. R., L. A. Toth, and D. R. Dudley. 1985. Fish communities of midwestern rivers: A history of degradation. *BioScience* 35: 90–95.

Karr, J. R., K. D. Fausch, P. L. Angermeier, P. R. Yant, and I. J. Schlosser. 1986. *Assessing biological integrity in running waters: A method and its rationale.* Illinois Natural History Survey Special Publication 5. Champaign, IL: Illinois Natural History Survey.

Kemf, E., and C. Phillips. 1998. "Whales in the Wild." World Wildlife Fund. Available at http://www.wwfus.org/species/whales/index.html; INTERNET.

Kempka, R., and P. Kollash. 1990. Comparison of national wetland inventory and a winter satellite inventory for the California central valley. Pages 179–187 in *Yosemite Centennial Symposium Proceedings* (October 13–20, Yosemite, California). Denver, CO: National Park Service.

Kentucky State Nature Preserves Commission. 1996. *Recommendations for the protection of biological diversity on the Daniel Boone National Forest.* Comment submitted toward the Revised Forest Land and Resource Management Plan for Daniel Boone National Forest, Kentucky. Frankfort, KY: Kentucky State Nature Preserves Commission.

Keown, M. P, E. A. Dardeau Jr., and E. M. Causey. 1981. *Characterization of the suspended-sediment regime and bedload gradation of the Mississippi River Basin.* Report 1, vols. 1 and 2. St. Paul, MN: U.S. Army Corps of Engineers, Environmental Lab, St. Paul District.

Kornfield I. L., and J. N. Taylor. 1983. A new species of polymorphic fish, *Cichlasoma minckleyi*, from Cuatro Ciénegas, Mexico (Teleostei: Cichlidae). *Proceedings Biological Society Washington* 96(2): 253–269.

Kostow, K. 1997. The status of steelhead in Oregon. Pages 145–178 in D. J. Stouder, P. A. Bisson, and R. J. Naiman (editors), *Pacific salmon and their ecosystems: Status and future options.* New York: Chapman Hall.

Krapu, G. L. 1974. Foods of breeding pintails in North Dakota. *Journal of Wildlife Management* 38: 408–417.

Krever, V., E. Dinerstein, D. Olson, and L. Williams (editors). 1994. *Conserving Russia's biological diversity: An analytical framework and initial investment portfolio.* Washington, DC: World Wildlife Fund.

Langecker, T. G., and G. Longley. 1993. Morphological adaptations of the Texas blind catfishes *Trogloglanis pattersoni* and *Satan eurystomus* (Siluriformes: Ictaluridae) to their underground environment. *Copeia* 1993: 976–986

Langner, L. L., and C. H. Flather. 1994. *Biological diversity: Status and trends in the United States.* USDA Forest Service General Technical Report RM–244. Fort Collins, CO: Rocky Mountain Forest and Range Experiment Station.

Laws, Joel. 1998. "Ozark Caving: Cave Pod People." Available at http://www.umsl.edu/~joellaws/ozark_caving/mss/pod.htm; INTERNET.

Lee, D. S., C. R. Gilbert, C. H. Hocutt, R. E. Jenkins, D. E. McAllister, and J. R. Stauffer Jr. 1980. *Atlas of North American freshwater fishes.* North Carolina Biological Survey Publication No. 1980-12. Raleigh, NC: North Carolina State Museum of Natural History.

León, J. R. 1969. The systematics of the frogs of the *Hyla rubra* group in Middle America. *University of Kansas Publications, Museum of Natural History* 18: 505–545.

Levin, T. 1996. Restoration of Connecticut River shows potential of Clean Water Act. *Sierra* 81(2): 78–79.

Light, T., D. C. Erman, C. Myrick, and J. Clarke. 1995. Decline of the shasta crayfish (*Pacifastacus fortis* Faxon) of northeastern California. *Conservation Biology* 9: 1567–1577.

Lincoln, R. J., G. A. Boxshall, and P. F. Clark. 1982. *A dictionary of ecology, evolution and systematics.* London: Cambridge University Press.

Lindsey, C. C., and J. D. McPhail. 1986. Zoogeography of fishes of the Yukon and Mackenzie basins. Pages 639–674 in C. H. Hocutt and E. O. Wiley (editors), *The zoogeography of North American freshwater fishes.* New York: John Wiley.

Liner, E. A. 1994. *Scientific and common names for the amphibians and reptiles of Mexico in English and Spanish.* Herpetological Circular No. 23. Lawrence, KS: Society for the Study of Amphibians and Reptiles.

Linton, J. 1997. *Beneath the surface: The state of water in Canada.* Ottawa, Ontario: Canadian Wildlife Federation.

Livingston, R. J. 1992. Medium-sized rivers of the Gulf Coastal Plain. Pages 351–385 in C. T. Hackney, S. M. Adams, and W. H. Martin (editors), *Biodiversity of the southeastern United States: Aquatic communities.* New York: John Wiley.

Lodge, T. E. 1994. *The Everglades handbook: Understanding the ecosystem.* Delray Beach, FL: St. Lucie Press.

Loiselle, P. V. 1993. Endangered islands and ancient lakes: Achievable conservation goals for the twenty-first century. Pages 68–89 in *Proceedings of the 3rd International Aquarium Congress.* Boston, MA: New England Aquarium.

Longley, G. 1981. The Edwards Aquifer: Earth's most diverse groundwater ecosystem? *International Journal of Speleology* 11: 123–128.

———. 1986. The biota of the Edwards Aquifer and the implications for paleozoogeography. Pages 51–54 in P. L. Abott and C. M. Woodruff Jr. (editors), *The Balcones Escarpment: Geology, hydrology, ecology, and social development in Central Texas.* San Antonio, TX: Geological Society of America.

Longley, G., and H. S. Karnei Jr. 1979a. Status of *Satan eurystomus* Hubbs and Bailey, the widemouth blindcat. *U.S. Fish and Wildlife Service Endangered Species Report* 5(2): 1–48.

Longley, G., and H. S. Karnei Jr. 1979b. Status of *Trogloglanis pattersoni* Eigenmann, the toothless blindcat. *U.S. Fish and Wildlife Service Endangered Species Report* 5(1): 1–54.

Lotspeich, F. B., and W. S. Platts. 1982. An integrated land-aquatic classification system. *North American Journal of Fisheries Management* 2: 138–149.

Lower Mississippi River Conservation Committee. 1998. "Environmental Impacts of the 1997 Mississippi flood in Louisiana." LMRCC Newsletter. Available at http://www.lmrcc.org/back_issues/flood.html; INTERNET.

Lozano-Vilano, M. D. L., and S. Contreras-Balderas. 1987. Lista zoogeografica y ecológica de la ictiofauna continental de Chiapas, México. *Southwestern Naturalist* 32: 223–236.

————. 1993. Four new species of *Cyprinodon* from southern Nuevo León, Mexico, with a key to the *C. eximius* complex (Teleostei: Cyprinodontidae). *Ichthyological Explorations of Freshwater* 4: 295–308.

Lundberg, J. G. 1982. The comparative anatomy of the toothless blindcat, *Troglolanis pattersoni* Eigenmann, with a phylogenetic analysis of the ictalurid catfishes. *Miscellaneous Publications of the Museum of Zoology, University of Michigan* 163: 1–85.

————. 1992. The phylogeny of ictalurid catfishes: A synthesis of recent work. Pages 392–420 in R. L. Mayden (editor), *Systematics, historical ecology, and North American freshwater fishes*. Stanford, CA: Stanford University Press.

Lydeard, C., and R. L. Mayden. 1995. A diverse and endangered aquatic ecosystem of the southeast United States. *Conservation Biology* 9: 800–805.

Lyons, J. 1989. Correspondence between the distribution of fish assemblages in Wisconsin streams and Omernik's ecoregions. *American Midland Naturalist* 122: 163–182.

MacIsaac, H. J. 1996. Potential abiotic and biotic impacts of zebra mussels on the inland waters of North America. *American Zoologist* 36: 287–299.

Maitland, P. S. 1985. Criteria for the selection of important sites for freshwater fish in the British Isles. *Biological Conservation* 31: 335–353.

Marsh, P. C. 1984. Biota of Cuatro Ciénegas, Coahuila, Mexico: Proceedings of a special symposium. Preface. *Journal of the Arizona-Nevada Academy of Science* 19(1): 1–2.

Martin, P. S. 1958. *A biogeography of reptiles and amphibians in the Gomez Farias region, Tamaulipas, Mexico*. Miscellaneous Publications No. 101. Museum of Zoology. Ann Arbor: University of Michigan.

Maser, C., and J. R. Sedell. 1994. *From the forest to the sea: The ecology of wood in streams, rivers, estuaries and oceans*. Delray Beach, FL: St. Lucie Press.

Master, L. 1991a. Aquatic animals: Endangerment alert. *Nature Conservancy*, March/April, pp. 26–27.

————. 1991b. Assessing threats and setting priorities for conservation. *Conservation Biology* 5: 559–563.

Master, L. E., S. R. Flack, and B. A. Stein (editors). 1998. *Rivers of life: Critical watersheds for protecting freshwater biodiversity*. Arlington, VA: The Nature Conservancy.

Master, L., G. Hammerson, M. Shwartz, D. Simberloff, and B. Stein. In prep. Chapter 6. Conservation status of U.S. species. In B. Stein, J. Adams, and L. Kutner (editors), *Precious heritage: Status of biodiversity in the United States*.

Matthews, W. J. 1998. *Patterns in freshwater fish ecology*. New York: Chapman Hall.

Mattson, R. A., J. H. Epler, and M. K. Hein. 1995. Description of benthic communities in karst, spring-fed streams of north central Florida. *Journal of the Kansas Entomological Society* 62(2): 18–41.

Maxwell, J. R., C. J. Edwards, M. E. Jensen, S. J. Paustian, H. Parrot, and D. M. Hill. 1995. *A hierarchical framework of aquatic ecological units in North America (Nearctic Zone)*. General Technical Report NC-176. St. Paul, MN: USDA Forest Service, North Central Forest Experiment Station.

May, E. 1998. *At the cutting edge. The crisis in Canada's forests*. Toronto, Ontario: Key Porter Books.

Mayden, R. L. 1988. Vicariance biogeography, parsimony, and evolution in North American freshwater fishes. *Systematic Zoology* 37: 329–355.

Mayden, R. L., B. M. Burr, L. M. Page, and R. R. Miller. 1992. The native freshwater fishes of North America. Pages 827–890 in R. L. Mayden (editor), *Systematics, historical ecology, and North American freshwater fishes*. Stanford, CA: Stanford University Press.

McAllister, D. E. 1990. *A list of the fishes of Canada*. Ottawa, Ontario: National Museum of Natural Sciences.

McAllister, D. E., B. J. Parker, and P. M. McKee. 1985. *Rare, endangered, and extinct fishes in Canada*. Syllogeus No. 54. Ottawa, Ontario: National Museum of Natural Sciences, National Museums of Canada.

McAllister, D. E., S. P. Platania, F. W. Schueler, M. E. Baldwin, and D. S. Lee. 1986. Ichthyofaunal patterns on a geographic grid. Pages 17–51 in C. H. Hocutt and E. O. Wiley (editors), *The zoogeography of North American freshwater fishes*. New York: John Wiley.

McAllister, D. E., F. W. Schueler, C. M. Roberts, and J. P. Hawkins. 1994. Mapping and GIS analysis of the global distribution of coral reef fishes on an equal-area grid. Pages 155–176 in R. I. Miller (editor), *Mapping the diversity of nature*. London: Chapman and Hall.

McAllister, D. E., A. L. Hamilton, and B. Harvey. 1997. Global freshwater biodiversity: striving for the integrity of freshwater ecosystems. *Sea Wind* 11: 1–140.

McCoy, C. J. 1982. Reptilia. Pages 515–520 in S. H. Hurlbert and A. Villalobos-Figueroa (editors), *Aquatic biota of Mexico, Central America and the West Indies*. San Diego, CA: San Diego State University.

McDonald, Sam. 1996. "Crayfish." Available at http://www.aqualink.com/fresh/z-crayfish1.html; INTERNET.

McIntyre, Peter B. 1997. "The Road to Rangeland Reform: A History, Review, and Prospectus." Available at http://fguardians.org/mcintyre/reform.htm; INTERNET.

McMahon, R. F. 1983. Ecology of an invasive pest bivalve, *Corbicula*. Pages 505–561 in W. D. Russel-Hunter (editor), *The mollusca*. Vol. 6, *Ecology*. New York: Academic Press.

McNab, W. H., and P. E. Avers (compilers). 1994. "Ecological Subregions of the United States." WO-WSA-5. U.S. Forest Service, ECOMAP Team. Available at http://www.fs.fed.us/land/pubs/ecoregions/index.html; INTERNET.

McPhail, J. D., and C. C. Lindsey. 1986. Zoogeography of the freshwater fishes of Cascadia (the Columbia system and rivers north to the Stikine). Pages 615–637 in C. H. Hocutt and E. O. Wiley (editors), *The zoogeography of North American freshwater fishes*. New York: John Wiley.

Meffe, G. K. 1989. Fish utilization of springs and cienegas in the arid southwest. Pages 475–485 in R. R. Shartiz and J. W. Gibbons (editors), *Freshwater wetlands and wildlife: Perspectives on natural, managed, and degraded ecosystems.* Department of Energy Symposium, Series 61. Oak Ridge, TN: Office of Science and Technology Information.

Mestre, J. E. 1997. "Integrated Approach to River Basin Management: Lerma-Chapala Case Study." Available at http://iwrn.ces.fau.edu/mestre.htm; INTERNET.

Miller, R. R., J. D. Williams, and J. E. Williams. 1989. Extinction of North American fishes during the past century. *Fisheries* 14(6):22–38.

Miller, R. R. 1958. Origin and affinities of the freshwater fish fauna of western North America. Pages 187–222 in C. L. Hubbs (editor), *Zoogeography.* Publication No. 51. Washington, DC: American Association for the Advancement of Science.

———. 1968. Two new fishes of the genus *Cyprinodon* from the Cuatro Ciénegas Basin, Coahuila, Mexico. *Occasional Papers of the Museum of Zoology, University of Michigan* 659: 1–5.

———. 1977. Composition and derivation of the native fish fauna of the Chihuahuan Desert region. Pages 365–382 in R. H. Wauer and D. H. Riskind (editors), *Transactions of the symposium on the biological resources of the Chihuahuan Desert Region, United States and Mexico,* Sul Ross State University, 1974. National Park Service Transactions and Proceedings Series No. 3. Washington, DC: U.S. Government Printing Office.

———. 1982. Pisces. Pages 486–501 in S. H. Hurlbert and A. Villalobos-Figueroa (editors), *Aquatic biota of Mexico, Central America and the West Indies.* San Diego: San Diego State University.

Miller, R. R., and M. L. Smith. 1986. Origin and geography of the fishes of central Mexico. Pages 487–517 in C. H. Hocutt and E. O. Wiley (editors), *The zoogeography of North American freshwater fishes.* New York: John Wiley.

Miller, R. R., J. D. Williams, and J. E. Williams. 1989. Extinctions of North American fishes during the past century. *Fisheries* 14(6): 22–38.

Minckley, W. L. 1969. Environments of the Bolson of Cuatro Ciénegas, Coahuila, Mexico, with special reference to the aquatic biota. *University of Texas at El Paso, Sci. Ser.* 2: 1–65.

———. 1973. *Fishes of Arizona.* Phoenix: Arizona Game and Fish Department.

———. 1977. Endemic fishes of the Cuatro Ciénegas Basin, northern Coahuila, Mexico. Pages 383–404 in R. H. Wauer and D. H. Riskind (editors), *Transactions of the symposium on the biological resources of the Chihuahuan Desert Region, United States and Mexico,* Sul Ross State University, 1974. National Park Service Transactions and Proceedings Series No. 3, Washington, DC: U.S. Government Printing Office.

Minckley, W. L. 1984. Cuatro Ciénegas fishes: Research review and a local test of diversity versus habitat size. Pages 13– 21 in Paul C. Marsh (editor), *Biota of Cuatro Ciénegas, Coahuila, Mexico: Proceedings of a special symposium.* Fourteenth Annual Meeting, Desert Fishes Council, Tempe, Arizona, 18–20 November 1983. *Journal of the Arizona-Nevada Academy of Science* 19(1).

Minckley, W. L. 1994a. A bibliography for natural history of the Cuatro Ciénegas Basin and environs, Coahuila,

Mexico. *Proceedings of the Desert Fishes Council* 25 (1993): 47–64.

Minckley, W. L. 1994b. Ecosystem conservation, with special reference to Bolsón de Cuatro Ciénegas, México. *Proceedings of the Desert Fishes Council* 25 (1993): 47 (abstract).

Minckley, W. L., and D. E. Brown. 1994. Wetlands. Pages 223–301 in D. E. Brown (editor), *Biotic communities: Southwestern United States and northwestern Mexico.* Salt Lake City: University of Utah Press.

Minckley, W. L., and G. A. Cole. 1968. Preliminary limnologic information on waters of the Cuatro Ciénegas basin, Coahuila, Mexico. *Southwestern Naturalist* 13(4): 421– 431.

Minckley, W. L., and J. E. Deacon (editors). 1991. *Battle against extinction. Native fish management in the American West.* Tucson: University of Arizona Press.

Minckley, W. L., and D. W. Taylor. 1969. A new world for biologists. *Pacific Disc.* 19: 18–22.

Minckley, W. L., D. A. Hendrickson, and C. E. Bond. 1986. Geography of western North American freshwater fishes: Description and relationships to intracontinental tectonism. Pages 519–613 in C. H. Hocutt and E. O. Wiley (editors), *The zoogeography of North American freshwater fishes.* New York: John Wiley.

Minckley, W. L., G. K. Meffe, and D. L. Soltz. 1991. Conservation and management of short-lived fishes: The cyprinodonts. Pages 247–282 in W. L. Minckley and J. E. Deacon (editors), *Battle against extinction: Native fish management in the American West.* Tucson: University of Arizona Press.

Moler, P. E., editor. 1992. *Rare and endangered biota of Florida.* Vol. 3, *Amphibians and reptiles.* Gainesville: University Press of Florida.

Mooney, H. A., and J. A. Drake (editors). 1986. *Ecology of biological invasions of North America and Hawaii.* New York: Springer-Verlag.

Morris, T. H., and M. A. Stubben. 1984. Geologic contrasts of the Great Basin and Colorado Plateau. Pages 9–25 in K. T. Harper, L. L. St. Clair, K. H. Thornes, and W. M. Hess (editors), *Natural history of the Colorado Plateau and Great Basin.* Niwot: University of Colorado Press.

Mosquin, T., P. G. Whiting, and D. E. McAllister. 1995. *Canada's biodiversity: The variety of life, its status, economic benefits, conservation costs and unmet needs.* Ottawa, Ontario: Canadian Museum of Nature.

Moyle, P. B. 1976. *Inland fishes of California.* Berkeley: University of California Press.

———. 1994. Biodiversity, biomonitoring, and the structure of stream fish communities. Pages 171–186 in S. L. Loeb and A. Spacie (editors), *Biological monitoring of aquatic systems.* Boca Raton, FL: Lewis Publishers.

Moyle, P. B., and R. A. Leidy. 1992. Loss of biodiversity in aquatic ecosystems: Evidence from fish faunas. Pages 127–169 in P. L. Fielder and S. K. Jain (editors), *Conservation biology: The theory and practice of nature, conservation, preservation, and management.* New York: Chapman and Hall.

Moyle, P. B., and J. E. Williams. 1990. Biodiversity loss in the temperate zone: Decline of the native fish fauna of California. *Conservation Biology* 4: 275–294.

Moyle, P. B., and R. M. Yoshiyama. 1992. *Fishes, aquatic diversity management areas, and endangered species: A plan to protect California's native aquatic biota.* The California Policy Seminar. Davis: University of California.

Mueller, L. 1993. *Winged maple leaf mussel and Higgins eye pearly mussel: Freshwater mussels threatened with extinction.* St. Paul: Minnesota Department of Agriculture.

Naiman, R. J., J. J. Magnuson, D. M. McKnight, and J. A. Stanford (editors). 1995a. *The freshwater imperative: A research agenda.* Washington, DC: Island Press.

Naiman, R. J., J. J. Magnuson, D. M. McKnight, J. A. Stanford, and J. R. Karr. 1995b. Freshwater ecosystems and their management: A national initiative. *Science* 270: 584–585.

National Audubon Society. 1992. *Saving wetlands: A citizen's guide for action in California.* New York: National Audubon Society.

Natural Heritage Central Databases, The Nature Conservancy, and the International Network of Natural Heritage Programs and Conservation Data Centers. November 1997. Available at http://www.consci.tnc.org/src/zoodata.htm; INTERNET.

Natural Resources Canada. 1998a. "Copper Redhorse (*Moxostoma hubbsi*)." Available at http://www-nais.ccrs.nrcan.gc.ca/schoolnet/issues/risk/fish/efish/coprdhrse.html; INTERNET.

———. 1998b. "Quebec Mining Facts." Available at http://www.nrcan.gc.ca/mms/efab/mmsd/facts/pq.htm; INTERNET.

The Nature Conservancy. 1994. "The Conservation of Biological Diversity in the Great Lakes Ecosystem: Issues and Opportunities." The Nature Conservancy, Great Lakes Program. Available at http://epaserver.ciesin.org/glreis/glnpo/docs/bio/divrpt.html; INTERNET.

———. 1996a. *Troubled waters: Protecting our aquatic heritage.* Arlington, VA: Conservation Science.

———. 1996b. "Preserve Profile: Understanding the Clinch Valley Bioreserve." Available at http://www.tnc.org; INTERNET.

———. 1996c. *Priorities for conservation: 1996 annual report card for U.S. plant and animal species.* Arlington, VA: The Nature Conservancy.

———. 1996d. *America's least wanted: Alien species invasions of U.S. ecosystems.* Arlington, VA: The Nature Conservancy.

———. 1996e. "Cheyenne Bottoms Preserve." Available at http://www.tnc.org/Kansas/bottoms.htm; INTERNET.

Nevada Division of Water Planning. 1998. "Water Words Dictionary." Available at http://www.state.nv.us/cnr/ndwp/dict-1/waterwds.htm; INTERNET.

Neves, R. J. 1991. Mollusks. Pages 251–320 in K. Terwilliger (editor), *Virginia's endangered species. Proceedings of a symposium.* Blacksburg, VA: McDonald and Woodward.

———. 1992. The status and conservation of freshwater mussels (Unionidae) in the United States. Paper pre-sented at the annual meeting of the Society for Conservation Biology, Blacksburg, VA, 30 June 1992.

Neves, R. J., A. E. Bogan, J. D. Williams, S. A. Ahlstedt, and P. W. Hartfield. 1997. Status of aquatic mollusks in the southeastern United States: A downward spiral of diversity. Pages 43–86 in G. A. Benz and D. E. Collins (editors), *Aquatic fauna in peril: The southern perspective.* Special Publication 1. Southeast Aquatic Research Institute. Decatur, GA: Lenz Design and Communications.

New Mexico Department of Game and Fish. 1997. "New Mexico Species List." Biota Information System of New Mexico, version 9/97. Available at http://www.fw.vt.edu/fishex/nmex_main/species; INTERNET.

New Mexico Ecological Services, U.S. Fish and Wildlife Service. 1997. "Rio Grande Silvery Minnow." Available at http://refuges.fws.gov/NWRSFiles/WildlifeMgmt/SpeciesAccounts/Fish/Rio_Grande_Silvery_Minnow.html; INTERNET.

Northcote, T. G., and D. Y. Atagi. 1997. Pacific salmon abundance trends in the Fraser River watershed compared with other British Columbia systems. Pages 199–219 in D. J. Stouder, P. A. Bisson, and R. J. Naiman (editors), *Pacific salmon and their ecosystems: Status and future options.* New York: Chapman Hall.

Northern River Basins Study. 1996. *Final report to the ministers.* Edmonton: Alberta Environmental Protection.

Noss, R. 1991. *Protecting habitats and biological diversity.* Part 1, *Guidelines for regional reserve systems.* New York: National Audubon Society.

Noss, R. F., and A. Y. Cooperider. 1994. *Saving nature's legacy: Protecting and restoring biodiversity.* Washington, DC: Defenders of Wildlife and Island Press.

Noss, R. F., and R. L. Peters. 1995. *Endangered ecosystems. A status report on America's vanishing habitat and wildlife.* Washington, DC: Defenders of Wildlife.

Nueces River Authority. 1997. "Basin Highlights Report Nueces River Authority." Available at http://robin.tamucc.edu/~nra/nrbrpt.html; INTERNET.

Nyman, L. 1991. *Conservation of freshwater fish: Protection of biodiversity and genetic variability in aquatic ecosystems.* Göteborg, Sweden: SWEDMAR.

O'Keeffe, J. H., D. B. Danilewitz, and J. A. Bradshaw. 1987. An "expert system" approach to the assessment of the conservation status of rivers. *Biological Conservation* 40: 69–84.

Obregón-Barboza, H., S. Contreras-Balderas, and M. L. Lozano-Vilano. 1994. The fishes of northern and central Veracruz, Mexico. *Hydrobiologia* 286: 79–85.

Oesch, R. D. 1984. *Missouri naiades: A guide to the mussels of Missouri.* Jefferson City, MO: Missouri Department of Conservation.

Olson, D. M., and E. Dinerstein. 1998. The Global 200: A representation approach to conserving the Earth's most biologically valuable ecoregions. *Conservation Biology* 12: 502–515.

Olson, D. M., E. Dinerstein, G. Cintrón, and P. Iolster (editors). 1996. *A conservation assessment of mangrove ecosystems of Latin America and the Caribbean. Report from a workshop.* Washington, DC: World Wildlife Fund.

Olson, D. M., B. Chernoff, G. Burgess, I. Davidson, P. Canevari, E. Dinerstein, G. Castro, V. Morisset, R. Abell, and E. Toledo (editors). 1997. *Freshwater biodiversity of Latin America and the Caribbean: A conservation assessment. Proceedings of a workshop.* Washington, DC: World Wildlife Fund.

Olson, D. M., I. Itoua, E. Underwood, and E. Dinerstein. In prep. *A conservation assessment of the biodiversity of Africa.* Conservation Science Program, Washington, DC: World Wildlife Fund–United States.

Omernik, J. M. 1987. Ecoregions of the conterminous United States. *Annals of the Association of American Geographers* 77: 118–125.

Ono, R. D., J. D. Williams, and A. Wagner. 1983. *Vanishing fishes of North America.* Washington, DC: Stone Wall Press.

Page, L. M. 1983. *Handbook of darters.* Neptune City, NJ: Tropical Fish Hobbyist Publications.

———. 1985. The crayfishes and shrimps (Decapoda) of Illinois. *Illinois Natural History Survey Bulletin* 33: 335–448.

Page, L. M., and B. M. Burr. 1991. *A field guide to freshwater fishes, North America north of Mexico.* Boston, MA: Houghton Mifflin Company.

Page, L. M., P. A. Ceas, D. L. Swofford, and D. G. Buth. 1992. Evolutionary relationships of the *Etheostoma squamiceps* complex (Percidae; subgenus *Catonotus*) with descriptions of five new species. *Copeia* 3: 615–646.

Parmalee, P. L., and M. H. Hughes. 1993. Freshwater mussels (Mollusca: Pelecypoda: Unionidae) of the Tellico Lake: Twelve years after impoundment of the Little Tennessee River. *Annals of Carnegie Museum* 62: 81–93.

Pearse, P. H., and F. Quinn. 1996. Recent developments in federal water policy, two steps back. *Canadian Water Resources Journal* 21(4): 329–340.

Peck, S. 1998. *Planning for biodiversity: Issues and examples.* Washington, DC: Island Press.

Pinkava, D. J. 1984. Vegetation and flora of the Bolson of Cuatro Ciénegas Region, Coahuila, Mexico: 4. Summary, endemism, and corrected catalogue. *Journal Arizona-Nevada Academy Sciences* 19: 23–47.

Pister, E. P. 1981. The conservation of desert fishes. Pages 411–445 in R. J. Naiman and D. L. Soltz (editors), *Fishes in North American deserts.* New York: John Wiley.

———. 1990. Desert fishes: An interdisciplinary approach to endangered species conservation in North America. *Journal of Fish Biology* 37: 183–187.

———. 1991. The Desert Fishes Council: Catalyst for change. Pages 55–68 in W. L. Minckley and J. E. Deacon (editors), *Battle against extinction: Native fish management in the American West.* Tucson: University of Arizona Press.

Poulson, T. L. 1992. The Mammoth Cave ecosystem. Pages 569–611 in A. I. Camacho (editor), *The natural history of biospeleology.* Madrid: Monagrafías Museo Nacional de Ciencias Naturales.

Poulson, T. L., and E. B. White. 1969. The cave environment. *Science* 165: 971–981.

Propst, D. L., and K. R. Bestgen. 1991. Habitat and biology of the loach minnow, *Tiaroga cobitis*, in New Mexico. *Copeia* 1991: 29–38.

Propst, D. L., and J. A. Stefferud. 1994. Distribution and status of the Chihuahua chub (Teleostei: Cyprinidae: *Gila nigrescens*), with notes on its ecology and associated species. *The Southwestern Naturalist* 39(3): 224–234.

Propst, D. L., M. D. Hatch, and J. P. Hubbard. 1985. White Sands pupfish (*Cyprinodon tularosa*). *Handbook of species endangered in New Mexico.* FISH/CT/CY/TU: 1–2. Santa Fe: New Mexico Department of Game and Fish.

Propst, D. L., K. R. Bestgen, and C. W. Painter. 1986. *Distribution, status, biology, and conservation of the spikedace* (Meda fulgida) *in New Mexico.* Endangered Species Report No. 15. Albuquerque, NM: U.S. Fish and Wildlife Service, Region 2.

———. 1988. *Distribution, status, biology, and conservation of the loach minnow* (Tiaroga cobitis) *in New Mexico.* Endangered Species Report No. 17. Albuquerque, NM: U.S. Fish and Wildlife Service, Region 2.

Propst, D. L., J. A. Stefferud, and P. R. Turner. 1992. Conservation and status of Gila trout (*Oncorhynchus gilae*). *The Southwestern Naturalist* 37: 117–125.

Proudlove, G. S. 1997a. A synopsis of the hypogean fishes of the world. In *Proceedings of the 12th International Congress of Speleology*, La Chaux-de-Fonds, Switzerland, 10–17 August 1997. Volume 3, Symposium 9: Biospeleology.

———. 1997b. The conservation status of hypogean fishes. Pages 77–81 in I. D. Sasowsky, D. W. Fong, and E. L. White (editors), *Conservation and protection of the biota of karst.* Karst Waters Institute Special Publication 3. Charles Town, WV: Karst Waters Institute.

Quammen, D. 1996. *The song of the dodo: Island biogeography in an age of extinction.* New York: Touchstone.

Rabeni, C. F. 1996. Prairie legacies—fish and aquatic resources. Pages 111–124 in F. B Samson and F. L. Knopf (editors), *Prairie conservation: Preserving North America's most endangered ecosystem.* Washington, DC: Island Press.

Rabinowitz, D., S. Cairns, and T. Dillon. 1986. Seven forms of rarity and their frequency in the flora of the British Isles. Pages 182–204 in M. E. Soulé (editor), *Conservation biology: The science of scarcity and diversity.* Sunderland, MA: Sinauer Associates.

Reddell, J. R. 1982. A checklist of the cave fauna of México. *Association of Mexican Cave Studies Bulletin* 8: 249–284.

———. 1994. The cave fauna of Texas with special reference to the western Edwards Plateau. Pages 31–49 in W. R. Elliot and G. Veni (editors), *The caves and karst of Texas.* Huntsville, AL: National Speleological Society.

Reisenbichler, R. R. 1997. Genetic factors contributing to declines of anadromous salmonids in the Pacific Northwest. Pages 223–244 in D. J. Stouder, P. A. Bisson, and R. J. Naiman (editors), *Pacific salmon and their ecosystems: Status and future options.* New York: Chapman and Hall.

Ricciardi, A., F. G. Whoriskey, and J. B. Rasmussen. 1995. Predicting the intensity and impact of *Dreissena* infestation on native unionid bivalves from *Dreissena* field density. *Canadian Journal of Fisheries and Aquatic Sciences* 52: 1449–1461.

Richter, B. D., D. P. Braun, M. A. Mendelson, and L. W. Master. 1997. Threats to imperiled freshwater fauna. *Conservation Biology* 11: 1081–1093.

Ricketts, T., E. Dinerstein, D. M. Olson, C. J. Loucks, W. M. Eichbaum, D. A. DellaSala, K. C. Kavanagh, P. Hedao, P. T. Hurley, K. M. Carney, R. A. Abell, and S. Walters. 1999a. *Terrestrial ecoregions of North America: A conservation assessment.* Washington, DC: Island Press.

Ricketts, T. H., E. Dinerstein, D. M. Olson, and C. Loucks. 1999b. Who's where in North America? Patterns of species richness and the utility of indicator taxa for conservation. *BioScience* 49: 369–382.

Robins, C. R., R. M. Bailey, C. E. Bond, J. R. Brooker, E. A. Lachner, R. N. Lea, and W. B. Scott. 1991. *Common and scientific names of fishes from the United States and Canada.* Bethesda, MD: American Fisheries Society.

Robison, H. W. 1986. Zoogeographic implications of the Mississippi River Basin. Pages 267–285 in C. H. Hocutt and E. O. Wiley (editors), *The zoogeography of North American freshwater fishes.* New York: John Wiley.

Robison, H. W., and T. M. Buchanan. 1988. *Fishes of Arkansas.* Fayetteville: The University of Arkansas Press.

Rohde, F. C., R. G. Arndt, D. G. Lindquist, and J. F. Parnell. 1994. *Freshwater fishes of the Carolinas, Virginia, Maryland, and Delaware.* Chapel Hill: University of North Carolina Press.

Rossman, D. A., N. B. Ford, and R. A. Seigel. 1996. *The garter snakes: Evolution and ecology.* Norman: University of Oklahoma Press.

Ryon, M. G. 1986. The life history and ecology of *Etheostoma trisella* (Pisces: Percidae). *American Midland Naturalist* 115: 73–86.

Sada, D. W., H. B. Britten, and P. F. Brussard. 1995. Desert aquatic ecosystems and the genetic and morphological diversity of Death Valley System speckled dace. *American Fisheries Society Symposium* 17: 350–359.

Saskatchewan Wetland Conservation Corporation (SWCC). 1997. "Saskatchewan Wetland Policy." Available at http://www.wetland.sk.ca/swccpolicy/policy.htm; INTERNET.

Schindler, D. W. 1998. Sustaining aquatic ecosystems in boreal regions. *Conservation Ecology* [online] 2(2):18. Available at http://www.consecol.org/vol2/iss2/art18/; INTERNET.

Schmidt, K. F., and D. W. Owens. 1944. Amphibians and reptiles of northern Coahuila, Mexico. *Zool. Ser., Field Museum Natural History* 29(6): 97–115.

Schmidt, R. E. 1986. Zoogeography of the northern Appalachians. Pages 137–159 in C. H. Hocutt and E. O. Wiley (editors), *The zoogeography of North American freshwater fishes.* New York: John Wiley.

Schmitz, D. A. 1994. The ecological impact of nonindigenous plants in Florida. Pages 10–28 in D. A. Schmitz and T. C. Brown (editors), *An assessment of invasive nonindigenous species in Florida's public lands.* Tallahassee: Florida Department of Environmental Protection.

Schoenherr, A. A. 1995. *A natural history of California.* Berkeley: University of California Press.

Scott, W. B., and E. J. Crossman. 1973. *Freshwater fishes of Canada.* Bulletin 184. Ottawa, Ontario: Fisheries Research Board of Canada.

Scudday, J. F. 1977. Some recent changes in the herpetofauna of the northern Chihuahuan Desert. Pages 513–522 in R. H. Wauer and D. H. Riskind (editors), *Transactions of the symposium on the biological resources of the Chihuahuan Desert Region, United States and Mexico,* Sul Ross State University, 1974. National Park Service Transactions and Proceedings Series No. 3. Washington, DC: U.S. Government Printing Office.

Scudder, G. G. E. 1996. *Terrestrial and freshwater invertebrates of British Columbia: Priorities for inventory and descriptive research.* British Columbia Working Paper 09/1996. Victoria, BC: Research Bureau, British Columbia Ministry of Forestry, and Wildlife Bureau, British Columbia Ministry of the Environment, Lands and Parks.

Sharp, J. M. Jr. 1990. Stratigraphic, geomorphic and structural controls of the Edwards aquifer, Texas, U.S.A. Pages 67–82 in E. S. Simpson and J. M. Sharp Jr. (editors), *Selected papers on hydrogeology.* Heise, Hannover, Germany: International Association of Hydrogeologists.

Sheehan, R. J. and J. L. Rasmussen. 1993. Large rivers. Pages 445–468 in C. C. Kohler and W. A. Hubert (editors), *Inland fisheries management in North America.* Bethesda, MD: American Fisheries Society.

Shepard, W. D. 1993. Desert springs—both rare and endangered. *Aquatic Conservation: Marine and Freshwater Ecosystems* 3: 351–359.

Sierra Club. 1998. "DuPont's Proposed Strip Mine Threatens Okefenokee National Wildlife Refuge." Available at http://tamalpais.sierraclub.org/wetlands/dupont.html; INTERNET.

Sigler, J. W., and W. F. Sigler. 1994. Fishes of the Great Basin and the Colorado Plateau: Past and present forms. Pages 163–208 in K. T. Harper, L. L. St. Clair, K. H. Thornes, and W. M. Hess (editors), *Natural history of the Colorado Plateau and Great Basin.* Niwot: University of Colorado Press.

Silveira, J. G. 1998. Avian uses of vernal pools and implications for conservation practices. Pages 92–106 in C. W. Witham, E. T. Bauder, D. Belk, W. R. Ferren Jr., and R. Ornduff (editors), *Ecology, conservation, and management of vernal pool ecosystems—Proceedings from a 1996 conference.* Sacramento: California Native Plant Society.

Slattery, R. N., J. T. Patton, and D. S. Brown. 1997. "Recharge to and Discharge from the Edwards Aquifer in the San Antonio Area, Texas, 1997." Available at http://tx.usgs.gov/reports/district/98/01/; INTERNET.

Smith, A. R., and G. Veni. 1994. Karst regions of Texas. Pages 7–12 in W. R. Elliott and G. Veni (editors), *The caves and karst of Texas.* Huntsville, AL: National Speleological Society.

Smith, H. M., and E. H. Taylor. 1950. *An annotated checklist and key to the reptiles of Mexico exclusive of the snakes.* Smithsonian Institution, United States National Museum Bulletin 199. Washington, DC: United States Government Printing Office.

——. 1966. *Herpetology of Mexico: Annotated checklists and keys to the amphibians and reptiles.* Ashton, MD: Eric Lundberg.

Smith, M. L., and R. R. Miller. 1986. The evolution of the Rio Grande Basin as inferred from its fish fauna. Pages 457–485 in C. H. Hocutt and E. O. Wiley (editors), *The zoogeography of North American freshwater fishes.* New York: John Wiley.

SOE. 1996. *The state of Canada's environment.* Ottawa, Ontario: Government of Canada.

Soltz, D. L., and R. J. Naiman. 1978. *The natural history of native fishes in the Death Valley system. Sci. Set.* 30: 1–76.

Sonoran Arthropod Studies Institute (SASI). 1998. "Crustaceans." Available at http://www.azstarnet.com/~sasi/arthnatr/arthzoo/crusts.htm; INTERNET.

Southeastern Fishes Council. "1997 Report for Region I–Northeast." Available at http://www.flmnh.ufl.edu/fish/organizations/SFC/regionalreports/R1Northeast97.htm; INTERNET.

Stager, J. C., and L. B. Cahoon. 1987. The age and trophic history of Lake Waccamaw, North Carolina. *The Journal of the Elisha Mitchell Scientific Society* 103: 1–13.

Stansbery, D. H. 1970. Eastern freshwater mollusks (1): The Mississippi and St. Lawrence River systems. *Malacologia* 10(1): 9–22.

Starnes, W. C., and D. A. Etnier. 1986. Drainage evolution and fish biogeography of the Tennessee and Cumberland rivers drainage realm. Pages 325–361 in C. H. Hocutt and E. O. Wiley (editors), *The zoogeography of North American freshwater fishes.* New York: John Wiley.

State Goals and Indicators Project. 1998. "Tennessee State of the Environment: Wildlife". U.S. Environmental Protection Agency and the Florida Center for Public Management. Available at http://www.fsu.edu/~cpm/segip/states/TN94/wild.html; INTERNET.

Stebbins, R. C. 1985. *A field guide to western reptiles and amphibians.* Boston, MA: Houghton Mifflin.

Stein, B. A., L. L. Master, L. E. Morse, L. S. Kutner, and M. Morrison. 1995. Status of U.S. species: Setting conservation priorities. Pages 399–401 in E. T. LaRoe, G. S. Farris, C. E. Puckett, P. D. Doran, and M. J. Mac (editors), *Our living resources: A report to the nation on the distribution, abundance, and health of U.S. plants, animals, and ecosystems.* Washington, DC: U.S. Department of the Interior, National Biological Service.

Steinhart, P. 1990. *California's wild heritage: Threatened and endangered animals in the Golden State.* Sacramento, CA: Sierra Club Books.

Stolzenburg, W. 1997. "Sweet Home Alabama." *The Nature Conservancy Magazine*, September/October. Available at http://www.tnc.org/news/magazine/septoct97/interior/eco.html; INTERNET.

Strain, P., and F. Engle. 1993. *Looking at Earth.* Atlanta, GA: Turner Publishing.

Stranahan, S. Q. 1995. *Susquehanna: River of dreams.* Baltimore, MD: Johns Hopkins University Press.

Streever, W. J. 1992. Report of a cave fauna kill at Peacock Springs cave system, Suwannee County, Florida. *Florida Scientist* 55(2):125–128.

Sumner, F. B., and M. C. Sargent. 1940. Some observations on the physiology of warm spring fishes. *Ecology* 21: 45–51.

Suttkus, R. D., and D. A. Etnier. 1991. *Etheostoma tallapossae* and *E. brevirostrum,* two new darters, subgenus *Ulocentra,* from the Alabama River drainage. *Tulane Studies in Zoology and Botany* 28: 1–24.

Suttkus, R. D., B. A. Thompson, and H. L. Bart Jr. 1994. Two new darters, *Percina (Cottogaster),* from the southeastern United States, with a review of the subgenus. *Occasional Papers Tulane University Museum of Natural History* 4: 1–46.

Swanson, G. A., M. I. Meyer, and J. R. Serie. 1974. Feeding ecology of breeding blue-winged teals. *Journal of Wildlife Management* 38: 396–407.

Swift, C. C., C. R. Gilbert, S.A. Bortone, G. H. Burgess, and R. W. Yerger. 1986. Zoogeography of the freshwater fishes of the southeastern United States: Savannah River to Lake Pontchartrain. Pages 213–265 in C. H. Hocutt and E. O. Wiley (editors), *The zoogeography of North American freshwater fishes.* New York: John Wiley.

Tagami, T. 1998. "A Lake Lands in Limbo." *Lexington Herald.* Available at http://www.kentuckyconnect.com/heraldleader/news/011998/fn1lake.html; INTERNET.

Tansley, A. G. 1935. The use and abuse of vegetational concepts and terms. *Ecology* 16: 284–307.

Taylor, C. A. 1997. Distributional data for North American crayfish, supplied under contract.

Taylor, C. A., and M. Redmer. 1996. Dispersal of the crayfish *Orconectes rusticus* in Illinois, with notes on species displacement and habitat preference. *Journal of Crustacean Biology* 16(3): 547–551.

Taylor, C. A., M. L. Warren, Jr., J. F. Fitzpatrick Jr., H. H. Hobbs III, R. F. Jezerinac, W. L. Pflieger, and H. W. Robison. 1996. Conservation status of crayfishes of the United States and Canada. *Fisheries* 21(4): 25–38.

Taylor, D. W. 1966. A remarkable snail fauna from Coahuila, Mexico. *Zool. Ser., Field Museum Natural History* 29(6): 97–115.

Tercafs, R. 1992. The protection of the subterranean environment: Conservation principles and management tools. Pages 481–524 in A. I. Camacho (editor), *The natural history of biospeleology.* Madrid: Monagrafias Museo Nacional de Ciencias Naturals.

Texas Natural Resource Conservation Commission. 1997a. "Summary of Water Quality in Texas by River Basin." Available at http://www.tnrcc.state.tx.us/admin/topdoc/sfr/046/summary_wq.html; INTERNET.

——. 1997b. "1996 Identified Water Quality Issues of the San Antonio-Nueces Coastal Basin." Available at http://www.tnrcc.state.tx.us/water/quality/data/wmt/sannue_assmt.html; INTERNET.

——. 1997c. "Identified Water Quality Issues of the San Antonio Lower Nueces-Frio River Subwatershed." Available at http://www.tnrcc.state.tx.us/water/quality/data/wmt/nra_assmt.html; INTERNET.

Thompson, B. A. 1995. Percina austroperca: *A new species of logperch (Percidae, subgenus* Percina*) from the Choctawhatchee and Escambia rivers in Alabama and Florida.* Occasional Papers of the Museum of Natural

Science No. 69. Baton Rouge: Louisiana State University.

Underhill, J.C. 1986. The fish fauna of the Laurentian Great Lakes, the St. Lawrence lowlands, Newfoundland and Labrador. Pages 105–136 in C. H. Hocutt and E. O. Wiley (editors), *The zoogeography of North American freshwater fishes*. New York: John Wiley.

University of Michigan. 1982. *Freshwater snails (Mollusca: Gastropoda) of North America*. Cincinnati, OH: Environmental Monitoring and Support Laboratory.

————. 1995. "An Electronic Distributional Atlas of Mexican Fishes." Web-based data from University of Michigan Fish Collection. Available at http://muse.bio.cornell.edu/mexico/; INTERNET.

U.S. Army Corps of Engineers. 1998. "Zebra mussel information." Ohio River Regional Headquarters. Available at http://www.ord-wc.usace.army.mil/zebra.html; INTERNET.

U.S. Congress, Office of Technology Assessment. 1993. *Harmful non-indigenous species in the United States*. Washington, DC: U.S. Government Printing Office.

U.S. Department of Agriculture. 1995. "National Resources Inventory Glossary." Available at http://www.ftc.nrcs.usda.gov/doc/nri/all_toc.html; INTERNET.

U.S. Environmental Protection Agency. 1997. *Index of watershed indicators*. Draft. Washington, DC: United States Environmental Protection Agency.

————. 1998. "Lakewide Management Plan for Lake Ontario." U.S. EPA Region 2 and GLIN. Available at http://www.epa.gov/glnpo/lakeont/; INTERNET.

————. 1995. *A Phase I Inventory of Current EPA Efforts to Protect Ecosystems*. EPA841-S-95-001. Washington, DC: Office of Water (4503F), United States Environmental Protection Agency.

U.S. Environmental Protection Agency. 1999. "Great Lakes Areas of Concern." Available at http://www.epa.gov/glnpo/aoc/; INTERNET.

U.S. Fish and Wildlife Service. 1986. *Chihuahua chub recovery plan*. Albuquerque, NM: U.S. Fish and Wildlife Service, Region 2.

————. 1990. *Endangered and Threatened Species of the Southeastern United States (the red book)*. Atlanta GA: U.S. Fish and Wildlife Service, Region 4.

————, Division of Endangered Species. 1993. Endangered and threatened wildlife and plants: Endangered status for eight freshwater mussels and threatened status for three freshwater mussels in the Mobile River drainage. FR Doc. 93–6162. *Federal Register*, March 17, 1993. Available at http://www.fws.gov/r9endspp/r/fr93495.html; INTERNET.

————. 1994. "Lake Erie water snake." Available at http://eelink.net/EndSpp/lake.html; INTERNET.

————. 1995. *Strategic plan for conservation of Fish and Wildlife Service trust resources in the Ohio River Valley ecosystem*. Cookville, TN: Cookville Ecological Services Field Office, USFWS Regions 5, 4, and 3, Ohio River Valley Ecosystem Team.

————. 1998. Endangered and threatened wildlife and plants: 90 day finding for a petition to list the robust

blind salamander, widemouth blindcat, and toothless blindcat. *Federal Register* 63: 48166–48167.

————. Division of Endangered Species. 1998. "Species accounts: Slackwater Darter." Available at http://www.fws.gov/r9endspp/i/e/sae1a.html; INTERNET.

U.S. Geological Survey. 1996. "First Detailed 'Report Card' on Mississippi River Shows Movement of Contaminants." Available at http://www.usgs.gov/public/press/public_affairs/press_releases/pr106m.html; INTERNET.

U.S. Geological Survey. 1999. *Ecological status and trends of the Upper Mississippi River System 1998: A report of the Long Term Resource Monitoring Program*. LaCrosse, WI: LTRMP 99-T001.U.S. Geological Survey, Upper Midwest Environmental Sciences Center.

Vazquez Gutierrez, F. 1997. Desarrollo urbano industrial de las cuencas en México. Pages 34–39 in lecture notes for "Curso de Limnologia Aplicada." Mexico City: Instituto de Ciencias Marinas del Mar y Limnologia, Universidad Autónoma Nacional de México (UNAM).

Veni, G. 1994. Hydrogeology and evolution of caves and karst in the southwestern Edwards Plateau, Texas. Pages 13–30 in W. R. Elliott and G. Veni (editors). *The caves and karst of Texas*. Huntsville, AL: National Speleological Society,

Vinyard, G. L. 1996. Distribution of a thermal endemic minnow, the desert dace (*Eremichthys acros*), and observations of impacts of water diversion on its population. *Great Basin Naturalist* 56(4): 360–368.

Vitousek, P. M., L. L. Loope, and C. P. Stone. 1987. Introduced species in Hawaii: Biological effects and opportunities for ecological research. *Trends in Ecology and Evolution* 2: 224–227.

Wake, D. B., and J. F. Lynch. 1976. The distribution, ecology, and evolutionary history of Plethodontid salamanders in tropical America. *Natural History Museum of Los Angeles County Science Bulletin* 25. Los Angeles, CA: Natural History Museum of Los Anglees County.

Wallace, J. B., J. R. Webster, and R. L. Lowe. 1992. High-gradient streams of the Appalachians. Pages 133–191 in C. T. Hackney, S. M. Adams, and W. H. Martin (editors). *Biodiversity of the southeastern United States: Aquatic communities*. New York: John Wiley.

Walsh, S. J. and Gilbert, C. R. 1995. New species of troglobitic catfish of the genus *Prietella* (Siluriformes: Ictalluridae) from northeastern Mexico. *Copeia* 1995: 850–860.

Walsh, S. J., N. M. Burkhead, and J. D. Williams. 1995. Conservation status of southeastern freshwater fishes. Pages 144–147 in E. T. LaRoe (editor), *Our living resources 1995: A report to the nation on the distribution, abundance, and health of U. S. plants, animals, and ecosystems*. Washington, DC: U.S. Department of Interior, National Biological Service.

Warren, M. L. Jr. and B. M. Burr. 1994. Status of freshwater fishes of the United States: overview of an imperiled fauna. *Fisheries* 19: 6–18.

Warren, M. L. Jr., P. L. Angermeier, B. M. Burr, and W. R. Haag. 1997. Decline of a diverse fish fauna: Patterns of imperilment and protection in the southeastern United States. Pages 105–164 in G. Benz and

D. E. Collins (editors), *Aquatic fauna in peril: The southern perspective.* Special Publication 1. Southeast Aquatic Research Institute. Decatur, GA: Lenz Design and Communications.

White, W. B., D. C. Culver, J. S. Herman, T. C. Kane and J. E. Mylroie. 1995. Karst lands. *American Scientist* 83(5): 450–459.

Wikramanayake, E., E. Dinerstein, P. Hedao, D. M. Olson, L. Horowitz, and P. Hurley. In press. *A conservation assessment of the terrestrial ecoregions of the Indo-Pacific region.* Washington, DC: World Wildlife Fund.

Wilcove, D. S., M. J. Bean, R. Bonnie, and M. McMillan. 1996. *Rebuilding the ark: Toward a more effective Endangered Species Act for private land.* Washington, DC: Environmental Defense Fund.

The Wilderness Society. 1998. "DuPont Mine Endangers Okefenokee." Available at http://www.wilderness.org/standbylands/refuges/dupont.htm; INTERNET.

The Wilderness Society. 1999. *15 most endangered wild lands 1999.* Washington, DC: The Wilderness Society.

Williams, J. D., and G. H. Clemmer. 1991. *Scaphirhynchus suttkusi,* a new sturgeon (Pisces: Acipenseridae) from the Mobile Basin of Alabama and Mississippi. *Bulletin Alabama Museum of Natural History* 10: 17–31.

Williams, J. D., S. L. H. Fuller, and R. Grace. 1992. Effects of impoundments on freshwater mussels (Mollusca: Bivalvia: Unionidae) in the main channel of the Black Warrior and Tombigbee rivers in western Alabama. *Bulletin of the Alabama Museum of Natural History* 13: 1–10.

Williams, J. D., M. L. Warren Jr., K.S. Cummings, J. L. Harris, and R. J. Neves. 1993. Conservation status of freshwater mussels of the United States and Canada. *Fisheries* 18(9): 6–22.

Williams, J. E. 1995a. Threatened fishes of the world: *Catostomus warnerensis* Snyder, 1908 (Catostomidae). *Environmental Biology of Fishes* 44 (4): 346.

———1995b. Threatened fishes of the world: *Gila boraxobius* Williams and Bond, 1980 (Cyprinidae). *Environmental Biology of Fishes* 43(3): 294.

Williams, J. E., D. B. Bowman, J. E. Brooks, A. A. Echelle, R. J. Edwards, D. A. Hendrickson, and J. J. Landye. 1985. Endangered aquatic ecosytems in North American deserts with a list of vanishing fishes of the region. *Journal of Arizona-Nevada Academy of Sciences* 20:1–62.

Williams, J. E., J. E. Johnson, D. A. Hendrickson, S. Contreras-Balderas, J. D. Williams, M. Navarro-Mendoza, D. E. McAllister, and J. E. Deacon. 1989. Fishes of North America endangered, threatened, or of special concern: 1989. *Fisheries* 14(6): 2–20.

Williams, S. C. 1968. Scorpions from northern Mexico: Five species of *Vejovis* from Coahuila, Mexico. *Occasional Papers of the California Academy of Sciences* 68: 1–24.

Wilson, E. O. 1992. *The diversity of life.* Cambridge, MA: Harvard University Press, Belknap Press.

——— (editor). 1988. *Biodiversity.* Washington, DC: National Academy Press.

Wood, R. M., and R. L. Mayden. 1993. Systematics of the *Etheostoma jordani* species group (Teleostei: Percidae), with descriptions of three new species. *Bulletin Alabama Museum of Natural History* 16: 31–46.

World Wildlife Fund and National Audubon Society. 1996. *Restoring the river of grass.* Washington, DC, and Miami, FL: World Wildlife Fund and National Audubon Society.

Wydoski, R. S. 1980. Potential impacts of alterations in streamflow and water quality on fish and macroinvertebrates in the Upper Colorado River basin. Pages 77–147 in W. O. Spofford Jr., A. L. Parker, and A. V. Kneese (editors), *Energy development in the Southwest: Problems of water, fish and wildlife in the Upper Colorado River Basin.* Vol. 2, *Research paper R-18.* Resources for the Future, Washington, DC: Resources for the Future.

Wynne-Edwards, V. C. 1952. *Freshwater vertebrates of the Arctic and Subarctic.* Bulletin No. 94. Ottawa, Ontario: Fisheries Research Board of Canada.

Zhao, E. M., and K. Adler. 1993. *Herpetology of China.* Contributions to Herpetology No. 10. Oxford, OH: Society for the Study of Amphibians and Reptiles.

AUTHORS

Robin A. Abell, M.S.
Freshwater Conservation Biologist
Conservation Science Program
World Wildlife Fund–United States

David M. Olson, Ph.D.
Senior Scientist
Conservation Science Program
World Wildlife Fund–United States

Eric Dinerstein, Ph.D.
Chief Scientist and Director
Conservation Science Program
World Wildlife Fund–United States

Patrick T. Hurley
Research Assistant
Conservation Science Program
World Wildlife Fund–United States

James T. Diggs, M.S.
Conservation Analyst
Conservation Science Program
World Wildlife Fund–United States

William Eichbaum, L.L.B.
Vice President
U.S. Program
World Wildlife Fund–United States

Steven Walters, M.S.
GIS Specialist
Conservation Science Program
World Wildlife Fund–United States

Wesley Wettengel, M.C.P.
GIS Specialist
Conservation Science Program
World Wildlife Fund–United States

Tom Allnutt, M.S.
Conservation Analyst/GIS Specialist
Conservation Science Program
World Wildlife Fund–United States

Colby J. Loucks, M.E.M.
Conservation Analysist/GIS Specialist
Conservation Science Program
World Wildlife Fund–United States

Prashant Hedao
GIS Specialist
Conservation Science Program
World Wildlife Fund–United States

CONTRIBUTORS

Richard Biggins
U.S. Fish and Wildlife Service
Ecological Services
Asheville Field Office
160 Zillicola Street
Asheville, NC 28801, USA

Noel M. Burkhead
Florida Caribbean Science Center
Biological Resources Division, USGS
7920 NW 71st Street
Gainesville, FL 32653-3071, USA

Brooks M. Burr
Zoology Department
Mail Code 6501
Southern Illinois University at Carbondale
Carbondale, IL 62901-6501, USA

Ronald Cicerello
Kentucky State Nature Preserves Commission
801 Schenkel Lane
Frankfort, KY 40601, USA

Salvador Contreras-Balderas
Bioconservación, A.C.
Loma Larga 2524, Col. Obispado
Monterrey, N.L., Mexico

Lynda Corkum
Department of Biological Sciences
University of Windsor
Windsor, Ontario N9B 3P4, Canada

Hector Espinoza
Instituto de Biología
U.N.A.M.
Ciudad Universitaria
04510 Mexico D.F., Mexico

Andrew Fahlund
Hydropower Reform Coalition
1025 Vermont Avenue NW
Suite 720
Washington, DC 20005, USA

Terrence Frest
Deixis
6842 24th Avenue NE
Seattle, WA 98115, USA

Christopher Frissell
Flathead Lake Biological Station
The University of Montana
311 BioStation Lane
Polson, MT 59860, USA

Dean A. Hendrickson
Texas Natural History Collections
LSF R4000
University of Texas, Austin
Austin, TX 78712, USA

Howard L. Jelks
Florida Caribbean Science Center
Biological Resources Division, USGS
7920 NW 71st Street
Gainesville, FL 32653-3071, USA

Jean Krejca
Department of Zoology
University of Texas at Austin
Austin, TX 78712, USA

James B. Ladonski
Department of Zoology
Southern Illinois University
Carbondale, IL 62901, USA

Tom Maloney
Connecticut River Watershed Council
1 Ferry Street
Easthampton, MA 01027, USA

Lawrence L. Master
Chief Zoologist
The Nature Conservancy
Eastern Conservation Science Division
201 Devonshire Street, 5th floor
Boston, MA 02110, USA

Don McAllister
Canadian Museum of Nature
Canadian Centre for Biodiversity
2086 Walkley Road
P.O. Box 3443, Station D
Ottawa, Ontario K1P 6P4, Canada
AND
Ocean Voice International
Box 37026
Ottawa, Ontario K1V 0W0, Canada

Gary K. Meffe
Department of Wildlife Ecology and Conservation
Newins-Ziegler Hall
University of Florida
Gainesville, FL 32611, USA

Patricia Melhop
New Mexico Natural Heritage Program
851 University Boulevard
Department of Biology
Albuquerque, NM 87131-1091, USA

W.L. Minckley
Department of Biology
Arizona State University
Tempe, AZ 85287-1501, USA

Peter B. Moyle
University of California–Davis
Department of Wildlife & Fisheries Biology
Davis, CA 95616, USA

John Pittenger
New Mexico Department of Game and Fish
Villagra Building
P.O. Box 25112
Santa Fe, NM 87504, USA

David Propst
New Mexico Department of Game and Fish
Villagra Building
P.O. Box 25112
Santa Fe, NM 87504, USA

Mark H. Stolt
Department of Natural Resources Science
University of Rhode Island
Kingston, RI 02881, USA

David L. Strayer
Institute for Ecosystem Studies
Box AB
Millbrook, NY 12545, USA

Christopher A. Taylor
Illinois Natural History Survey
Center for Biodiversity
172 Natural Resources Building
607 East Peabody Drive
Champaign, IL 61820, USA

Craig Tenbrink
Missouri Department of Conservation
Fish and Wildlife Research Center
1110 South College Avenue
Columbia, MO 65201, USA

Peter J. Unmack
Department of Biology
Arizona State University
Tempe, AZ 85287-1501, USA

Barbara Vlamis
Butte Environmental Council
116 W. Second Street, Suite 3
Chico, CA 95928, USA

Steven J. Walsh
Florida Caribbean Science Center
Biological Resources Division, USGS
7920 NW 71st Street
Gainesville, FL 32653-3071, USA

G. Thomas Watters
Ohio Biological Survey and Aquatic Ecology
   Laboratory
Ohio State University
1315 Kinnear Road
Columbus, OH 43212, USA

James D. Williams
Florida Caribbean Science Center
Biological Resources Division, USGS
7920 NW 71st Street
Gainesville, FL 3653-3071, USA

Freshwater Ecoregions of North America

Acid-mine runoff, 210. *See also* Mining
Acid precipitation, 231, 232, 236
ACT compact (water-allocation plan), 216
Agricultural pollution, 203, 207, 219, 220
    from agrochemicals, 84–85, 200, 205–6
    sedimentation and, 196, 210, 258, 262
Agriculture
    in Canada, 84–85
    eutrophication from, 225, 227, 233
    hog farming, 225, 226
    spring habitat and, 52
    threat from, 202, 241, 257
    water quality and, 250, 253, 259
    water use by, 185, 191
    See also Irrigation
Alabama Natural Heritage Program, 199, 215, 217
Alabama Rivers Alliance, 217, 220, 221
Alabama Wilderness Association, 217
Alaska Boreal Forest Council, 244
Alberta Wilderness Association, 201, 239, 240
Alien species, in Canada, 84. *See also* Exotic species
Altacal Audubon, 54
Alvord chub (*Gila alvordensis*), 176, 177
American Fisheries Society, 74
American Rivers, 174, 183
    Columbia River basin, 167, 168
    large temperate river MHTs, 181, 186, 199
    Pacific coastal MHTs, 166, 169, 170
    Southeast region, 217, 220, 221, 226
    temperate headwaters and lakes MHTs, 203, 204, 210, 211
Amphibians, 1, 121
    deformities in, 84–85
    endemic, 225, 258
    species richness in, 152
    See also Herpetofauna
Amphipods, aquatic, 189–90. *See also* Snails
Anacostia Watershed Society, 227
Anadromous fish species, 245
    Atlantic, 218, 225, 226–27
    dams and, 227, 228, 236
    decline in, 167
    Pacific, 46–47, 103, 170, 171, 172
    See also Salmon
Anthony's river snail (*Atherania anthonyi*), 213

Anthropogenic impact, 36, 85, 206–7
    See also Human activity
Apache trout (*Oncorhynchus apache*), 179, 180
Apalachicola [37], 62, 81, 218–20
Appalachian region, 30, 208–9
Aquaculture industry, 84, 205
Aquarium fish, 53, 67–68
Aquatic invertebrates, 9
Aquifers
    over-pumping of, 208
    subterranean, 192
Arctic rivers and lakes MHT, 63, 96–97
    Arctic Islands [60], 25, 248–49
    East Arctic [59], 247–48
    East Hudson [54], 243–44
    hydrographic integrity, 64
    imperiled species in, 78
    integration matrix for, 156
    lack of introduced species in, 66
    Lower MacKenzie [56], 244–45
    North Arctic [58], 246–47
    South Hudson [53], 242–43
    Upper MacKenzie [57], 245–46
    Yukon [55], 244
Arizona State University, 96
Arkansas Natural Heritage Commission, 199, 204, 205
Army Corps of Engineers. *See* U.S. Army Corps of Engineers
Artesian springs, 221. *See also* Spring habitat
Ash Meadows (Nevada), 52, 53, 177, 178
Asian clam (*Corbicula fluminea*), 68, 192, 217
Assessment criteria and methods
    biodiversity and, 15
    for biological distinctiveness, 23, 121–26
    for conservation status, 74, 127–30
    Global 200, 6, 101, 102, 103
    overview of, 2–6
    scale-dependent, 15
    threat, 70
    See also Snapshot conservation status
Atlantic Salmon Federation, 229
Atlantic salmon (*Salmo salar*), 84, 228, 236
Atlantic-Ungava [48], 62
Balsas [70], 258–59
Banff/Bow Valley Naturalists, 239

Bear Lake, 173, 174
Beluga whales (*Delphinapterus leucas*), 234, 236
Beta-diversity, 16, 126
Bioconservacion, A.C., 189, 192, 193
Biodiversity
    in Appalachian region, 208
    in Canada, 83–85
    defined, 4
    and ecoregion size, 150–52
    globally outstanding, 2, 87–88, 101
    indicators for, 45
    invasive species threat to, 67–69
    latitude and, 84
    patterns of, 9, 149–50
    in South Hudson [53], 242
    in springs, 53
    terrestrial vs. freshwater, 7
    See also Species richness
Biodiversity conservation, 6–7
    agenda for, 103–4, 105, 108
    sites for, 24
Biological distinctiveness, 4–5, 58
    assessment methods for, 23, 121–26
    category, 49
    conservation status and, 22–24, 87–94, 96–99
    data and scores, 48, 58, 122, 131–39
    final, 57, 126
    index (BDI), 12, 121, 126
    matrix for, 6
    overview of, 15–17
    ratings, 28–29
    See also Appendix C; Endemism
Birdlife International, 124
Birds, habitat for, 201, 239
    waterfowl, 54, 55, 56
Blue tilapia (*Oreochromis aureus*), 251, 252, 255, 260, 261
Bonneville [8], 62, 173–74
Bonytail chub (*Gila elegans*), 68
Borax lake chub (*Gila boraxobius*), 176, 177
Boulder darter (*Etheostoma wapiti*), 213
British Columbia Ministry of Environment, Lands, and
    Parks, 166
Burr, B. M., 210
Butte Environmental Council, 54

Cahaba River, 217
Cahaba River Society, 217
California, 47
    habitat modification in, 171–72
    vernal pools in, 53–55
California Department of Fish and Game, 54, 172
California Native Grass Association, 172
California Native Plant Society, 170, 172
Canada, 4
    freshwater biodiversity in, 83–85
Canadian Arctic Resources Committee, 239, 241, 244,
    245, 246, 247, 248, 249
Canadian Nature Federation, 201, 238, 240, 245
Canadian Parks and Wilderness Society, 201, 237, 239,
    245, 246
Canadian Rockies [49], 238–39
Canyon Preservation Trust, 190

Carp, exotic, 34, 67, 182, 197
Catchment (watershed), 32
    alteration, 19, 62, 63, 129
    characterization data, 79–83
    as conservation priority, 119
    vulnerability of, 80–81
Catemaco [72], 259–60
Cattle (livestock) grazing, 53, 211, 261. *See also* Pasturage
Cave fauna, 95, 106–8, 202, 203
    crayfish, 221, 222
Cave (subterranean) habitat, 106–8, 192, 263
Central Cascades Alliance, 168
Central Prairie [28], 202–3
Centre de Donnees sur le Patrimoine Naturel du Québec,
    235, 237
Centro de Investigationes Biológicas del Noreste, S.C.
    (CIBNOR), 173
Centro de Investigación y de Estudios Avanzados (CIN-
    VESTAV), 264
Centro Interdisciplinario de Investigacion para el Desarrollo
    Integral Regional (CIIDIR), 191, 261, 262
Channelization, 199, 219
    habitat loss and, 214, 216
    hydrographic integrity and, 64
Chapala [65], 93, 101, 253–54
Chattahoochee Riverkeeper, 220
Chesapeake Bay [41], 81, 226–27
Chesapeake Bay Foundation, 227
Cheviot Project proposal, 238
Chihuahua chub (*Gila nigrescens*), 187, 188
China, 208
Chinese carp (*Cyprinus carpio*), 197
Chinook salmon (*Oncorhynchus tshawytscha*), 46, 47
Chub fish (*Gila* spp.), 68, 182, 183–84, 187
Cichlid fish (*Cichlasoma minckleyi*), 95, 194
Ciscos (*Coregonus* spp.), 66, 231, 240
Clean Water Act, 55, 56, 118
Clearcutting, 85, 191. *See also* Logging
Clinch River area, 211
Coal mining, 210, 211, 217, 238
Coastal Plains Institute and Land Conservancy, 226
Coatzacoalcos [73], 260–61
Coho salmon (*Oncorhynchus kisutch*), 46, 47
Colegio de la Frontera Sur (ECOSUR), 264
Colorado [12], 75, 179–81
Colorado Rivers Alliance, 181
Colorado squawfish (*Ptychocheilus lucius*), 179, 180
Columbia Environmental Research Center, 197, 203, 206
Columbia Glaciated [2], 166–67
Columbia River Inter-Tribal Fish Commission, 167
Columbia Unglaciated [3], 167–68
Comanche Springs pupfish (*Gamubusia elegans*), 189, 190
Comisión Nacional del Agua, 253, 256, 257
Comisión Nacional para el Conocimiento y Uso de la Bio-
    diversidad (CONABIO), 7, 73, 83, 119, 189, 191,
    192, 250
    subtropical regions and, 251, 252, 256, 258, 259, 261,
    262, 263, 264
    xeric regions and, 193, 195, 196, 254
Committee on Restoration of Aquatic Ecosystems, 18
Committee on the Status of Endangered Wildlife, 84
Confederated Tribes of the Umatilla Indian Reservation,
    167

Connecticut River Watershed Council, 229
Conservation
  assessment, 2
  biodiversity and, 6–7, 24, 103–4, 105, 108
  priority setting matrix for, 87–94, 96–99
Conservation Council of New Brunswick, 229
Conservation groups, 7, 119
  See also specific groups
Conservation intervention
  dams and, 112, 116–19
  site-specific, 108–12
Conservationists, 3
Conservation status
  biological distinctiveness and, 22–24, 87–94, 96–99
  criteria for, 15, 17–20
  of ecoregions, 4–5, 61–62
  final, 22, 70, 72, 88, 90
  global ranks, 73, 74–75
  index (CSI), 14, 72, 127
  projected, 130
  snapshot, 17, 20–22, 59–60, 62–66, 72, 88–89
Conservation status assessment, 6, 20
  methods for, 127–30
Conservation status data
  and assessment, 141–47
  Canadian freshwater biodiversity, 83–85
  imperiled fauna, 70–79
  national watershed characterization, 79–83
Continentally outstanding ecoregions, 90
Contreras-Arquieta, A., 95, 96
Coosa River Basin Initiative, 217
Coosa River Society, 218
Crawford, A. B., 179
Crayfish, 3, 33, 203
  ecoregion delineation by, 9
  endemic, 35, 149, 150, 198, 204, 213, 221, 224
  range data for, 124
  relict species, 255
  rusty, 36–37, 197, 210, 235
  species richness, 25, 34–35, 149–50
  subterranean, 221, 222
  taxonomic diversity, 36–37
Critically imperiled ecoregions, 87, 88
Critical status, for habitat, 20–21
Crustaceans, 104. *See also* Crayfish; Snails
Cryptic fish species, 31–32
Cuatro Ciénegas [22], 88, 94–96, 101, 194–95
  beta-diversity in, 16
  endemism in, 75, 95, 194
  spring habitat in, 52, 53, 194
Cui-ui (*Chasmistes cujus*), 175
Culver, D. C., 107
Cutthroat trout (*Oncorhynchus clarkii*), 46, 47
Cyclid (*Herichthys carpintis*), 255

Dams, 64, 240
  decommissioning/modification of, 112, 116–19
  fish migrations and, 168, 227, 228, 236
  habitat fragmentation and, 64
  impact of, 84, 195, 216, 246, 250
  river impoundment by, 1, 212, 214, 218–19, 225
Daniel Boone National Forest (DBNF), 210, 211, 214, 215

Darters (*Percidae* spp.), 31–32, 33
Data precision, 5
Death Valley [11], 25, 44, 75, 177–79
Deepwater fishes, 230, 231
Defenders of Wildlife, 7, 181
Deforestation, 262. *See also* Logging
Desert dace (*Erimichthys acros*), 175
Desert Fishes Council, 95, 173, 174, 176, 179, 181, 183,
    184, 186, 188, 195
Desert pupfish (*Cyprinodon macularius*), 180
Deserts. *See* Xeric-region rivers, lakes and springs
Devils Hole pupfish (*Cyprinodon diabolis*), 178
Deyrup, M., 222
Discriminators, conservation status, 14
Distinctiveness, biological. *See* Biological distinctiveness;
    Endemism
Disturbances, of habitat, 13. *See also* Habitat modification
Diversity, 212. *See also* Beta-diversity; Biodiversity
Drainage basins, 32. *See also* Catchments
Drengson, A. R., 85
Duck habitat, 55, 56
Ducks Unlimited Canada, 201, 240
DuPont Corporation, 223

East Arctic [59], 247–48
Eastern U.S., conservation sites in, 113
East Hudson [54], 63, 64, 243–44
East Texas Gulf [32], 36, 82, 206–7
Ecological phenomena. *See* Rare phenomena
Ecology Action Centre, 229
Ecology North, 241, 244, 245, 246, 247, 248, 249
Ecoregion-based conservation (ERBC), 109–12
Ecoregions, freshwater, 2–4, 10–11
  assessment of, 4–6, 59
  biological distinctiveness and, 16–17
  biota of, 3–4
  conservation status of, 4–5, 9, 17, 61–62
  continentally outstanding, 16, 90
  critically imperiled, 87, 88
  defined, 2
  globally outstanding, 16, 48, 58, 98, 101
  and island ecosystems compared, 68
  major habitat types for, 12–13
  priority-setting matrix for, 87–94, 96–99
  rare habitat in, 126
  size of, and biodiversity scores, 150–51
  and terrestrial compared, 7, 87
  water quality-vulnerabilty corelation in, 80–82
  See also Snapshot conservation status
Ecosystem Recovery Program for the Bay-Delta Region, 47
Edwards Aquifer, 208
Edwards Dam, 117, 118
Edwards Plateau, 206, 207
Eglin Air Force Base, 221
Endangered habitat, 21. *See also* Habitat loss
Endangered species, 55, 75. *See also* Imperiled fauna
Endangered Species Act, 3, 74, 118, 186
Endemism, 26, 182, 212, 252
  assessment of, 15, 16
  biological distinctiveness and, 16, 48, 121
  in Canadian fauna, 83–84
  in Cuatro Ciénegas, 94–95, 194

Endemism (continued)
  decision rules for, 124–25
  in endorheic basins, 173, 187, 191, 254, 257
  in karst biota, 106
  in Mississippi River species, 34
  Mobile Bay [36], 216
  in mollusks, 168, 169, 218, 248
  and species imperilment, 75
  species richness and, 149–50
  spring habitat and, 52, 182
  Teays-Old Ohio MHT, 209
  See also Crayfish; Fish species; Herpetofauna, endemic
Endorheic rivers, lakes, and springs MHT, 58, 98
  Bonneville [8], 62, 173–74
  Death Valley [11], 25, 44, 75, 177–79
  endemism in, 173, 187, 191, 254, 257
  Guzmán [16], 98, 187–88
  imperiled species in, 78, 79
  integration matrix for, 160
  Lahontan [9], 98, 174–76
  Lerma [69], 44, 93, 98, 101, 256–58
  Llanos el Salado [66], 25, 98, 254–55
  Mapimí [19], 98, 190–91
  Oregon Lakes [10], 25, 176–77
English-Winnipeg Lakes [52], 241–42
Environmental Defense Fund, 181, 244
Environment Canada, 83, 166
Environment North, 231
EPA. See U.S. Environmental Protection Agency
Erie [45], 70, 232–33
Estación Ecológica Chapala (Universidad Autónoma de
    Guadalajara), 254, 257
Etowah River, 217
Eulachon (Thaleichthys pacificus), 46, 47
Eutrophication, 107, 246, 260
  agricultural runoff and, 225, 227, 233
  in Great Lakes, 229
Everglades, 98, 222–23
Everglades Coalition, 223
Evolution, of fish, 33
Evolutionary phenomena. See Rare phenomena
Exotic species, 192, 197
  aquaculture industry and, 205
  control of, 105
  disturbed habitat and, 254
  fish species, 175, 180, 181
  future threats from, 66, 103
  in Mississippi River, 34
  native species and, 185–86, 251, 252
  in northern ecoregions, 235
  in subtropical ecoregions, 260, 261
  in xeric regions, 182, 184, 190, 193, 194, 195
  See also Introduced/invasive species; and specific species
Expert assessment, 5, 81, 83
  workshop, 103–4, 105
Exploitation, species, 65, 66, 70, 130
  See also Mussels, overharvesting of
Extinction, 1, 180
  causes of, 17–18
  conservation status and, 20
  human activities and, 33, 216, 225
  imperiled species and, 31, 73
  in invertebrates, 104

  of island species, 68
  of mussels, 37
  of native species, 171–72
  secondary vs. global, 19
  threat of, 191
Extractive industries, 238. See also Logging; Mining

Fairy shrimp (Branchinecta spp.), 53
Federal Energy Regulatory Commission (FERC), 117, 118
Federation of Alberta Naturalists, 201, 240
Federation of Nova Scotia Naturalists, 229
Federation of Ontario Naturalists, 231, 232, 233, 235, 237,
    243
Final biological distinctiveness categorization, 57, 126
Final conservation status, 22, 70, 72
  priority class of ecoregions and, 88, 90
Fish and Wildlife Service. See U.S. Fish and Wildlife Ser-
    vice
Fish migrations. See Anadromous fish; Migratory fish
Fish species, 3
  causes for extinction of, 17–18
  cryptic, 31–32
  deepwater, 230, 231
  ecoregion delineation by, 9
  effect of dams on, 118
  exotic/non-native, 67–69, 175, 180, 181
  exploitation of, 66, 70
  game, 30, 66, 67
  globally rare, 231
  imperilment of, 1, 30–32, 75, 76, 79
  range maps for, 121
  richness of, 13, 25, 27, 33, 149, 150
  spring-inhabiting, 52–53
  See also Anadromous fish species and specific species
Fish species, endemic, 44, 164, 206, 218
  in endorheic ecoregions, 173, 187, 191, 254, 257
  in Florida ecoregions, 220–22
  imperilment of, 76
  Pacific coastal, 168, 169, 170, 171
  pupfish, 178, 180, 185, 186, 254, 263
  South Atlantic [40], 224–25
  and species richness, 27, 149, 150, 212
  in subtropical ecoregions, 252, 256, 259, 260, 261
  in temperate headwaters and lakes, 209, 212–13
  in xeric ecoregions, 172, 188, 193, 194, 195, 253
Floods, 33, 246
  pulses, 199
  simulation of, at dams, 118
Flores-Villela, Oscar, 121
Florida [39], 93, 101, 221–23
  Everglades preservation and, 98, 222–23
Florida Audubon Society, 220, 221, 223
Florida Defenders of the Environment, 223
Florida Gulf [38], 220–21
Florida Natural Areas Inventory, 223
Fondo Mexicano para la Conservacion de la Naturaleza
    (FMCN), 191
Foreign species. See Exotic species
Forest Guardians, 186, 190
Forestry, 84, 85, 235
Franz, R., 222
Freshwater biota, 3–4. See also Species

Freshwater ecoregions. *See* Ecoregions, freshwater
Friends of the Animas River, 181
Friends of the Boundary Waters Wilderness, 242
Friends of the Earth, 166
Friends of the Los Angeles River, 173
Friends of the White Salmon, 168
Frogs, 39, 84–85, 225, 258
Fundación Ecologica de Cuixmala, 252
Future threats, 101, 103, 130

Game fish species, 30, 66, 67. *See also* specific species
Georgia Natural Heritage Program, 215, 218, 223, 226
Giant stickleback (*Gasterosteus* sp.), 166
"Gifts to the Earth" campaign, 112
Gila [14], 183–84
Gila trout (*Salmo gilae*), 183, 187
Gilbert, C. R., 222
Glaciation
    endemism and, 229, 241
    saltwater pools and, 248
    speciation and, 13
Glen Canyon Dam, 118
Global conservation status ranks, 73, 74–75
Global 200 Ecoregions analysis, 6, 101, 102, 103
Global importance, of ecoregions, 23
Globally outstanding
    biodiversity, 2, 87–88
    ecoregions, 16, 48, 58, 98, 101
    rating, 252, 257
Global warming, 247
Glochidia, 37–39
Grand Canyon Trust, 181, 183
Grand Council of the Crees, 238, 243, 244
Greater Yellowstone Coalition, 201
Great Lakes, 48, 229–35
    hydrographic integrity of, 63, 64
    pollution of, 229, 230, 232, 233, 234
    priority class for, 97
    species exploitation in, 66
    water-quality degradation of, 62
    zebra mussel in, 66, 68, 230, 232, 233, 234
    See also Large temperate lakes; *and specific lake*
Great Lakes Commission, 231, 232, 233, 235
Great Lakes Water Quality Agreement, 229
Great Recycling and Northern Development (GRAND), 243
Green sturgeon (*Acipenser medirostris*), 46, 47
Grijalva-Usumacinta [75], 262–63
Groundwater
    contamination, 107, 205–6
    extraction, 188, 257
    prairie potholes and, 55–56
Guadalupe River, 206
Gulf of St. Lawrence, 229
Guzmán [16], 98, 187–88
Gypsum extraction, 96

Habitat
    critical status for, 20–21
    destruction, 214
    disturbances, 13

fragmentation, 64, 65, 129
loss, 5, 18, 66, 129, 216
modification, 171–72, 184, 196–97, 222
quality, 17
rarity of, 16, 48, 126
representation, 2
subterranean, 106–8, 192, 263
Hammond, H., 85
Hanford Reach, 167
Headwaters Environmental Center, 171
Hellbender (*Cryptobranchus alleganiensis*), 208
Herpetofauna, 203, 218
    amphibians, 1, 84–85, 121, 152, 225, 258
    frogs, 39, 84–85, 225, 258
    imperilment of, 75, 77, 78–79
    salamanders, 39, 196, 208, 209, 213, 225
    species richness, 25, 34, 149–50, 152
Herpetofauna, endemic, 44, 209
    imperilment of, 77
    in Mexican ecoregions, 152, 188, 194, 258, 259, 261, 262
    species richness of, 149, 150
Hershler, R., 95
Hill Country Federation, 207
Hocutt, C. H., 237
Hog farming, 225, 226
Holsinger, J. R., 107
Human activities, 261–62
    anthropogenic impact, 36, 85, 206–7
    extinction and, 36, 216, 225
    impact on native biota by, 188, 213
    native mussels and, 39
    pressure on habitat from, 256
    water quality degradation and, 52, 62
    See also specific activity
Human health, non-native species and, 67
Humboldt State University, 171
Hybridization, 18, 37, 190, 255
Hydrobiid snails, 52
Hydroelectric dams, 84, 168, 250. *See also* Dams
Hydrographic integrity, 63–64, 129
Hydrological modification, 198–99
    to caves, 107
    in Everglades, 222, 223
    of Mississippi River basin, 33–34
    See also Channelization
HydroQuebec, 243

Ichthyofauna. *See* Fish species
Idaho Rivers United, 168, 169
Illinois Department of Natural Resources, 197, 232
Illinois Natural Heritage Division, 199, 211
Illinois Natural History Survey, 124
Imperiled fauna, 1–2, 19, 70–79, 99, 152–53
    in Canada, 84
    categorizing, 73–79
    fish, 30–32
    water quality and, 81
Impoundment, of rivers, 1, 212, 214, 216, 218–19, 225. *See also* Dams
Index of Watershed Indicators (IWI) project, 79, 81
Indiana Department of Natural Resources, 197

Indiana Natural Heritage Data Center, 211
Industry
    mussel overharvesting by, 39, 197, 199, 211, 214, 225
    pollution from, 210, 219, 220, 232, 244, 262
    water-quality degradation by, 253
Insects, 104
Instituto de Ecología A.C., 259, 260
Instituto Nacional de Ecología, 191, 195, 258
Instituto Tecnologico Regional de Chetumal, 264
Instituto Tecnologico y de Estudios Superiores de Monter-
    rey (ITESM), 193, 196, 250, 251
Integration matrices, for MHTs, 155–63
Introduced/invasive species, 67–69, 172, 253
    aquarium fish, 53, 67–68
    in Great Lakes, 229, 230, 232, 233
    impact of, 65, 66, 129, 217
    non-native, 36–37, 214, 225
    See also Exotic species
Invertebrates, aquatic, 3, 9, 185, 223
    species richness of, 3, 104
    vulnerability of, 235
Irrigation, 185, 191, 202, 257
Island species, extinction of, 68
IUCN Red Data Book series, 20, 127

James Bay Project, 243
June sucker (Chasmistes liorus), 173, 174

Kansas Natural Heritage Inventory, 202, 203, 206
Karst biotas, 106–8. See also Cave fauna
Karst Waters Institute, 108
Kentucky Natural Heritage Program, 199, 211, 215
Kiamichi River, 204
Klamath Forest Alliance, 171

Labrador Inuit Association, 238
Lacustrine marine relict species, 248
Lahontan [9], 98, 174–76
Lakes
    impacts on, 18
    in northern ecoregions, 235
    See also Great Lakes; Large Temperate Lakes; and specif-
      ic lake
Lake sturgeon (Acipenser fulvescens), 230
Lake Tahoe, 174
Lamprey (family Lampetra), 46, 47
Land-cover. See Catchment; Watershed
Land use
    freshwater biota and, 217
    freshwater habitat and, 17, 21
Large temperate lakes MHT
    Erie [45], 232–33
    imperiled species in, 78, 79
    integration matrix for, 157
    Michigan-Huron [44], 75, 97, 231–32
    Ontario [46], 234–35
    priority class for, 97
    Superior [43], 97, 230–31
    See also Great lakes
Large temperate rivers MHT, 97–98
    Colorado [12], 75, 179–81
    integration matrix for, 159

Lower Rio Grand/Bravo [20], 192–93
Middle Missouri [27], 201–2
Mississippi [24], 37, 75, 196–97
Mississippi Embayment [25], 23, 33, 34, 93, 97–98,
    101, 198–99
Upper Missouri [26], 200–201
Upper Rio Grande/Río Bravo del Norte [15], 62, 75, 98,
    185–86
Latitude, species richness and, 12, 84, 152
Leon Springs pupfish (Gamubusia bovinus), 189, 190
Lerma [69], 44, 93, 98, 101, 256–58
Lerma-Chapala River Basin Council, 254, 258
Livebearer (Poeciliopsis sp.), 252
Livestock (cattle) grazing, 53, 211, 216. See also Pasturage
Llanos el Salado [66], 25, 98, 254
Loach minnow (Tiaroga cobitis), 183, 184
Local conservation, 105
Logging, 228, 235, 242
    clearcutting, 85, 191
    sedimentation and, 170, 213, 246, 257, 262
    water quality and, 202, 238
Longfin smelt (Spirinchus thaleichthys), 46, 47
Louisiana Natural Heritage Program, 199, 205
Lower MacKenzie [56], 244–45
Lower Mississippi River Conservation Committee, 199
Lower Rio Grand/Bravo [20], 192–93
Lower Saskatchewan [51], 240–41
Lower St. Lawrence [47], 62, 64, 75, 235–37

Mackenzie river ecoregions, 64, 244–46
MacMillan-Bloedel, 85
Maine Natural Areas Program, 229
Major habitat types (MHTs), 18
    assignment of, 12–13
    integration matrices for, 155–63
    latitude and species richness in, 152
    patterns of biodiversity in, 149–50
    See also Habitat; and specific major habitat types
Manantlan-Ameca [64], 252–53
Manitoba Naturalists Society, 242
Mapimí [19], 98, 190–91
Marsh, E. P., 94
Master, L., 225
May, E., 85
McAllister, D. E., 83, 85
Mean biodiversity indicators, 45
Medialuna, La, 254–55
Mexico, 4, 83
    conservation sites in, 115
    habitat loss in, 66
    herpetofauna in, 152, 188, 194, 258, 259, 261, 262
    imperiled species in, 72, 78
    mussels in, 15
    priority hydrologic basins in, 119
    species presence in, 121, 124
    species richness in, 34
    See also specific ecoregions
Michigan Environmental Council, 231
Michigan-Huron [44], 75, 97, 231–32
Michigan Land Use Institute, 232
Michigan Natural Areas Council, 231
Mid-Atlantic region, species richness in, 25
Middle Missouri [27], 201–2
Migratory birds, 201

Migratory fish, 47, 168, 222, 227. *See also* Anadromous
     fish; Salmon
Mill Creek, 210
Minckley, W. L., 59, 94, 95, 96, 107
Mining, 202, 244
     coal, 210, 211, 217, 238
     pollution from, 180, 181, 217, 248, 250, 258
     uranium, 242, 246
     water quality and, 217, 238, 247
Minnesota Department of Natural Resources, 197
Mississippi [24], 33–34, 37, 75, 196–97
Mississippi Embayment [25], 93, 97–98, 101, 198–99
     species richness in, 25, 33, 34, 198
Mississippi Natural Heritage Program, 199, 215, 218
Mississippi Wildlife Federation, 199
Missouri Department of Conservation, 203
Missouri Department of Natural Resources, 197
Missouri Natural Heritage Database, 199
Missouri River Coalition, 201
Missouri River ecoregions, 200–202
Moapa speckled dace (*Moapa coriacea*), 182
Mobile Bay [36], 79, 93, 98, 101, 215–18
     species richness in, 25, 37, 216
Mollusks, 68, 104, 216
     endemism in, 168, 169, 218, 248
Mono Lake, 174, 175
Mono Lake Committee, 176
Monotypic species, 30
MoRAP, 197, 199, 203, 204
Mosquin, T., 83
Moyle, P. B., 171
Mussels, 1, 9, 124
     endangered species, 213, 216, 228
     endemic, 44, 149, 220, 224
     larval stage of, 37–39
     overharvesting of, 39, 197, 199, 211, 214, 225
     species richness of, 3, 34, 149, 213, 218, 231–32
     threats to, 15, 39, 68, 214, 219
     See also Zebra mussels

Nashville crayfish (*Orconectes shoupi*), 213
National Audubon Society, 167, 172, 193, 223
National Cattlemen's Beef Association, 206
National Forum on Biodiversity (1986), 1
National Park Service, 167
National Watershed Characterization, 79–81, 153
National Wildlife Federation, 199, 231, 232, 233, 235
Nationwide Permit 26, 54–55
Native biota, invasive species and, 67
Native Plant Society, 54
Natural Heritage database, 72
Natural Resources Conservation Service, 201
Natural Resources Council of Maine, 229
Natural Resources Defense Council, 227
Nature Conservancy, 7, 99, 108, 119, 124
     Mexican ecoregions and, 250, 251, 252, 254, 255, 256,
          258, 259, 260, 261, 262, 263, 264
     Natural Heritage database, 72
Nature Conservancy of Canada
     Alberta office, 239, 240, 246
     Atlantic region, 238
     British Columbia office, 166, 167, 245, 246
     Headquarters, 245, 246, 248
     Manitoba office, 241, 242, 246, 248

Ontario office, 233, 242, 243
     Québec office, 235, 238, 243, 244
     Saskatchewan office, 240, 241, 242, 246
Nature Conservancy (U.S.) field offices
     Alabama, 199, 215, 218, 220, 221
     Alaska, 166, 244
     Arizona, 181, 184
     Arkansas, 199, 203, 204, 205
     California, 171, 172, 173, 176, 179, 181
     Colorado, 181, 186, 188, 202, 206
     Connecticut/Delaware, 229
     Florida, 220, 221, 223
     Georgia, 218, 220, 223, 226
     Idaho, 167, 168, 169, 174
     Illinois, 197, 199, 211
     Indiana, 197, 211, 232, 233
     Iowa, 202
     Kansas, 202, 203
     Kentucky, 199, 211, 215
     Louisiana, 199, 205, 207
     Maine, 229
     Maryland, 211, 227, 229
     Minnesota, 197, 231, 242
     Mississippi, 199, 218
     Missouri, 197, 199, 202, 203, 204
     Montana, 167, 168, 201
     Nebraska, 202, 206
     Nevada, 168, 169, 174, 179, 183, 1176
     New Mexico, 181, 184, 186, 188, 190, 206, 207
     New York, 211, 227, 229, 233, 235, 237
     North Carolina, 215, 226
     North Dakota, 201, 242
     Ohio, 211, 233
     Oklahoma, 203, 205, 206
     Oregon, 168, 171, 177
     Pennsylvania, 211, 227, 229, 233
     South Carolina, 226
     South Dakota, 201
     Tennessee, 199, 215
     Texas, 186, 188, 190, 193, 205, 206, 207, 208
     Utah, 174, 181, 183
     Virginia, 215, 226, 227
     Washington, 166, 167
     West Virginia, 211, 227
     Wisconsin, 197, 231, 232
     Wyoming, 169, 181, 201, 202
Nature Conservancy (U.S.) regional offices
     Midwest, 211, 215
     Southeast, 211, 215, 216, 218, 220, 221, 223
Nature Saskatchewan, 240, 242
Nebraska Natural Heritage Program, 202
New Brunswick Federation of Naturalists, 229
New Hampshire Natural Heritage Inventory, 229
New Mexico Department of Game and Fish, 184, 187, 188
New York Natural Heritage Program, 211, 233, 235, 237
Nonindigenous Aquatic Nuisance Prevention and Control
          Act of 1990, 67
Non-native species, 36–37, 214, 225
     See also Exotic species; Introduced/invasive species
North Arctic [58], 246–47
North Atlantic [42], 228–29
North Atlantic-Ungava [48], 237–38
North Carolina Coastal Federation, 226
North Carolina Natural Heritage Program, 215, 226
Northcoast Environmental Center, 171

Northern Appalachian Restoration Project, 229
Northern ecoregions, 235
    See also Arctic rivers, lakes and springs
Northern River Basins Study, 245–46
North Pacific Coastal [1], 83, 165–66
Northwatch, 231, 243
Northwest Power Planning Council, 167
Nuclear power plants, 227
Nunavit Wildlife Management Board, 248, 249

Ohio Natural Heritage Database, 211
Ohio River ecosystem, 209–11
    See also Teays-Old Ohio [34]
Ohio Valley Environmental Coalition, 211
Oklahoma Natural Heritage Inventory, 205
Ono, R. D., 197
Ontario [46], 234–35
Oregon Lakes [10], 25, 176–77
Oregon Lakes Association, 177
Oregon Natural Desert Association, 169
Oregon Natural Resources Council, 169
Oregon Trout, 177
Ouachita Highlands [30], 203, 204–5
    endemic species in, 33, 36, 204
    water quality in, 81, 82
Ouellet, Martin, 84
Ozark cavefish (*Amblyopsis rosae*), 202
Ozark Highlands [29], 25, 59, 82, 203–4
    endemic species in, 33, 36, 203
Ozark Society, 204

Pacific Central Valley [6], 44, 171–72
Pacific coast
    anadromous fish on, 46–47, 103
    Columbia ecoregions, 166–68
    fish exploitation in, 66
    North [1] MHT, 83, 165–66
    South [7] MHT, 82, 172–73
Pacific Mid-Coastal [5], 170–71
Pacific Rivers Council, 169
Paddlefish (*Polyodon spathula*), 30
Page, L. M., 210
Pahranagat roundtail chub (*Gila robusta jordani*), 182
Pahrump poolfish (*E. latos*), 178
Pallid sturgeon (*Scaphirhyncus albus*), 200
Papaloapan [71], 44, 79, 259
Parasitic life cycles, 37–39
"Partners in Flight," 74
Pasturage, 196, 203, 207. *See also* Cattle grazing
Pearl industry, overharvesting by, 39, 197, 199, 211, 214, 225
Pearson correlation, 150, 151
Pecos [18], 189–90
Pecos gambusia (*Gambusia nobilis*), 189, 190
Pennsylvania Natural Diversity Inventory-West, 211
Pesticides, 85, 200
    See also Agricultural pollution
Peterson, D. F., 179
Phenomena, rare. *See* Rare phenomena
Pigmy whitefish (*Prosopium coulteri*), 230
Pink salmon (*Oncorhynchus gorbuscha*), 46, 47
Pirate perch (*Aphredoderus sayanus*), 30
Plant species, wetland, 54

Pollution, 39, 107, 208, 236
    in Everglades, 223
    in Great Lakes, 229, 230, 232, 233, 234
    industrial, 210, 219, 220, 232, 244, 262
    from mining, 180, 181, 248, 250, 258
    from population expansion, 193, 222
    of Rio Grande, 185, 186, 192
    of spring habitat, 190
    urban development and, 219, 228
    water quality and, 210, 217
    See also Agricultural pollution
Population fragmentation, species decline and, 31
Population growth
    threat from, 31, 204, 210, 219
    water resources and, 193, 254
Prairie Conservation Forum, 201, 240
Prairie potholes, 55–56, 196, 200, 201
Priority class, ecoregion distribution and, 91–94
Priority hydrologic basins, 119
Priority-setting matrix, for conservation, 87–94, 96–99
PROFAUNA, 193, 195
Pronatura Peninsula de Yucatan, 264
Proserpine shiner (*Cyprinella proserpina*), 189, 190
Protection Area for Flora and Fauna, 96
Proyecto UNAM-PEMEX, 261
Pupfish (*Cyprinodon* spp.), 52, 53, 68
    endemism in, 178, 180, 185, 186, 254, 263
Putah Creek Council, 47
Pygmy sunfish (*Elassoma* spp.), 224–25
Pyramid Lake Paiute Tribe, 176

Railroad Valley, 174, 175
Range maps, for species presence, 121
"Range rank," 73
Rare habitat type, 16, 48, 126
Rare phenomena, 50–51, 58
    ecological/evolutionary, 15–16, 46, 48, 49, 106, 125–26
Rare species
    imperilment of, 71
    ranking, 75
Rating, of ecoregions. *See* Assessment
Razorback sucker (*Xyrauchen texanus*), 68, 179, 180
Red Data Book series (IUCN), 20, 127
Regional conservation, 105
Relatively intact/stable habitat, 21–22
Relicensing, of dams, 118, 119
Relict species, 95, 248, 255
Reptiles, 121
Reservoirs, fish species and, 118
Resource extraction, 202. *See also* Logging; Mining
Restore the Northwoods, 229
Rio Conchos [17], 93, 101, 188–89
Rio Grande Alliance, 186, 190
Rio Grande ecoregions, 62, 75, 98, 185–86, 192–93
Rio Grande silvery minnow (*Hybognathus amarus*), 185, 186
Río Lerma [69], 78
Río Salado [21], 193
Río San Juan [23], 93, 101, 195
Río Tamesí, 256
Río Verde Headwaters [67], 75, 254–55
Riparian habitat, 105
    vegetation loss, 170, 207
Rivers
    dam removal from, 117–19

impacts on, 18–19
impoundment of, 212, 216, 218–19, 225
See also Arctic rivers; Large temperate rivers; Temperate
coastal rivers; Subtropical coastal rivers; *and specific
rivers*
Roosevelt, Theodore, 119
Roundtail chub (*Gila robusta*), 184
Rusty crayfish (*Orconectes rusticus*), 36–37, 197, 210, 235

Sacramento perch (*Archoplites interruptus*), 171
Sacramento River Preservation Trust, 172
St. Lawrence River ecoregion, 62, 64, 75, 229, 236–37
Salamanders, 39, 196, 208, 209, 213, 225
Salmon, 46, 47, 66, 168
    Atlantic, 84, 228, 236
    decline in, 167
    hatcheries, 170
Saltwater pools, 248
Santiago [63], 251–52
Saskatchewan River ecoregion, 239–41
Saskatchewan Wetland Conservation Corporation, 240
Saskatchewan Wildlife Federation, 240
Save Barton Creek Association, 207
Save Our Streams, 227
Save Our Wild Salmon, 169
Scale-dependent assessment, 15. *See also* Assessment
Schindler, David, 85, 235
Seasonal wetlands, 53–56
Secretaria de Medio Ambiente Recursos Naturales y Pesca
    (SEMARNAP), 263
Sedimentation, 34, 196, 210, 214, 258
    logging and, 170, 213, 246, 257, 262
Shortjaw cisco (*Coregonus zenithicus*), 240
Shrimp species, 53
Sierra Club, 54, 172, 174, 176, 202, 207, 208, 211, 215,
    218, 223, 226, 227, 239, 244
    Cascade Chapter, 166
    Rocky Mountain Chapter, 181
Sinaloan Coastal [62], 250–51
Sinkholes, 106
Siskiyou Regional Education Project, 171
Site-selection tools, 119
Site-specific interventions, 108–12, 113–15
    dams, 112, 116–19
Slackwater darter (*Etheostoma boschungi*), 214
Snails, aquatic, 52, 216, 221
    endemic, 176, 213
    springsnails, 182, 189–90
Snake River ecoregions, 75, 169
Snapshot conservation status, 20–22, 59–60, 72
    criteria for, 17, 62–66
    ecoregion priority classes and, 88–89
Society for Range Management, 206
Society for the Protection of New Hampshire Forests, 229
Society of Grassland Naturalists, 201, 240
Soil erosion, 214. *See also* Sedimentation
Sonoran [61], 249–50
Sonoran Institute, 184
South African rivers, 19
South Atlantic [40], 34, 98, 101, 224–26
    species richness in, 25, 93, 224–25
South Carolina Heritage Trust, 226
Southeast Alaska Conservation Council, 166
Southeast (U.S.)

dam relicensing in, 119
fish species imperilment in, 30–32
freshwater alteration in, 197
species richness in, 25
Southern Plains [31], 62, 63, 64, 205–6
Southern Utah Wilderness Alliance, 174, 183
South Hudson [53], 242–43
South Pacific Coastal [7], 82, 172–73
Southwest Center for Biodiversity, 181, 184, 186, 190
Southwest (U.S.)
    conservation sites in, 115
    non-native fish in, 68
    water quality in, 81
Species, imperilment of. See Imperiled fauna
Species, spatial scales for, 3–4
Species exploitation, 65, 66, 70, 130
    See also Mussels, overharvesting of
Species loss, conservation status and, 20
Species presence/absence
    biological distinctiveness and, 121
Species richness, 12–13, 25–27, 48, 98
    assessment of, 15, 16
    endemism and, 149–50
    in invertebrates, 104
    latitude and, 12, 84, 152
    measure of, 125
    in Mobile Bay MHT, 25, 216
    in South Hudson MHT, 242
    in Teays-Old Ohio MHT, 25, 92, 209
    See also Biodiversity
Spring habitat, 52–53, 223
    artesian, 221
    endangerment of, 190, 192, 194
    endemic species in, 52, 106, 182, 223
State GAP analysis programs, 7
State of the Environment (SOE) reports, 83
Steelhead trout (*Oncorhynchus mykiss*), 46, 47
Stenothermy, 52
Strategies Saint-Laurent, Inc., 237
Stream capture, 170
Sturgeon (*Acipenser* spp.), 236
Subterranean aquifer, 192
Subterranean habitat, 106–8, 192, 263
Subtropical coastal rivers, lakes, and springs MHT, 79
    Balsas [70], 258
    Catemaco [72], 259–60
    Coatzacoalcos [73], 260–61
    Florida [39], 93, 98, 101, 221–23
    Grijalva-Usumacinta [75], 262–63
    herpetofauna in, 150
    integration matrix for, 163
    Manantlan-Ameca [64], 252
    Papaloapan [71], 44, 79, 259
    Santiago [63], 251–52
    Sinaloan Coastal [62], 250–51
    Tamaulipas-Veracruz [68], 34, 79, 94, 101, 256–57
    Tehuantepec [74], 39, 44, 79, 261–62
    West Texas Gulf [33], 82, 207–8
    Yucatán [76], 263–64
Suburban sprawl, 214. *See also* Urbanization
Superior [43], 66, 97, 230–31
Surface mining, 210. *See also* Mining
Surface water, degradation of, 63
Suwannee River, 223
Swales (vernal wetlands), 53–55

Tadpole shrimp (*Triops* sp.), 249
Tamaulipas-Veracruz [68], 34, 79, 94, 101, 256–57
Taylor, Christopher, 124
Taylor, D. W., 85, 95
Teays-Old Ohio [34], 34, 37, 44, 101, 209–11
    species richness in, 25, 92, 209
Tehuantepec [74], 39, 44, 79, 261–62
Temperate coastal rivers, lakes, and springs MHT, 98
    Apalachicola [37], 62, 81, 218–20
    Chesapeake Bay [41], 81, 226–27
    Columbia Glaciated [2], 166–67
    Columbia Unglaciated [3], 167–68
    East Texas Gulf [32], 36, 82, 206–7
    Florida Gulf [38], 220–21
    integration matrix for, 162
    Lower St. Lawrence [47], 62, 64, 75, 236–37
    Mobile Bay [36], 25, 37, 79, 93, 98, 101, 215–18
    North Atlantic [42], 228–29
    North Atlantic-Ungava [48], 237–38
    North Pacific Coastal [1], 83, 165–66
    Pacific Central Valley [6], 44, 171–72
    Pacific Mid-Coastal [5], 170–71
    South Atlantic [40], 25, 34, 93, 98, 101, 224–26
    Upper Snake [4], 75, 169
Temperate headwaters and lakes MHT, 97
    Canadian Rockies [49], 238–39
    Central Prairie [28], 202–3
    English-Winnipeg Lakes [52], 241–42
    integration matrix for, 158
    Lower Saskatchewan [51], 240–41
    Ouachita Highlands [30], 33, 36, 81, 82, 203, 204–5
    Ozark Highlands [29], 25, 33, 36, 59, 82, 203–4
    Southern Plains [31], 62, 63, 64, 205–6
    Teays-Old Ohio [34], 25, 34, 44, 92, 101, 209–11
    Upper Saskatchewan [50], 239–40
    See also Tennessee-Cumberland [35]
Tennessee Aquarium, 215
Tennessee-Cumberland [35], 25, 92–93, 101, 211, 212–15
    endemism in, 44, 212, 213
    fish species imperilment in, 79, 213
    mussel species in, 34, 37, 213, 214
Tennessee Division of Natural Heritage, 199, 211, 215
Tennessee-Tombigbee Waterway, 216
Tennessee Valley Authority (TVA), 214, 215
Terrestrial ecoregions, and freshwater compared, 7, 87
Texas Center for Policy Studies, 208
Texas Conservation Data Center, 205
Texas Parks and Wildlife, 207, 208
Threat
    assessment, 70
    future, 101, 103. 130
Threatened ranking, 75
Threespine stickleback (*Gasterosteus aculeatus*), 46, 47
Tourism, 238–39, 261–62
Toxic chemicals, 234, 235, 236, 244.
    See also Agrochemicals
"Triage" logic, 23
Troglobitic organisms, 106, 107
Trout Unlimited, 229
Tui chub (*Gila bicolor*), 175

Umack, Peter J., 107
Unionid mussels. *See* Mussels
Union Québecoise pour la Conservation de la Nature
    (UQCN), 235, 237, 238, 244

Universidad Autónoma Agraria Antonio Narro (UAAAN),
    193
Universidad Autónoma de Aguascalientes, 252
Universidad Autónoma de Baja California Sur (UABCS),
    173
Universidad Autónoma de Baja California (UABC), 173
Universidad Autónoma de Chihuahua (UACH), 189, 250,
    262
Universidad Autónoma de Guerrero (UAG), 262
Universidad Autónoma de Nayarit (UAN), 252
Universidad Autónoma de Nuevo León (UANL), 96, 189,
    191, 193, 195, 196, 250
Universidad Autónoma de Sinaloa (UAS), 250, 251
Universidad del Mar, 262
Universidad del Noreste, 255, 256
Universidad de Occidente, 250, 251
Universidad de Quintana Roo, 264
Universidad de Sonora (UNISON), 184, 250, 251
Universidad de Tampico, 255
Universidad Juarez Autónoma de Tabasco (UJAT), 261,
    263
Universidad Michoacana, 258
Universidad Nacional Autónoma de México (UNAM),
    173, 250, 251, 252, 262
    Instituto de Biología, 259, 260
Universidad Veracruzana, 259, 260
University of Arizona, 250
University of Wisconsin Sea Grant Program, 231, 232, 233,
    235
Upper Chattahoochee Riverkeeper, 220
Upper Mackenzie [57], 64, 245–46
Upper Missouri [26], 200–201
Upper Rio Grande/Río Bravo del Norte [15], 62, 75, 98,
    185–86
Upper Saskatchewan [50], 239–40
Upper Snake [4], 75, 169
Uranium mining, 242, 246
Urbanization, 98, 210, 219, 228
    See also Population growth
U.S. Army Corps of Engineers, 54–55, 117, 199, 219
    Everglades and, 223
U.S. Bureau of Land Management, 172
U.S. Environmental Protection Agency (EPA), 2, 54, 79,
    234
    water quality data from, 81–82, 153
U.S. Fish and Wildlife Service (USFWS), 56, 74, 203, 210,
    213, 216
    mussel species endangerment and, 39, 211, 216
U.S. Forest Service, 9, 211, 215
U.S. Geological Survey (USGS), 79, 99
Utah Lake, 173, 174
Utah Rivers Council, 174

Vegas-Virgin [13], 75, 181–83
Vermont Nongame and Natural Heritage Program, 229
Vernal pools, 53–55
Viability, of imperiled species, 73
Virginia Division of Natural Heritage, 211, 215, 226
Vulnerability, watershed, 80–81
Vulnerable habitat, 21
Vulnerable species. *See* Imperiled fauna; Rare species

Warner sucker (*Catostomus warnerensis*), 176
Washington Environmental Council, 167
Water-allocation plans, 216

Waterfowl, 54, 55, 56
Water quality, 153, 198
    and conservation status, 21
    extractive industries and, 202, 238
    pollution and, 217
    threats to, 213
    vulnerability corelation, 81–82
Water-quality degradation, 62–63, 129, 210
    indicators for, 19
    in Mexican ecoregions, 191, 195, 249–50, 251, 253, 257
Watersheds
    characterization data, 79–83
    as conservation priorities, 119
    vulnerability of, 80–81
    See also Catchment
Water use
    for irrigation, 185, 191, 202, 257
    in xeric areas, 105
West Central Research and Extension Station, University of Nebraska, 202
Western Pennsylvania Conservancy, 211, 233
Western spadefoot toad (*Scaphiopus hammondii*), 53
West Texas Gulf [33], 82, 207–8
West (U.S.)
    conservation sites in, 114
    habitat loss in, 66
    imperiled species in, 78
    water quality in, 81
West Virginia, mining impact in, 210
West Virginia Highlands Conservancy, 211
West Virginia Natural Heritage Program, 211
Wetlands, 239
    assessment of, 9
    Canadian, 83, 84
    destruction of, 54–55, 56, 199
    prairie potholes, 55–56, 196, 200, 201
    vernal (swales), 53–55
White river spinedace (*Lepidoneda albivallis*), 182
White Sands pupfish (*Cyprinodon tularosa*), 185, 186
Wild Earth, 232, 235, 237
Wilderness Society, 166, 176, 179
    Southeast Region, 215
Wildlands League, 231, 233, 235, 243
Wildlands Project, 171, 211, 215
Wiley, 237
Williams, J. E., 255
Williams, James D., 124

Wisconsin Department of Natural Resources, 197, 232
World Conservation Union, 74
World Wildlife Fund, 99
    Global 200 assessment, 6, 101, 102, 103
World Wildlife Fund Canada, 201
    arctic rivers and lakes MHTs, 243, 244, 245, 246, 247, 248, 249
    Great Lakes region, 231, 232, 233, 235
    temperate coastal MHTs, 166, 167, 229, 237, 238
    temperate headwaters and lakes MHTs, 239, 240, 241, 242 Endangered Spaces Campaign, 7
World Wildlife Fund México, 181, 250
    endorheic ecoregions, 191, 254, 258
    subtropical ecoregions, 251, 252, 256, 259, 260, 262, 263, 264
    xeric ecoregions, 173, 181, 184, 188, 189, 193, 195, 196, 250, 254, 255
World Wildlife Fund (U.S.), 2, 7, 14
    "Gifts to the Earth" campaign, 112
Woundfin (*Plogopterus argentissimus*), 183

Xeric-region rivers, lakes, and springs MHT, 52, 58, 98
    Chapala [65], 93, 101, 253–54
    Cuatro Ciénegas [22], 16, 75, 88, 94–96, 101, 194–95
    Gila [14], 183–84
    imperiled species in, 78
    integration matrix for, 161
    Pecos [18], 189–90
    Río Conchos [17], 93, 101, 188–89
    Río Salado [21], 193
    Río San Juan [23], 93, 101, 195
    Río Verde Headwaters [67], 75, 254–55
    Sonoran [61], 249–50
    South Pacific Coastal [7], 172–73
    Vegas-Virgin [13], 75, 181–83

Yellow perch (*Perca flavescens*), 233
Yellowstone River, 200
Yucatán [76], 263–64
Yukon [55], 244

Zebra mussel (*Dreissena polymorpha*)
    native mussels and, 39, 84, 197, 211, 217 in Great Lakes, 66, 68, 230, 232, 233 in Northern ecoregions, 235, 236 threat from, 66, 214, 215
Zona Carbonifera, 192